MCBU
Molecular and Cell Biology Updates

Series Editors:

Prof. Dr. Angelo Azzi
Institut für Biochemie
und Molekularbiologie
Bühlstr. 28
CH - 3012 Bern
Switzerland

Prof. Dr. Lester Packer
Dept. of Molecular
and Cell Biology
251 Life Science Addition
Membrane Bioenergetics Group
Berkeley, CA 94720
USA

Alpha-Keto Acid Dehydrogenase Complexes

Edited by   M.S. Patel
             T.E. Roche
             R.A. Harris

Birkhäuser Verlag
Basel · Boston · Berlin

Editors

Prof. M.S. Patel
Department of Biochemistry
School of Medicine and Biomedical
Sciences
State University of New York
140 Farber Hall
3435 Main Street
Buffalo, New York 14214
USA

Dr. T.E. Roche
Department of Biochemistry
Kansas State University
104 Willard Hall
Manhattan, KS 66506
USA

Dr. R.A. Harris
Department of Biochemistry and
Molecular Biology
Indiana University School of Medicine
635 Barnhill Drive
Indianapolis, IN 46202
USA

Library of Congress Cataloging-in-Publication Data

Alpha-keto acid dehydrogenase complexes / edited by M.S. Patel, T.E.
Roche, R.A. Harris.
        p.   cm. - (Molecular and cell biology updates)
Includes bibliographical references and index.
ISBN 3-7643-5181-0 (alk. paper). - ISBN 0-8176-5181-0 (alk.
paper)
1. Alpha-keto acid dehydrogenase complexes. I. Patel, Mulchand
S. II. Roche, Thomas E. III. Harris, Robert A. (Robert Allison),
1939– . IV. Series.
QP603.A37A47 1996
574.19'258 - dc20
                                        95-25649
                                        CIP

Deutsche Bibliothek Cataloging-in-Publication Data

Alpha-keto acid dehydrogenase complexes / ed. by M. S. Patel - Basel ;
Boston ; Berlin : Birkhäuser, 1996
(Molecular and cell biology updates)
ISBN 3-7643-5181-0
NE: Patel, Mulchand [Hrsg.]

© 1996 Birkhäuser Verlag, PO Box 133, CH-4010 Basel, Switzerland
Printed on acid-free paper produced from chlorine-free pulp. TCF ∞
Printed in Germany

ISBN 3-7643-5181-0
ISBN 0-8176-5181-0

9 8 7 6 5 4 3 2 1

Special Recognition

The 1995-96 Academic Year is being celebrated as the sesquicentennial year of the School of Medicine and Biomedical Sciences at the State University of New York at Buffalo, Buffalo, New York. During this auspicious celebration, the Editors of this volume would like to congratulate and to recognize this Medical School for its continuing commitment to excellence in Biomedical Research and to wish the School success in making significant research contributions in life sciences in the years to come.

Contents

Preface

This volume brings together wide ranging findings of a group of international researchers investigating the α-keto acid dehydrogenase complexes. This ubiquitous family of related but very diverse multienzyme systems function at strategic points in carbohydrate metabolism (pyruvate dehydrogenase complex), in the citric acid cycle (α-ketoglutarate dehydrogenase complex), in amino acid catabolism (branched-chain α-keto acid dehydrogenase complex). These systems are among the largest and most complicated of multienzyme complexes with molecular weights ranging upward to 9 million daltons. Although an intense research target for more than 30 years, the α-keto acid dehydrogenase complexes have been particularly rewarding subjects of investigation in recent years due to advances using molecular biology and biophysical approaches. Recombinant forms of components have been produced and their functions dissected by reconstruction approaches. Past and recent results can be comprehensively interpreted and tested based on insights from three-dimensional structures of components and component domains. The possibility that the organization of these complexes yields a molecular process not attained in isolated enzyme systems is continually supported by findings in this volume. Advanced explanations have been gained into the exquisite mechanisms that facilitate the integrated channeling of reactants between component enzymes. Unique mitochondrial protein kinases and protein phosphatases, dedicated to the regulation of these complexes, are shown to participate in unprecedented mechanisms in responding to a wide variety of metabolites and hormone signals. The chapters reveal how the specific metabolic demands of bacteria, yeast, plants, lower and higher animals are met by the interplay of specialized molecular features imposed on a framework of common structural motifs. The delineation of gene structures has begun to reveal how expression at varied sites operates to yield integrated, hormone responsive expression of components at the desired level. With this enhanced understanding, novel studies are increasingly successful in explaining medical problems involving inherited defects in components, component-directed autoimmune responses, and altered regulation in diseased states. The reader is presented with an overview of recent advances in these areas as well as exciting questions that remain to be answered.

The Editors wish to thank each contributor for their authoritative and insightful reviews. Their combined effort has created an important resource for both beginning students and established investigators in this and other fields. For some time, it has been apparent that highly organized systems, in which several enzymes operate in concert, are prevalent throughout cells. We believe that the broad scope of the progress on these very complex enzyme systems will influence future research in diverse areas.

Mulchand S. Patel
Thomas E. Roche
Robert A. Harris   August 1995

Alpha-Keto Acid Dehydrogenase Complexes
M.S. Patel, T.E. Roche and R.A. Harris (eds)
© 1996 Birkhäuser Verlag Basel/Switzerland

Interaction of protein domains in the assembly and mechanism of 2-oxo acid dehydrogenase multienzyme complexes

R.N. Perham

Cambridge Centre for Molecular Recognition, Department of Biochemistry, University of Cambridge, Tennis Court Road, Cambridge CB2 1QW, England, UK

Summary. The structures of the lipoyl domain and the peripheral subunit-binding domain of the dihydrolipoyl acyltransferase component of 2-oxo acid dehydrogenase multienzyme complexes are described. Unravelling the interactions of these domains with the other enzymes in the complexes is crucial to our understanding of the multi-step catalytic mechanism and is throwing new light on the underlying processes of biomolecular assembly.

Introduction

The 2-oxo acid dehydrogenase multienzyme complexes constitute an almost ubiquitous family of enzymes that catalyse the oxidative decarboxylation of various 2-oxo acid substrates in glycolysis, the citric acid cycle and branched-chain amino acid metabolism, generating the corresponding acyl CoA and NADH (for previous reviews, see Patel and Roche, 1990; Perham, 1991 and Mattevi et al., 1992a). In the pyruvate dehydrogenase (PDH) complex, the three enzymes that make up the assembly are: pyruvate decarboxylase [pyruvate dehydrogenase (lipoamide); E1p; EC 1.2.4.1], dihydrolipoyl acetyltransferase (E2p, EC 2.3.1.12) and dihydrolipoyl dehydrogenase (E3; EC 1.8.1.4). For the 2-oxoglutarate dehydrogenase (2OGDH) complex, the corresponding enzymes are 2-oxoglutarate decarboxylase (E1o, EC 1.2.4.2), dihydrolipoyl succinyltransferase (E2o, EC 2.3.1.61) and dihydrolipoyl dehydrogenase (E3, EC 1.8.1.4). The three short branched-chain 2-oxo acids produced by transamination of the amino acids leucine, isoleucine and valine are oxidatively decarboxylated by a single multienzyme complex (BCDH) of comparable structure.

The structural core of each complex is formed from multiple copies of the relevant E2 polypeptide chain arranged with octahedral (24-mer) or icosahedral (60-mer) symmetry, depending on the source (Reed and Hackert, 1990; Perham, 1991). The peripheral components, E1 and E3, are also present in multiple copies, and are bound tightly but non-covalently to the surface of the E2 core. The E2 chain has a pronounced domain-and-linker structure, (Reed and Hackert, 1990; Perham, 1991) indicated schematically in Figure 1. A large (approximately 28 kDa) C-terminal inner core domain houses the active site responsible for the acyltransferase activity and its aggregation dictates the octahedral or icosahedral symmetry of the assembled complex. The structure of the octahedral inner core of the PDH complex from *Azotobacter vinelandii* has recently been solved to 2.6 Å resolution by means of X-ray crystallography (Mattevi et al., 1992b). On the N-terminal side of the inner core (acyltransferase) domain and separated from it by an inter-domain segment of polypeptide chain that is sensitive to proteolysis is a smaller domain, responsible at least in part for binding the peripheral E1 and E3 subunits. The structures of this peripheral subunit-binding domain (remarkably small at approximately 35 amino acid residues) from the E2o chain

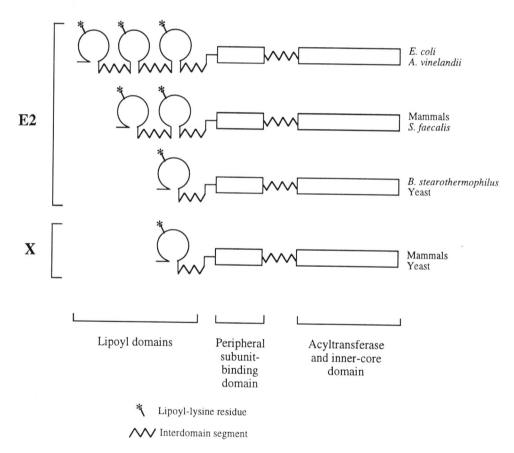

Figure 1. The segmented structure of the E2 polypeptide chain. The N-terminal lipoyl domains (approximately 80 residues) are followed by the peripheral subunit-binding domain (approximately 35 residues) and the C-terminal acyltransferase (inner-core) domain. The domains are connected by flexible but extended linker regions (approximately 25–30 residues). The number of lipoyl domains varies with the origin of the E2 chain. The segmented structure of protein X is shown for comparison.

of the 2-oxoglutarate dehydrogenase complex of *Escherichia coli* (Robien et al., 1992) and the E2p chain of the PDH complex from *Bacillus stearothermophilus* (Kalia et al., 1993) have been solved by means of NMR spectroscopy.

The N-terminal portion of the E2 chain comprises one or more lipoyl domains, each approximately 80 amino acids long and each carrying a lipoic acid prosthetic group. The number of lipoyl domains per E2 chain, like the symmetry of the assembled E2 core, depends on the source, varying from one in the icosahedral PDH complexes of yeast and the Gram-positive *Bacillus* species, to two in the icosahedral PDH complexes from mammalian mitochondria and the Gram-positive *Streptococcus faecalis*, to three in the octahedral PDH complexes of Gram-negative bacteria (Perham, 1991; Perham and Packman, 1989). The lipoyl group is attached in amide linkage to the N^6-amino group of a lysine residue and acts as a 'swinging arm', ferrying the substrate between the active sites of the complex (Reed, 1974). The three-dimensional structures of the lipoyl domain from the *B. stearothermophilus* (Dardel et al., 1993) and *E. coli* (Green et al., 1995) PDH complexes have been determined by means of NMR spectroscopy. The structure appears to form a new protein module and has been predicted to recur as the biotinylated domain of ATP-dependent carboxylases (Brocklehurst and Perham, 1993). The lipoyl domains are separated from one another and from the peripheral subunit-binding domain by further long (20–30 amino acid residues) segments of proteolytically sensitive polypeptide chain.

The octahedral E2 cores of the PDH complexes of eukaryotes, yeast and mammals, are further complicated by the presence of a few (about 6) copies of an additional protein, designated protein X, which contains a lipoyl domain and a peripheral subunit-binding domain, but whose C-terminal domain has no acetyltransferase activity or sequence similarity with any other known protein (Reed and Hackert, 1990; Mattevi et al., 1992a). Protein X is required for the proper association of the E3 component in yeast and mammalian PDH complexes, and its lipoyl group can participate in the catalytic reaction.

In the assembled 2-oxo acid dehydrogenase complexes, the lipoyl domains and the peripheral subunit-binding domain extend outwards from the inner E2 core, interdigitating between the E1 and E3 components (Reed and Hackert, 1990; Perham, 1991; Hale et al., 1992; Roche et al., 1993). The initial decarboxylation of the 2-oxo acid by the E1 component and the final reoxidation of the dihydrolipoyl group by the E3 component take place outside the E2 core, whereas the acyltransferase reaction takes place in the inner E2 core. The inter-domain segments in the E2 chain appear to be extended yet flexible, as judged by the results of NMR spectroscopy (Perham, 1991; Green et al., 1992; Machado et al., 1993). This flexibility is likely to play an important part in the mechanism of the 2-oxo acid dehydrogenase complexes, notably in the system of active site coupling by means of which the substrate is transferred between the three active sites that must act successively in the overall complex reaction (Reed and Hackert, 1990; Perham, 1991).

Thus there are interesting questions to be answered about how the peripheral subunits come to be bound by the E2 core, how the substrate is ferried between the physically separated active sites, whether there is any system of substrate chanelling in the complex, and if so how this is achieved. Ultimately such questions will only be answered in full by the determination of the complete three-dimensional structure of one or more 2-oxo acid dehydrogenase complexes. So far, probably because of their structural complexity and inherent conformational flexibility, particularly in the E2 core, the intact complexes have resisted all attempts to crystallize them. The domain-and-linker structure of the E2 chain and the application of the techniques of molecular biology and protein engineering open up another approach, that of piecing together the structure and mechanism from a study of isolated components and sub-assemblies.

In the present article I review the results of recent experiments that have thrown new light on the mechanism and assembly of these multienzyme complexes, which have long served as paradigms of self-assembling biomolecular structures. The holy grail of a complete three-dimensional structure can now reasonably be contemplated.

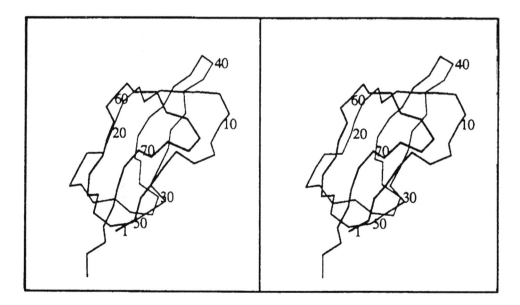

Figure 2. Stereo representation of the structure of a lipoyl domain from the *E. coli* E2p chain. The N- and C-terminal amino acids are at the bottom, the β-turn with the lipoyl-lysine residue (position 40 in the *E. coli* domain) at the top. From Green et al. (1995).

Structure of the lipoyl domain

A sub-gene encoding the N-terminal 80 or so amino acids (Ala1-Phe85) of the native E2p polypeptide chain from the *B. stearothermophilus* PDH complex has been created and over-expressed in *E. coli*, permitting the determination of the structure by means of NMR spectro-scopy (Dardel et al., 1993). A similar approach has subsequently permitted the determination of the structure of a lipoyl domain from the *E. coli* E2p chain (Green et al., 1995). The domain is composed of two four-stranded β-sheets making a flattened β-barrel, with a core of well-defined hydrophobic residues. The polypeptide chain weaves backwards and forwards between the two sheets, except for a single type-I β-turn in one sheet which incorporates the lipoyl-lysine residue, position 42 in the *B. stearothermophilus* domain and 40 in the *E. coli* domain, at its tip (Fig. 2). The N- and C-terminal amino acids (Ala1 and Gly79 in the *B. stearothermophilus* domain and Ala1 and Gly77 in the *E. coli* domain) of the structured region are found close together in space

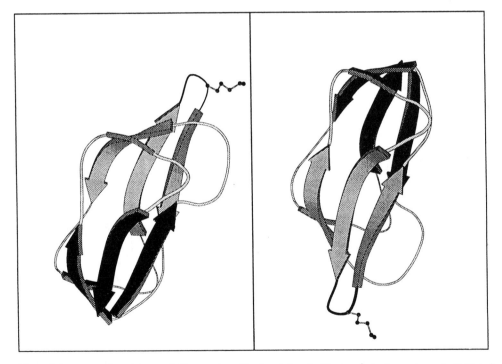

Figure 3. Representation of the two-fold symmetry of the lipoyl domain. The two β-sheets are shown in light and dark shading, the β-turn containing the lipoyl-lysine residue is shown in black. The orientation of the left-hand structure is the same as that of the stereo representation in Figure 2. Strands of the light sheet: residues 14 to 19, 23 to 28, 41 to 47 and 62 to 64; strands of the dark sheet: residues 1 to 6, 25 to 27, 51 to 57 and 69 to 75. The figure was prepared using the program MOLSCRIPT (Kraulis, 1991). From Green et al. (1995).

on the opposite side of the molecule, in the other β-sheet; the remaining 6 – 8 amino acids at the C-terminal end of the molecule are not localized in the structure and appear to form part of the first long flexible region that links the lipoyl domain to the peripheral subunit-binding domain in the intact E2p chain (Green et al., 1992). Another important feature of the barrel structure is its two-fold axis of quasi-symmetry (Fig. 3). This reflects a weak similarity between the N- and C-terminal halves of the amino acid sequence noticed earlier (Spencer et al., 1984).

The amino acid sequences of all lipoyl domains thus far published share certain key residues required for the folding, suggesting that they will likewise adopt very similar β-barrel structures (Dardel et al., 1993). Further support for this comes from the derivation of the secondary structure of the N-terminal lipoyl domain from the *A. vinelandii* PDH complex (Berg et al., 1994), also by means of NMR spectroscopy. As predicted (Brocklehurst and Perham, 1993), the lipoylated H-protein of the multifunctional glycine decarboxylase system turns out to have a closly comparable crystal structure (Pares et al., 1994), despite the low level of amino acid sequence identity between lipoyl domains and H-proteins. Moreover, it turns out that it is possible to build a plausible structure for a putative biotinyl domain from the amino acid sequence of yeast pyruvate carboxylase, based on that of the lipoyl domain, in which the biotinyl-lysine residue is correspondingly perched at the tip of an exposed β-turn (Brocklehurst and Perham, 1993). This mode of displaying the prosthetic group may be important to the biological function of these 'swinging arms' and can be expected to play a significant part in the mechanism of active-site coupling (see below).

Interactions of the lipoyl domain

The lipoyl domain is recognised *in vivo* specifically by two enzymes: the lipoylating enzyme(s) of the cell and the E1 component of the 2-oxo acid dehydrogenase complexes. The lipoylating enzyme, a lipoate protein ligase, attaches the lipoic acid to the domain (Brookfield et al., 1991; Morris et al., 1994), selecting both the domain and the specific lysine residue that is to be lipoylated. Lipoylation causes no detectable change of conformation, as judged by NMR spectroscopy of the lipoylated and unlipoylated domain from the *B. stearothermophilus* (Dardel et al., 1990) and *A. vinelandii* (Berg et al., 1994) PDH complexes, in keeping with the function of the lipoyl-lysine residue as a swinging arm which is effectively in free solution (Perham, 1991).

This makes it all the more surprising that the E1p component of the *E. coli* PDH complex requires the presence of the entire lipoyl domain for the reductive acetylation of the pendant lipoyl group, despite the fact that the dithiolane ring at the end of the swinging arm is approximately

1.4 nm from the backbone of the E2p polypeptide chain. Thus, E1p cannot use free lipoic acid, lipoamide, or a lipoylated decapeptide with an amino acid sequence identical to that surrounding the lipoyl-lysine residue as substrate, whereas an intact lipoyl domain functions well, the value of k_{cat}/K_m being raised by a factor of 10^4 (Graham et al., 1989). Moreover, the E1o component from the *E. coli* 2-oxoglutarate dehydrogenase complex cannot recognise the lipoyl domain from the *E. coli* PDH complex and vice versa, implying a specific molecular recognition between the lipoyl domain and the E1 component of the parent 2-oxo acid dehydrogenase complex (Graham et al., 1989). The reactions catalysed by E2 and E3 appear to involve no such specific interaction, since they can both take place with free lipoic acid as substrate (Reed et al., 1958).

In lipoyl domains from many different sources, the amino acid residues on either side of the lipoylated lysine are highly conserved as the motif DKA (Dardel et al., 1993). These residues form part of the exposed turn containing the lipoyl-lysine residue (Fig. 1). Site-directed mutagenesis of a sub-gene encoding the *B. stearothermophilus* domain indicates that they can be varied substantially (Asp41 changed to alanine, glutamate and lysine and Ala43 to methionine, glutamate and lysine) without obvious effect on the ability of the domain to be lipoylated *in vivo* in *E. coli* (Wallis and Perham, 1994). The conserved aspartic acid and alanine residues are therefore not important for recognition of the lipoyl domain by the lipoate protein ligase. The D41K and A43K mutations place a second lysine residue on the N- and C-terminal sides, respectively, of the original target lysine, in each case generating two potentially lipoylatable side-chains in adjacent positions. However, only one lipoyl group is attached *in vivo* and that is found on the natural lysine residue (Lys42). Similarly, if the target lysine residue is moved one place to the N-terminal or C-terminal side of its native position, it fails to become lipoylated. These experiments imply that there is an exquisitely specific interaction between the apo-lipoyl domain and lipoate protein ligase, such that only the N^6-amino group of a lysine residue at the very tip of the β-turn in the lipoyl domain is selected for post-translational modification (Wallis and Perham, 1994).

The recognition of the domain by the E1 component of the PDH complex does, however, appear to be dependent on the side-chains surrounding the lipoyl-lysine residue. Replacement of the aspartic acid and alanine residues in the DKA motif causes a decrease in the rate of reductive acetylation of the *B. stearothermophilus* PDH domain by the cognate E1 (Wallis and Perham, 1994). The aspartic acid residue is at the mouth of a cleft in the structure (Dardel et al., 1993) and substitution of large residues at this position may obstruct access to other residues within the cleft which are recognised by E1. Replacement of the alanine residue by residues with charged side-chains, lysine and glutamate, leads to a decrease in the rate of reductive acetylation whereas substitution of the larger methionine residue is without effect, suggesting that the alanine residue may be involved in a hydrophobic interaction with E1 (Wallis and Perham, 1994).

The E1 component

The E1 component of 2-oxo acid dehydrogenase complexes comes in two forms. In the octahedral PDH and the 2OGDH (all of which thus far are octahedral) complexes, the E1 component exists as an α_2 homodimer, the subunit M_r of which is about 100 000 Da. In contrast, the icosahedral PDH and the BCDH (all of which thus far are octahedral) complexes, E1 is an $\alpha_2\beta_2$ heterotetramer, with subunit M_r values of ca. 41 000 Da and 36 000 Da, respectively (Patel and Roche, 1990; Perham, 1991). In the heterotetrameric form, the α-chain is believed to be involved in binding the essential cofactor, thiamin diphosphate (ThDP), whereas the β-chain is responsible for binding the E1 component to the E2 core (Stepp and Reed, 1985; Wynn et al., 1992; Lessard and Perham, 1995).

We know very little yet about the structure of E1 or that of its active site, although a common amino acid sequence motif has been detected in all ThDP-dependent enzymes, including the E1 (E1α) chains of all 2-oxo acid dehydrogenase complexes, and predicted to be involved in ThDP binding (Hawkins et al., 1989). This prediction has been substantiated and extended by the determination of the three-dimensional structures of transketolase (Lindqvist et al., 1992), pyruvate oxidase (Muller and Schulz, 1993) and pyruvate decarboxylase (Dyda et al., 1993), and by the results of site-directed mutagenesis of pyruvate decarboxylase (Diefenbach et al., 1992), all ThDP-dependent enzymes. Chemical modification experiments have identified a cysteine residue in the ox E1α chain (Ali et al., 1993) and a tryptophan residue in the human E1β chain (Ali et al., 1995) that are also associated with thiamin diphosphate binding.

Further structural studies of E1 and of its interaction with the lipoyl domain will be needed to uncover the detail of the protein-protein interactions involved in the reductive acylation reaction. The availability of large amounts of E1 component from the cloning and over-expression in *E. coli* of the structural genes for the E1α and E1β chains of the mammalian BCDH complex (Davie et al., 1992), and the *B. stearothermophilus* (Lessard and Perham, 1994) and human (Korotchkina et al., 1995) PDH complexes should greatly facilitate this.

In mammalian PDH complexes, there is a regulatory mechanism that involves reversible inactivation by phosphorylation and dephosphorylation of the E1 component (Patel and Roche, 1990). The ATP-dependent kinase is complex-specific and appears to be tightly bound by the inner of the two lipoyl domains in the E2p chain of such complexes, and the binding may involve the lipoyl group itself (Liu et al., 1995). This is referred to again below in the section devoted to enzyme mechanism and assembly, and is dealt with in detail in this volume (Roche et al.; Motokawa et al.).

Interaction of protein domains in the assembly and mechanism of . . .

9

Structure of the peripheral subunit-binding domain

The structures of synthetic peptides representing the peripheral subunit-binding domains of the *E. coli* E2o chain (Robien et al., 1992) and *B. stearothermophilus* E2p chain (Kalia et al., 1993) have been solved by means of NMR spectroscopy. The folded structure consists of approximately 35 residues in each instance, with the N-terminal and C-terminal regions of the polypeptide chain (5 – 10 amino acid residues) that lie outside the folded structure appearing more or less disordered in solution and presumably forming part of the proteolytically sensitive linker regions that connect the peripheral subunit-binding domain to the lipoyl domain(s) and acyltransferase domain, respectively (Fig. 1).

The folded region consists of two parallel α-helices of 8 – 10 amino acids, separated by a short extended strand, a short 3, 10 helix-like turn, and an irregular and more disordered loop. Despite its small size, the smallest yet described for any well-folded protein devoid of stabilizing

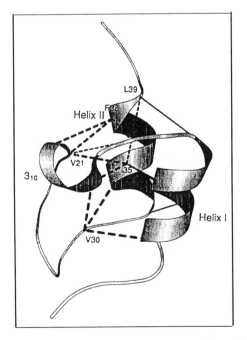
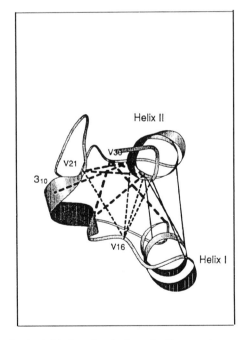

Figure 4. Representations of the structure of the peripheral subunit-binding domain from the *B. stearothermophilus* E2p chain. Some of the major hydrophobic contacts are depicted, to show how the hydrophobic core can be constructed despite the small size (approximately 35 amino acid residues) of the folded region. From Kalia et al. (1993).

disuphide bridges or bound metal ions, the domain has a significant close-packed hydrophobic core (Fig. 4). Indeed, the domain may be approaching the lower size limit for a three-dimensional structure that possesses such features characteristic of larger structures, which have hitherto been thought to require at least 50 amino acid residues to achieve (Privalov and Gill, 1988). The organization of the side-chains in the hydrophobic core are of particular interest. Close inspection of the peripheral subunit-binding domain of the B. stearothermophilus E2p chain suggests that two buried and highly conserved hydrophilic residues, Asp34 and Thr24 in Figure 4, participate in a set of side-chain-main-chain hydrogen bonds that are crucial to the domain stability (Kalia et al., 1993).

The peripheral subunit-binding domain has also been prepared by limited proteolysis (Hipps et al., 1994) from a di-domain (lipoyl domain and binding domain joined together by their natural linker region) encoded by a sub-gene over-expressed in E. coli (Hipps and Perham, 1992). NMR spectroscopy of the di-domain and its constituent domains indicate that the two domains fold independently and do not interact detectably (Hipps and Perham, 1992; Hipps et al., 1994).

Interactions of the peripheral subunit binding domain

As judged from the results of limited proteolysis and genetic reconstruction, in the octahedral BCDH (Wynn et al., 1992) and icosahedral PDH (Perham, 1991; Mattevi et al., 1992) complexes, the E1 component binds principally to the peripheral subunit-binding domain of the E2 chain. In contrast, in the octahedral 2OGDH and PDH complexes of E. coli, a major part at least of the binding site for E1 resides in the acyltransferase (inner-core) domain of the E2 chain (Perham and Packman, 1989; Perham, 1991). Likewise, mutagenesis and reconstruction of the gene encoding the A. vinelandii E2p chain suggests that in this octahedral PDH complex the E1 component is bound jointly by the peripheral subunit-binding domain and the acetyltransferase domain, perhaps between two E2p chains (Schulze et al., 1993).

The peripheral subunit-binding domain is also responsible for binding the E3 component to the E2 core in almost all 2-oxo acid dehydrogenase complexes, irrespective of their symmetry (Perham, 1991). The only significant exceptions are the icosahedral PDH complexes of eukaryotic origin, in which the E3-binding domain is located in protein X, an additional subunit in their E2 cores (Patel and Roche, 1990; Reed and Hackert, 1990), and the octahedral mammalian OGDH complex in which the E3-binding domain is found as an additional domain at the N-terminus of the E1 component (Rice et al., 1992). Thus, despite its very small size, the peripheral subunit-binding domain has a major part to play in binding two large enzymes (E1 and E3, both with M_r values of over 100 000) to the E2 core in most 2-oxo acid dehydrogenase complexes. This also poses some interesting problems of protein-protein interaction.

The cloning and expression in *E. coli* of a sub-gene encoding a di-domain comprising the *B. stearothermophilus* lipoyl domain attached by means of its native linker region to the peripheral subunit-binding domain (Hipps and Perham, 1992) has made it possible to tackle these problems directly for the *B. stearothermophilus* PDH complex. Thus it has been found that the *B. stearo-thermophilus* E1 and E3 components both bind tightly to the binding domain *in vitro* but with an unexpected stoichiometry of 1 mol of binding domain per mol of heterotetrameric ($\alpha_2\beta_2$) E1 (Lessard and Perham, 1995) or per mol of homodimeric (α_2) E3 (Hipps et al., 1994). The 1:1 stoicheiometry may be due to some form of steric hindrance or an anti-cooperative conformational change in the E3 dimer or E1 heterotetramer that renders the association of a second binding domain impossible when the first is in place. The simplest and most economical explanation (Lessard and Perham, 1995; Brocklehurst et al., 1994) is that in both E1 and E3 the binding site for the peripheral subunit-binding domain should lie on or close to the C_2 symmetry axis at the subunit interface (the E1β-E1β interface in the case of E1).

Another important feature of the interaction *in vitro* is that the binding of E1 and E3 to the *B. stearothermophilus* binding domain is mutually exclusive (Lessard and Perham, 1995). It is not yet clear how widely these observations apply to other icosahedral 2-oxo acid dehydrogenase complexes, in which the peripheral subunit-binding domain is responsible for binding both E1 and E3 (see above). However, competition in the binding of E1 and E3 to the intact octahedral E2 component of rat liver BCDH complex (Mori et al., 1994) and *E. coli* PDH complex (Reed et al., 1975) has been reported and ascribed potentially to competition for space on the surface of the E2 core. In the *B. stearothermophilus* PDH complex it may well be that there is insufficient space on the peripheral subunit-binding domain for E1 and E3 to bind simultaneously. Once more, it will be necessary to pursue detailed structural studies of the interaction of the domain with E1 and E3 to obtain fuller answers to these questions.

Implications for the assembly and mechanism of 2-oxo acid dehydrogenase complexes

The interactions of the lipoyl domain and peripheral subunit-binding domain with other component enzymes of the 2-oxo acid dehydrogenase complexes that have been described above have important implications for the assembly and the catalytic mechanism. The specificity of the interaction of the lipoyl domain with its cognate E1 component provides for recognition of the lipoyl group by the enzyme, E1, catalysing the first committed step (the irreversible decarboxylation of the 2-oxo acid) in the three-step reaction. Thus we have the molecular basis of an elegant system of substrate channelling (Perham, 1991). Part of the underlying structural explanation will almost certainly be found in the prominent display of the pendant lipoyl group at the very tip of the β-turn in the lipoyl domain, an unusual feature unexpectedly shared by the biotinyl-lysine swinging arm in the biotinyl domain of ATP-dependent carboxylases (Brocklehurst and Perham, 1993).

The catalytic mechanism of the 2-oxo acid dehydrogenase complexes involves the transfer of substrate between the three successive active sites by means of a network of interacting lipoyl groups on mobile lipoyl domains (Perham, 1991). Distant active sites can be coupled in this way, relieving the enzyme complex of the need to have the three active sites closely juxtaposed or symmetrically arranged. There is clear evidence of an elongated arrangement of the lipoyl and peripheral subunit-binding domains away from the C-terminal inner core domain to make this possible (Reed and Hackert, 1990; Perham, 1991; Hale et al., 1992; Roche et al., 1993). The lipoyl domains interdigitate between the E1 and E3 subunits where the first and third steps of the overall reaction are catalysed. The structure of the E2 inner core will dictate the way in which the lipoyl domains approach the acyltransferase active sites in the inner core domains for the middle reaction to take place and this is dealt with in detail elsewhere in this volume.

Given the competitive binding of E1 and E3 for the same peripheral subunit-binding domain in the *B. stearothermophilus* E2p chain (Lessard and Perham, 1995), it is difficult to see how the E1 and E3 molecules can be distributed with strict precision and symmetry over the surface of the E2 core unless there are structural constraints imposed by the three-dimensional structure of the inner core on the selection of E1 and E3 molecules for interaction with the peripheral subunit-binding domains. This will need to be carefully examined. Similarly, there are questions of the distribution of the protein X component in the E2 core of eukaryotic PDH complexes that need to be reconciled. For example, cryoelectron microscopy of frozen-hydrated specimens of ox heart and kidney PDH complex and E2 core suggest that the approximately six copies of protein X do not appear to be clustered and are not likely to be symmetrically distributed (Wagenknecht et al., 1991). In contrast, reassembly experiments with the yeast PDH complex *in vitro* indicate that one molecule of protein X carrying one E3 dimer is capable of binding to each of the 12 faces of the pentagonal dodecahedral E2 core (Maeng et al., 1994).

It is clear that the past few years have seen rapid strides in our understanding of the structure and assembly of the 2-oxo acid dehydrogenase complexes, multifunctional assemblies that continue to provide new insights into enzyme catalysis and biomolecular assembly processes. The stage is set for further interesting advances.

Acknowledgements
I thank The Wellcome Trust and the Biotechnology and Biological Sciences Research Council (formerly the Science and Engineering Research Council) for financial support. I am grateful to many colleagues for their valued contributions to the work from my laboratory and to Mr C. Fuller for skilled technical assistance.

References

Ali, M.S., Roche, T.E. and Patel, M.S. (1993) Identification of the essential cysteine residue in the active site of bovine pyruvate dehydrogenase. *J. Biol. Chem.* 268: 22353–22356.

Ali, M.S., Shenoy, B.C., Eswaran, D., Andersson, L.A., Roche, T.E. and Patel, M.S. (1995) Identification of the tryptophan residue in the thiamin pyrophosphate binding-site of mammalian pyruvate dehydrogenase. *J. Biol. Chem.* 270: 4570–4574.

Berg, A., De Kok, A. and Vervoort, J. (1994) Sequential ^{1}H and ^{15}N nuclear magnetic resonance assignments and secondary structure of the N-terminal lipoyl domain of the dihydrolipoyl transacetylase component of the pyruvate dehydrogenase complex from *Azotobacter vinelandii*. *Eur. J. Biochem.* 221: 87–100.

Brocklehurst, S.M. and Perham, R.N. (1993) Prediction of the three-dimensional structures of the biotinylated domain from the pyruvate carboxylase of yeast and of the lipoylated H-protein from the glycine cleavage system of pea leaf: a new, automated method for the prediction of protein tertiary structures. *Protein Sci.* 2: 626–639.

Brocklehurst, S.M., Kalia, Y.N. and Perham, R.N. (1994) Protein-protein recognition mediated by a mini-protein domain: possible evolutionary significance. *Trends Biochem. Sci.* 19: 360–361.

Brookfield, D.E., Green, J., Ali, S.T., Machado, R.S. and Guest, J.R. (1991) Evidence for two protein-lipoylation activities in *Escherichia coli*. *FEBS Lett.* 295: 13–16.

Dardel, F., Packman, L.C. and Perham, R.N. (1990) Expression in *Escherichia coli* of a sub-gene encoding the lipoyl domain of the pyruvate dehydrogenase complex of *Bacillus stearothermophilus*. *FEBS Lett.* 264: 206–210.

Dardel, F., Davis, A.L., Laue, E.D. and Perham, R.N. (1993) The three-dimensional structure of the lipoyl domain from *Bacillus stearothermophilus* pyruvate dehydrogenase multienzyme complex. *J. Mol. Biol.* 229: 1037–1048.

Davie, J.R., Wynn, R.M., Cox, R.P. and Chuang, D.T. (1992) Expression and assembly of a functional E1 component ($\alpha_2\beta_2$) of mammalian branched-chain α-keto acid dehydrogenase complex in *Escherichia coli*. *J. Biol. Chem.* 267: 16601–16606.

Diefenbach, R.J., Candy, J.M., Mattick, J.S. and Duggleby, R.G. (1992) Effects of substitution of aspartate-440 and tryptophan-487 in the thiamin diphosphate binding region of pyruvate decarboxylase from *ZymoMonas mobilis*. *FEBS Lett.* 296: 95–98.

Dyda, F., Furey, W., Swaminathan, S., Sax, M., Farrenkopf, B. and Jordan, F. (1993) Catalytic centers in the thiamin diphosphate dependent enzyme pyruvate decarboxylase at 2.4 Å resolution. *Biochemistry* 32: 6165–6170.

Graham, L.D., Packman, L.C. and Perham, R.N. (1989) Kinetics and specificity of reductive acylation of lipoyl domains from 2-oxo acid dehydrogenase multienzyme complexes. *Biochemistry* 28: 1574–1581.

Green, J.D.F., Perham, R.N., Ullrich, S.J. and Appella, E. (1992) Conformational studies of the inter-domain linker peptides in the dihydrolipoyl acetyltransferase component of the pyruvate dehydrogenase multienzyme complex of *Escherichia coli*. *J. Biol. Chem.* 267: 23484–23488.

Green, J.D.F., Laue, E.D., Perham, R.N., Ali, S.T. and Guest, J.R. (1995) Three-dimensional structure of a lipoyl domain from the dihydrolipoyl acetyltransferase component of the pyruvate dehydrogenase multienzyme complex of *Escherichia coli*. *J. Mol. Biol.* 248: 328–343.

Hale, G., Wallis, N.G. and Perham, R.N. (1992) Interaction of avidin with the lipoyl domains in the pyruvate dehydrogenase multienzyme complex: three-dimensional location and similarity to biotinyl domains in carboxylases. *Proc. R. Soc. Lond. B* 248: 247–253.

Hawkins, C.F., Borges, A. and Perham, R.N. (1989) A common structural motif in thiamin pyrophosphate-binding enzymes. *FEBS Lett.* 255: 77–82.

Hipps, D.S. and Perham, R.N. (1992) Expression in *Escherichia coli* of a sub-gene encoding the lipoyl and peripheral subunit-binding domains of the dihydrolipoyl acetyltransferase component of the pyruvate dehydrogenase complex of *Bacillus stearothermophilus*. *Biochem. J.* 283: 665–671.

Hipps, D.S., Packman, L.C., Allen, M.D., Fuller, C., Sakaguchi, K., Appella, E. and Perham, R.N. (1994) The peripheral subunit-binding domain of the dihydrolipoyl acetyltransferase component of the pyruvate dehydrogenase complex of *Bacillus stearothermophilus*: preparation and characterization of its binding to the dihydrolipoyl dehydrogenase component. *Biochem. J.* 297: 137–143.

Kalia, Y.N., Brocklehurst, S.M., Hipps, D.S., Appella, E., Sakaguchi, K. and Perham, R.N. (1993) The high resolution structure of the peripheral subunit-binding domain of dihydrolipoamide acetyltransferase from the pyruvate dehydrogenase multienzyme complex of *Bacillus stearothermophilus*. *J. Mol. Biol.* 230: 323–341.

Korotchkina, L.G., Tucker, M.M., Thekkumkara, T.J., Madhusudhan, K.T., Pons, G., Kim, H.J. and Patel, M.S. (1995) Overexpression and characterization of human tetrameric pyruvate dehydrogenase and its individual subunits. *Protein Expression and Purification* 6: 79–90.

Kraulis, P.J. (1991) MOLSCRIPT: a program to produce both detailed and schematic plots of protein structures. *J. Appl. Cryst.* 24: 946–950.

Lessard, I.A.D. and Perham, R.N. (1994) Expression in *Escherichia coli* of genes encoding E1α and E1β subunits of the pyruvate dehydrogenase complex of *Bacillus stearothermophilus* and assembly of a functional E1 component (α₂β₂) *in vitro. J. Biol. Chem.* 269: 10378–10383.

Lessard, I.A.D. and Perham, R.N. (1995) Interaction of component enzymes with the peripheral subunit-binding domain of the pyruvate dehydrogenase multienzyme complex of *Bacillus stearothermophilus*: stoicheiometry and specificity in self-assembly. *Biochem. J.* 306: 727–733.

Lindqvist, Y., Schneider, G., Ermler, U. and Sundström, M. (1992) Three-dimensional structure of transketolase, a thiamine diphosphate dependent enzyme, at 2.5 Å resolution. *EMBO J.* 11: 2373–2379.

Liu, S.J., Baker, J.C. and Roche, T.E. (1995) Binding of the pyruvate dehydrogenase kinase to recombinant constructs containing the inner lipoyl domain of the dihydrolipoyl acetyltransferase component. *J. Biol. Chem.* 270: 793–800.

Machado, R.S., Guest, J.R. and Williamson, M.P. (1993) Mobility in pyruvate dehydrogenase complexes with multiple lipoyl domains. *FEBS Lett.* 323: 243–246.

Maeng, C.Y., Yazdi, M.A., Niu, X.D. Lee, H.Y. and Reed, L.J. (1994) Expression, purification and characterization of the dihydrolipoamide dehydrogenase-binding protein of the pyruvate dehydrogenase complex from *Saccharomyces cerevisiae. Biochemistry* 33: 13801–13807.

Mattevi, A., de Kok, A. and Perham, R.N. (1992a) The pyruvate dehydrogenase complex. *Curr. Opin. Struct. Biol.* 2: 277–287.

Mattevi, A., Obmolova, G., Schulze, E., Kalk, K.H., Westphal, A., de Kok, A. and Hol, W.G.J. (1992b) Atomic structure of the cubic core of the pyruvate dehydrogenase multienzyme complex. *Science* 255: 1544–1550.

Mori, T., Oshima, Y. and Kochi, H. (1994) Binding properties of the components of branched-chain 2-oxo acid dehydrogenase complex on ELISA. *J. Biochem.* 116: 1111–1116.

Morris, T.W., Reed, K.E. and Cronan, J.E., Jr. (1994) Identification of the gene encoding lipoate-protein ligase A of *Escherichia coli. J. Biol. Chem.* 269: 16091–16100.

Muller, Y.A. and Schulz, G.E. (1993) Structure of the thiamine- and flavin-dependent enzyme pyruvate oxidase. *Science* 259: 965–967.

Pares, S., Cohen-Addad, C., Sieker, L., Neuburger, M. and Douce, R. (1994) X-ray structure determination at 2.6 Å resolution of a lipoate-containing protein: The H-protein of the glycine decarboxylase complex from pea leaves *Proc. Natl. Acad. Sci. USA* 94: 4850–5853.

Patel, M.S. and Roche, T.E. (1990) Molecular biology and biochemistry of pyruvate dehydrogenase complexes. *FASEB J.* 4: 3224–3233.

Perham, R.N. (1991) Domains, motifs and linkers in 2-oxo acid dehydrogenase multienzyme complexes: a paradigm in the design of a multifunctional protein. *Biochemistry* 30: 8501–8512.

Perham, R.N. and Packman, L.C. (1989) 2-oxo acid dehydrogenase multienzyme complexes: domains, dynamics and design. *In:* T.E. Roche and M.S. Patel (eds): *Alpha-Keto Acid Dehydrogenase Complexes: Organization, Regulation and Biomedical Ramifications,* Vol. 573. *Ann. N.Y. Acad. Sci.*, pp 1–20.

Privalov, P.L. and Gill, S.J. (1988) Stability of protein structure and hydrophobic interaction. *Adv. Prot. Chem.* 39: 191–234.

Reed, L.J., Koike M., Levitch, M.E. and Leach F.R. (1958) Studies on the nature and reactions of protein bound lipoic acid. *J. Biol. Chem.* 232: 143–158.

Reed, L.J. (1974) Multienzyme complexes. *Acc. Chem. Res.* 7: 40–46.

Reed, L.J., Pettit, F.J., Eley, M.H., Hamilton, L., Collins, J.H. and Oliver, R.M. (1975) Reconstitution of the *Escherichia coli* pyruvate dehydrogenase complex. *Proc. Natl. Acad. Sci. USA* 72: 3068–3072.

Reed, L.J. and Hackert, M.L. (1990) Structure-function relationships in dihydrolipoamide acyltransferases. *J. Biol. Chem.* 265: 8971–8974.

Rice, J.E., Dunbar, B. and Lindsay, J.G. (1992) Sequences directing dihydrolipoamide dehydrogenase (E3) binding are located on the 2-oxoglutarate dehydrogenase (E1) component of the mammalian 2-oxoglutarate dehydrogenase multienzyme complex. *EMBO J.* 9: 3229–3235.

Robien, M.A., Clore, G.M., Omichinski, J.G., Perham, R.N., Appella E., Sakaguchi K. and Gronenborn, A.M. (1992) Three-dimensional solution structure of the E3-binding domain of the dihydrolipoamide succinyltransferase core from the 2-oxoglutarate dehydrogenase multienzyme complex of *Escherichia coli. Biochemistry* 31: 3463–3471.

Roche, T.E., Powers Greenwood, S.L., Shi, W.F., Zhang, W.B., Ren, S.Z., Roche, E.D., Cox, D.J. and Sorensen, C.M. (1993) Sizing of bovine heart and kidney pyruvate dehydrogenase complex and dihydrolipoyl transacetylase core by quasi-elastic light scattering. *Biochemistry* 21: 5629–5637.

Schulze, E., Westphal, A.H., Hanemaaijer, R. and DeKok, A. (1993) Structure-function-relationships in the pyruvate dehydrogenase complex from *Azotobacter vinelandii*. Role of the linker region between the binding and catalytic domain of the dihydrolipoyl transacetylase component. *Eur. J. Biochem.* 211: 591–599.

Spencer, M.E., Darlison, M.G., Stephens, P.E., Duckenfield, I.K. and Guest, J.R. (1984) Nucleotide sequence of the sucB gene encoding the dihydrolipoamide succinyltransferase of *Escherichia coli* K12 and homology with the corresponding acetyltransferase. *Eur. J. Biochem.* 141: 361–374.

Stepp, L.R. and Reed, L.J. (1985) Active-site modification of mammalian pyruvate dehydrogenase by pyridoxal 5'-phosphate. *Biochemistry* 24: 7187–7191.

Wagenknecht, T., Grassucci, R., Radke, G.A. and Roche, T.E. (1991) Cryoelectron microscopy of mammalian pyruvate dehydrogenase complex. *J. Biol. Chem.* 266: 24650–24656.

Wallis, N.G. and Perham, R.N. (1994) Structural dependence of post-translational modification and reductive acetylation of the lipoyl domain of the pyruvate dehydrogenase multienzyme complex. *J. Mol. Biol.* 236: 209–216.

Wynn, R.M., Chuang, J.L., Davie, J.R., Fischer, C.W., Hale, M.A., Cox, R.P. and Chuang, D.T. (1992) Cloning and expression in *Escherichia coli* of mature E1β subunit of bovine mitochondrial branched-chain α-keto acid dehydrogenase complex. *J. Biol. Chem.* 267: 1881–1887.

Probing the active site of mammalian pyruvate dehydrogenase

L.G. Korotchkina, M.S. Ali and M.S. Patel

Department of Biochemistry, School of Medicine and Biomedical Sciences, State University of New York at Buffalo, Buffalo, NY 14214, USA

Introduction

The pyruvate dehydrogenase (E1) is the first catalytic component of the multienzyme pyruvate dehydrogenase complex (PDC). E1 catalyzes the two partial reactions: the thiamin pyrophosphate (TPP)-dependent decarboxylation of pyruvic acid to 2-hydroxyethylidene-TPP (HETPP) [Eq. 1] and reductive acetylation of lipoic acid residues covalently linked to the second catalytic component – dihydrolipoamide acetyltransferase (E2) [Eq. 2] (Reed, 1974):

$$CH_3COCO_2^- + E1\text{–}TPP \longrightarrow CH_3C(OH)=TPP\text{–}E1 + CO_2 \qquad [1]$$

$$CH_3C(OH)=TPP\text{–}E1 + E2\text{–}lipoate(S_2) \longrightarrow E1\text{–}TPP + E2\text{–}lipoate(SH)SCOCH_3 \quad [2]$$

Mammalian E1 is a heterotetramer ($\alpha_2\beta_2$) containing two 41 kDa α subunits and two 36 kDa β subunits. The role of individual subunits is not clear. It was proposed that E1α can catalyze the first partial reaction and E1β the second (Roche and Reed, 1972). E1α is thought to be involved in TPP binding as it contains TPP motif found by sequence comparison of several TPP requiring enzymes (Hawkins et al., 1989). Recent evidence indicates that both subunits are involved in TPP binding (Robinson and Chun, 1993; Ali et al., 1995; Korotchkina et al., 1995).

Eukaryotic E1 has two active sites with equal catalytic efficiency but with different affinity for the coenzyme as was determined by spectral analysis (Khailova and Korochkina, 1982) and by the E1 titration with transition state analog and equilibrium dialysis (Butler et al., 1977). The two active sites of E1 are proposed to operate in "alternating site mechanism", interacting with each other during catalysis (Khailova et al., 1989). The oxidation of HETPP takes place only in one active site at a time. The second site can catalyze this reaction after deacetylation of the first site is complete. The intersubunit interaction is essential for the regulation of E1 activity. This interaction is also displayed in cooperative behavior of E1 *versus* TPP and pyruvate; in different reactivity of essential amino acid residues belonging to the two active sites (Khailova et al., 1990) and during phosphorylation as phosphorylation of one active site in E1 is enough for complete inactivation of the enzyme (Yeaman et al., 1978). Three phosphorylation sites are identified in the E1α subunit and phosphorylation of the first site causes major inactivation (Yeaman et al., 1978). However, the role of the sites 2 and 3 is not well understood (Sale and Randle, 1982; Teague et al., 1979; Kerbey et al., 1981).

In this chapter, we will discuss the recent developments on (i) the identification of the active site residues of mammalian E1 using chemical modification procedure, (ii) characterization of the individually overexpressed subunits of E1 as well as tetrameric ($\alpha2\beta2$) E1 component, and (iii) the role of the three phosphorylation sites in the regulation of E1 using site-directed mutagenesis.

Identification of the active site residues of mammalian pyruvate dehydrogenase

Extensive studies from the laboratories of Dr. Khailova (Khailova et al., 1989) and Dr. Reed (Stepp and Reed, 1985) have identified by chemical modification several catalytically important amino acid residues such as tryptophan, cysteine, arginine, histidine and lysine in eukaryotic E1 (Tab. 1). However, the exact location of any one of these residues either in the α or β subunit of E1 was not known until very recently. From chemical modification studies of pigeon breast muscle (Khailova et al., 1983) and *Escherichia coli (E. coli)* E1 (Schwartz and Reed, 1970), a highly reactive cysteine residue located within or near the active site was indicated. Using sulfhydryl specific reagent N-ethylmaleimide, a reactive cysteine residue (Cys-62 in E1α) that is rapidly and exclusively modified with complete loss of E1 activity has been identified in bovine kidney E1 (Ali et al., 1993). Since TPP and pyruvate protect this inactivation, it is suggested that essential

Table 1. Identification by chemical modification of amino acid residues involved in catalysis of E1

Amino acid	Chemical used	Location	Predicted function
Cysteine	N-Ethylmaleimide 5, 5'-dithiobis (nitrobenzoate)	E1α Cys-62	Catalysis
Tryptophan	N-bromosuccinimide	E1β Trp-135	TPP-binding
Arginine	Pyreneglyoxal Phenyglyoxal 2, 3-butanedione	E1β Arg-239	Pyruvate-binding
Lysine	Pyridoxal phosphate	E1α (?)	TPP-binding
Histidine	Diethylpyrocarbonate		catalysis

cysteine is protected by the HETPP intermediate. The possibility of this cysteine to be localized at the active site was indicated by spectroscopic analysis for an acyl intermediate formed by pigeon breast E1 (Khailova et al., 1985). Flournoy and Frey (1989) have proposed the role of thiol group in *E. coli* E1 as chemical catalyst. The formation of HETPP and eventual decomposition of the adduct by E1 involves general acid-base catalysis and cysteinyl thiol group could contribute to catalytic process.

Tryptophan has been implicated in TPP binding from spectral analysis and chemical modification studies in a number of TPP-dependent enzymes like pigeon breast E1 (Khailova et al., 1989). To explore the specific localization of this tryptophan in the E1 component of mammalian PDC, an approach involving differential peptide mapping, reverse phase high performance liquid chromatography followed by sequence analysis was used. An essential tryptophan residue involved in TPP binding was identified at position 135 in the human E1β subunit (Ali et al., 1995). Both natural and magnetic circular dichroism studies indicate that TPP interacts with Trp of the E1β subunit, supporting the observation that there is formation of charge transfer complex between the thiazolium ring of TPP and indole ring of tryptophan (Korochkina et al., 1984).

Previously it has been reported that one arginine residue is important for the binding and decarboxylation of pyruvate but is not essential for further oxidative conversion of the HETPP intermediate (Nemerya et al., 1984). Using pyreneglyoxal, a fluoresecent analog of phenylglyoxal, and pyruvate and TPP as protective agents, it was possible to specifically modify an arginine residue of bovine E1 (Eswaran et al., 1995, in press). Sequence analysis of the modified peptide revealed that either Arg-239 or Arg-242 of E1β subunit is the essential arginine residue but not both. The results of Stepp and Reed (1985) provided evidence that lysine residues are important for coenzyme binding in bovine E1. This interaction apparently occurs on the α subunit. Using photooxidation and chemical modification, Khailova et al. (1989) observed that histidine residues are essential for catalysis. The most probable function of histidine residue in E1 catalysis is proton transfer during catalysis and the stabilization of the covalent intermediate between substrate and cofactor (Lindqvist et al., 1992). But the specific positions of critical lysine and histidine residues in the primary sequence of E1 have not yet been identified.

Genetic defects provide the possibility to identify essential amino acid residues and study their roles in the structure-function relationships (Dahl et al., 1992; Wexler et al., 1991). Approximately 20 point mutations in the α subunits have recently been reported, but their specific roles have largely remained undetermined.

Table 2. Properties of recombinant human E1, E1α and E1β

Assay or property	Bovine E1	Human E1	Human E1α	Human E1β
Recons. assay[1], U/mg	15.8	14.4	0	0
DCPIP reduction[2], mU/mg	n.d.	26.9	0.2	0
[$^{14}CO_2$] assay[3] (+ferrycianide), mU/mg	141.6	187.0	0.3	0
[$^{14}CO_2$] assay[3] (-ferricyanide), mU/mg	63.7	60.6	0.2	0
Acetoin production[4], mU/mg	46.5	44.0	0.1	1.4
TPP-binding measured by circular dichroism	n.d.	normal	none	none
Phosphorylation by E2-X-kinase	normal	normal	~12% of E1	none

Activity is measured by:
[1]NAD^+ reduction after reconstitution of E1 with purified E2-X-kinase subcomplex and E3;
[2]Reduction of 2, 6-dichlorophenolindophenol (DCPIP);
[3]Decarboxylation of [1-^{14}C]-pyruvate;
[4]Colorimetric determination of the products of acyloin condensation.
n.d.: not determined.

Characterization of overexpressed E1α, E1β and E1 (α2β2) proteins

During the last 5 years substantial progress has been achieved in cloning the α and β subunits of tetrameric (α2β2) E1 in the PDC from eukaryotes and Gram positive bacteria (Dahl et al., 1987; Ho and Patel, 1990; Hawkins et al., 1990). Availability of these clones have presented the opportunity to overexpress and purify E1 in bacterial systems. So far this has been achieved for PDC E1 of *Bacillus stearothermophilus* (Lessard and Perham, 1994) and human E1 (Korotchkina et al., 1995; Jeng et al., 1994).

For expression of human E1 and its individual subunits, pQE-9 was used which added a polyhistidine extension at the NH_2-terminus of the recombinant protein to rapidly purify by the affinity Ni-nitrilotriacetic-agarose chromatography (Korotchkina et al., 1995). The polyhistidine extension did not affect the E1 activity. Coexpression of the two subunits resulted in the formation of a functionally active and tetrameric E1 as determined by gel filtration on either Sepharose 6B-CL or TSKG 3000 PW gel filtration column. However, the individually expressed and purified subunits were less stable and were found mostly in inclusion bodies. Recombinant human E1 catalyzed all five reactions (including oxidative and nonoxidative decarboxylation of pyruvic acid) as efficiently as highly purified bovine kidney E1 (Tab. 2). However, the individually expressed E1α and E1β subunits did not display any significant activity in these reactions (Tab. 2). There was no reconstitution of functional E1 tetramers after mixing preparations of individually purified E1α and E1β subunits.

Recently, Jeng et al. (1994) have observed that coexpression of both subunits is necessary to produce an active enzyme. E1α subunit expressed alone did not possess catalytic activity. Coexpression of E1α and E1β was not necessary for production of the *Bacillus stearothermophilus* E1 (Lessard and Perham, 1994). Mixing the individual subunits expressed separately resulted in the formation of the tetrameric E1 as judged by gel filtration and activity recovery indicating that no chaperonin is needed to promote the assembly of individual E1α and E1β subunits of E1 from *B. stearothermophilus*. However, chaperonins did help in the assembly of mammalian branched-chain α-keto acid dehydrogenase when expressed in a bacterial system (Wynn et al., 1992). The individually expressed subunits of this protein were also unstable and formed aggregates which lacked decarboxylation activity when determined with separate or combined subunits (Davie et al., 1992). It was suggested that the folding and assembly of one or both subunits was dependent upon factors in the bacterial cytoplasm and on the other subunit of the tetramer. In other studies it has been observed that patients with genetic defects resulting in a loss of one subunit also show a loss of the other subunit (Wexler et al., 1992).

TPP binding to the pigeon breast muscle E1 is accompanied by the formation of a charge transfer complex with a tryptophan residue in the active site of E1 which causes broad band in the circular dichroism spectra at 330 nm (Khailova and Korochkina, 1982; Korochkina et al., 1991). Pyruvate binding results in the disappearance of this band and it is restored after protein deacetylation with dithiothreitol (Khailova et al., 1989). Recombinant human E1 displays the same spectral characteristics, i.e., binding of TPP results in formation of a charge transfer complex whose intensity changes during the catalytic reaction in the same way as is shown for the pigeon breast muscle enzyme (Korotchkina et al., 1995). The spectra of individual E1α or E1β differed from that for the E1 and there was no change of the spectra during the incubation of the subunits with TPP, suggesting that these subunits did not interact with TPP to form a charge transfer complex (Tab. 2).

The ability to undergo phosphorylation of the recombinant human E1α, E1β, and E1 by E2-X-kinase subcomplex provided additional insight (Korotchkina et al., 1995). Recombinant human E1 was phosphorylated in the expected way (Sugden and Randle, 1978) by incorporating nearly three phosphoryl groups per E1 molecule. E1α subunit alone also was phosphorylated by the E2-X-kinase subcomplex. However, the amount of radioactivity incorporated into E1α subunit alone was much less (about 12%) than that observed with E1 (Tab. 1). This is most likely due to a different conformation of E1α from the one present in tetrameric E1. Jeng et al. (1994) have also observed phosphorylation of the recombinant E1 by the E1-kinase.

E1 catalyzes two partial reactions: (i) decarboxylation of pyruvate and (ii) reductive acetylation of lipoic acid residues bound to E2. Earlier it was proposed that the decarboxylation reaction was catalyzed by the α subunit while the second partial reaction was performed by β (Roche and Reed, 1972). This suggestion was based on the observation that phosphorylation of E1 by E1-kinase inhibited only the first reaction and not the second. However, it was not possible to separate functionally active subunits under nondenaturing conditions. E1α subunit is thought to bind TPP, as the common motif of 30 amino acid residues is found in the sequence of E1α subunit for several TPP-binding enzymes (Hawkins et al., 1989), suggesting a similarity in cofactor binding. Further investigation on the crystal structures of three TPP-requiring enzyme, namely transketolase (TK) (Lindqvist et al., 1992; Nikkola et al., 1994), pyruvate decarboxylase (PD) (Dyda et al., 1993), pyruvate oxidase (POX) (Muller et al., 1994) which show no obvious primary sequence similarity confirms the conserved structure of TPP binding region as a common feature of other TPP enzymes (Fig. 1). Two different domains are required to bind the coenzyme, one binding the pyrophosphate group and the other the pyrimidine ring. All of these three enzymes are composed of identical subunits (2 for TK; 4 for PD and POX each). Each subunit of the enzymes contains both domains, however in all cases TPP is bound in a cleft between subunits so that the pyrophosphate domain belongs to one subunit, and the pyrimidine domain to the other subunit. Therefore, a dimer structure is the minimal functional unit for all three enzymes. Such a mode of binding was suggested by Robinson and Chun (1993) for human E1 based on alignment of amino acid sequences of human E1 with TK. Amino acid residues localized in the TPP-binding region belonging to the E1α and E1β sequences have been predicted (Fig. 1).

TPP-binding motif found by Hawkins et al. (1989) serves a role in anchoring TPP by binding the pyrophosphate on a divalent cation-dependent interaction. Thus they identified a critical part of the pyrophosphate-binding domain for TK, PD, POX. Structural alignment of TK, PD and

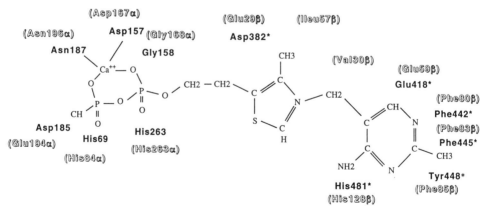

Figure 1. The predicted model for human E1 TPP binding site, based on sequence alignment (Robinson and Chun, 1993) and TK crystal structure (Lindqvist et al., 1992). The amino acid residues for human E1 are shown in parenthesis. Asterisk (*) denotes the residues of the other TK subunit.

POX (Muller et al., 1993) showed 10 conserved residues in the pyrophosphate domain: Gly68, Gly119, Gly156, Gly158, Asp157, Asn187, Leu177, Val217, Asp 202, Ile244 (numbers for TK sequence). Gly156 (Gly166α) (numbers for TK sequence, in parentheses for human E1 sequence), Gly158 (Gly168α), Asp157 (Asp167α), Asn187 (Asn196α) are the part of the TPP binding motif. Gly158 forms hydrogen bond to an oxygen of the diphosphate group. Asp157 and Asn187 in TK form the part of the metal-binding site. Mutagenic studies on *Zymomonas mobilis* PD and *E. coli* E1 confirmed the importance of several amino acid residues of TPP motif: Asp440 for PD (Asp157 for TK) (Diefenbach et al., 1992), Gly231 for *E. coli* E1 (Gly158 for TK) (Russell et al., 1992), Asn467 for PD (Asn187 for TK) (Candy and Duggleby, 1994). Two histidines: His69 (His84 of human E1α) and His263 (His263 of human E1α) of TK are shown to form salt bridges to the phosphates (Lindqvist et al., 1992). Other residues surrounding pyrophosphate moiety found in TK and predicted for human E1 are shown on Figure 1. The thiazolium ring bound between the pyrophosphate and pyrimidine domains has no conserved interaction with the proteins. The pyrimidine ring is buried in a hydrophobic pocket, none of these hydrophobic residues are conserved, for TK and predicted for E1 residues are shown on Figure 1. A very important conserved residue is Glu 418 of TK (Glu59 of E1β) that is involved together with His481 of TK (His128 of human E1β) in deprotonation of a C2 atom of thiazolium ring of TPP. The mechanism was tested and confirmed by site-directed mutagenesis of Glu418 on TK (Wikner et al., 1994). Based on their crystal structure, the thiazole C2 atom is the only one accessible to solution in the active site cleft of TK, PD, and POX. There is a narrow channel leading to C2 which is lined with conserved residues in TK including a cluster of histidines (30, 69, 103, 263, 481*; *denotes a residue from the other subunit) and hydrophobic side-chains (Ile191, Leu383*, Phe442*, Phe445*) at the bottom, Asp477*, Ser386*, His469* in the middle, and Arg359*, Arg528* near the entrance of the active site. The conformation of E1 can be different, as E1 does not bind TPP so tightly as TK, but E1 may reach the conformation revealed by the crystal structure for TK after the coenzyme binding. A large conformational change following TPP binding has been shown for E1 from pigeon breast muscle (Korochkina et al., 1991).

Based on (i) the study of the properties of the individual subunits (Korotchkina et al., 1995; Jeng et al., 1994; Davie et al., 1992), (ii) localization of the essential amino acid residues on both E1α and E1β subunits (Ali et al., 1993; Ali et al., 1995; Stepp and Reed, 1985), and (iii) prediction of the structure of the TPP-binding site (Robinson and Chun, 1993), one can conclude that both subunits are involved in active site formation. E1α subunit is the substrate for the E1-kinase regulating the PDC activity by phosphorylation; however, phosphorylation in the absence of E1β subunit is less effective (Korotchkina et al., 1995).

The role of the phosphorylation sites in the regulation of mammalian pyruvate dehydrogenase

PDC in many higher eukaryotes is regulated by phosphorylation-dephosphorylation carried out by specific regulatory enzymes: E1-kinase and phospho-E1-phosphatase, respectively (Reed, 1974; Patel and Roche, 1990; Linn et al., 1969a; Khailova et al., 1977; Randall et al., 1981; Thissen and Komuniecki, 1988). Three serine residues [site 1 (Ser-264), site 2 (Ser-271), and site 3 (Ser-203)] belonging to the E1α subunit of mammalian PDC are phosphorylated (Yeaman et al., 1978). Figure 2 shows the comparison of the sequences of PDC E1 from several different species. As can be seen, the sequences surrounding the three phosphorylation sites are almost the same for mammalian and nematode E1s (Yeaman et al., 1978; Sugden et al., 1979; Wexler et al., 1991; Cullingford et al., 1993; Fitzgerald et al., 1993; Johnson et al., 1992). *Neurospora crassa* E1 is also regulated by phosphorylation-dephosphorylation by concomitant kinase and phosphatase (Wieland et al., 1972). *Saccharomyces cerevisiae* PDC has no specific kinase and phosphatase; however, phosphorylation and dephosphorylation of the baker's yeast E1 of at least one site present in a homologous sequence can occur *in vitro* using bovine E1-kinase and phosphatase

Species (PDC-E1)	Sequence of sites 1 and 2		Sequence of site 3	Presence of kinase and phosphatase	Phospho-rylation *in vitro*
	1	2	3		
Human (s)	Y H G H **S** M S	D P G V **S** Y R	Y G M G T **S** V E R	+	+
Human (t)	Y H G H **S** M S	D P G V **S** Y R	Y G M G T **S** T E R	+	+
Pig (s)	Y H G H **S** M S	D P G V **S** Y R	Y G M G T **S** V E R	+	+
Bovine (s)	Y H G H **S** M S	N P G V **S** Y R	Y F M G T **S** V E R	+	+
Mouse (s)	Y H G H **S** M S	D P G V **S** Y R	Y G M G T **S** V E R	+	+
Mouse (t)	Y H G H **S** M S	D P G I **S** Y R	Y G M G T **S** N E R	+	+
Rat (s)	Y H G H **S** M S	D P G V **S** Y R	Y G M G T **S** V E R	+	+
Rat (t)	Y H G H **S** M S	D P G I **S** Y R	Y G M G T A I E R	+	+
Marsupials (s)	Y H G H **S** M S	D P G V **S** Y R	Y G M G T **S** V E R		
A. suum	Y S G H **S** M S	D P G T **S** Y R	Y G M G T **S** V E R	+	+
Plant	Y H G H **S** M S	D P G S T Y R	Y G M G T A E W R	+	+
S. cerevisiae	Y G G H **S** M S	D P G T T Y R	Y G M G T A A S R	-	+
B. stearother-mophilus	Y G P H T M S G D D P T R Y R		F A I S T P E V K	-	-
N. crassa	-		-	+	+
E. coli	F K S K D G A Y V R E H F F		Q R L D G P V T G	-	-

Figure 2. Amino acid sequences surrounding the phosphorylation sites in PDC-E1 from different species. (s) somatic form of E1α, (t) testis-specific form of E1α.

(Uhlinger et al., 1986; Behal et al., 1989). For bacterial E1 phosphorylation and the presence of the corresponding kinase and phosphatase were not found for *Bacillus stearothermophilus* which E1 is also has $\alpha_2\beta_2$ structure (there is no serine residues in position of sites 1, 2, 3 in this sequence) (Hawkins et al., 1990). The corresponding sequence for *Escherichia coli* E1 is completely different (it has structure of α_2 dimer) and there is no regulation by covalent modification. Plants have PDC in mitochondria and plastid, and both of them are regulated by phosphorylation-dephosphorylation; this regulation also depends upon light (Budde and Randall, 1990).

The role of individual phosphorylation sites in regulation of mammalian E1 has not been well defined. Three sites undergo phosphorylation in the presence of E2-X-kinase subcomplex at different rates as determined earlier using tryptic digestion and high-voltage paper electrophoresis (Yeaman et al., 1978; Kerbey and Randle, 1985). Phosphorylation of site 1 correlated with major inactivation of the enzyme, and half-of-the-site reactivity exhibited as phosphorylation of only one serine residue of one of the two α-subunits was sufficient for complete inactivation of the enzyme (Yeaman et al., 1978). However, it was difficult to determine the role of sites 2 and 3 in enzyme inactivation in the presence of the highly reactive site 1 (Sugden et al., 1979; Sale and Randle, 1982). It was suggested that site 2 phosphorylation results in only 0.7–6.4% of inactivation, and phosphorylation at site 3 was considered to be non-inactivating (Sale and Randle, 1982). Using thiophosphorylation of the sites 2 and 3, Teague et al. (1979) showed that these sites may be also inactivating. Also, it was not clear whether phosphorylation of the sites 2 and 3 can be catalyzed only after phosphorylation of site1 or the mechanism of phosphorylation is random. Sites 2 and 3 were suggested to be important for dephosphorylation and reactivation of E1 by the phospho-E1-phosphatase (Kerbey et al., 1981; Sugden et al., 1978).

Recombinant human E1 mutants having only site 1 (*mutated site MS2, 3*), site 2 (MS1, 3), site 3 (MS1, 2) or all possible combinations of two sites (MS1, MS2, MS3) generated by site-directed mutagenesis have allowed us to study the linkage between phosphorylation and inactivation and between dephosphorylation and activation separately for each site and in combination with the other sites (Korotchkina and Patel, 1995). Our recent findings can be summarized as follows:

(1) Each site present individually or in combination with the other sites can be phosphorylated by the E1-kinase bound to E2. The findings show that site 1 phosphorylation is not required for phosphorylation of sites 2 and 3.

(2) Phosphorylation of all mutants including enzymes with only site 1, site 2 or site 3 available results in enzyme inactivation (Fig. 3). The half-of-the-site reactivity during inactivation is revealed not only for site 1 but also for sites 2 and 3 as phosphorylation of one Ser residue (Ser-264, Ser-271 or Ser-203) per E1 tetramer causes complete inactivation of the enzyme assayed in reconstitution PDC assay. The maximal level of ^{32}P incorporation in our experiments was not

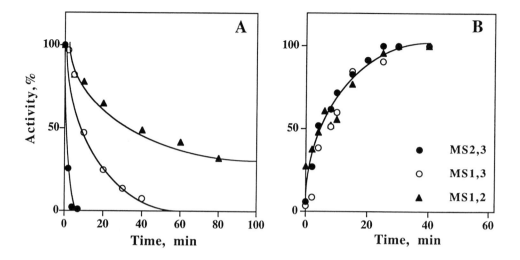

Figure 3. (A) Inactivation of E1 mutants during phosphorylation by the E2-X-kinase in the presence of 40 μM ATP. (B) Reactivation of E1 mutants during dephosphorylation by phospho-E1-phosphatase. Activity is measured by NAD$^+$ reduction after reconstitution of E1 with E2-X-kinase and E3. One hundred percent (100%) activity corresponds to the activity of each E1 mutant protein before phosphorylation. 10 μg of E1, 8 μg of E2 and 1 μg of phospho-E1-phosphatase (during dephosphorylation) were used per each reaction.

more than 1 mol ^{32}P/mol E1 for mutants with only one site, not more than two for mutants with two sites present while the maximum number of phosphoryl groups incorporated in E1 is not more than three, which corresponds to the previously reported data on highly purified porcine PDC (Sugden and Randle, 1978). As suggested previously (Sugden and Randle, 1978), the phosphorylation of one α-subunit can cause a conformational change making the sites on the other subunit unavailable or less available for the E1-kinase.

(3) The rates of phosphorylation and inactivation are site-specific. Site 1 has the highest rate of phosphorylation and inactivation, the rates are lower for site 2, the lowest for site 3 (Fig. 3). However phosphorylation of sites 2 and 3 is important for the E1 activity as it inactivates the enzyme.

(4) The role of the three sites in dephosphorylation and reactivation has been investigated using the site specific human E1 mutants. In previous studies, the rate of dephosphorylation was found to be higher for site 2 than site 3, and also higher for site 3 than site 1 (Teague et al., 1979; Kerbey et al., 1981). The mechanism of reactivation was not clear; it was proposed to be random (Teague et al., 1979) or that dephosphorylation of site 1 in fully phosphorylated E1 is not independent of that of site 2 (Kerbey et al., 1981). The E1 mutants are dephosphorylated by the phospho-E1-phosphatase with sites 1, 2, or 3, present individually or in combination with the

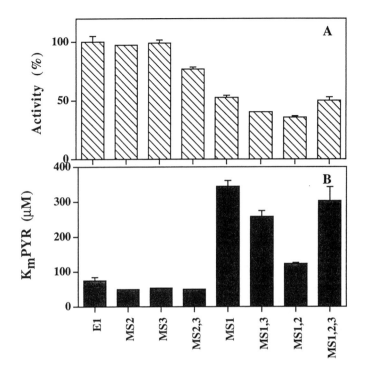

Figure 4. Activity of E1 mutants (A) and K_m for pyruvate (B) determined by NAD^+ reduction after the reconstitution of E1 with E2-X-kinase and E3. Values are mean ± S.E. of 2–5 measurements.

other sites. This finding supports a random (Teague et al., 1979) and not a sequential (Kerbey et al., 1981) mechanism for dephosphorylation of the three sites. The rates of dephosphorylation of sites 1, 2, and 3 in human E1 are similar. The dephosphorylation of all mutant E1 enzymes results in complete reactivation (Fig. 3). The enzymes having two and three sites needed longer incubation for complete reactivation than enzymes with only one site present because of the presence of two instead of one site (or three instead of two sites) being required to be dephosphorylated to activate the enzyme. The similar rates of reactivation for the mutants with sites 1, 2, and 3 (Fig. 3) clearly indicate that sites 1, 2, and 3 can be dephosphorylated independently of each other.

(5) The E1 mutants with the substitution of Ser271 or Ser203 to alanine has nearly the same specific activity as the wild type E1. The same change in position 264 (site 1) decreases the specific activity of the enzyme to 40–50% of the wild type enzyme (Fig. 4) and causes increase in K_ms for TPP and pyruvate for mutants with substitution in site 1 (Fig. 4).

The mechanism of E1 inactivation by phosphorylation is not well understood. Roche and Reed (1972) observed that phosphorylation of bovine E1 affected the decarboxylation, but not the oxidation step of the reaction. Also, phosphorylation increased K_D for TPP by about 12-fold and TPP in turn reduced the rate of the enzyme phosphorylation. Later it was suggested that phosphorylation produced a conformational change in E1 that displaced a catalytic group (or groups) at the active site (Butler et al., 1977). Study of the pig heart E1 revealed that phosphorylation inhibited all the partial reactions (forward and backward) leading to the formation of HETPP (Walsh et al., 1976). The recent spectral studies of pigeon breast muscle phospho-E1 showed that it binds TPP (with decreased affinity) in the active conformation with the formation of a charge transfer complex and is able to interact with HETPP by the alternating site mechanism, but could not bind the substrate pyruvate (Korochkina et al., 1993). It is suggested that a negative phosphoryl residue may compete for the active site arginine and thereby block the substrate binding.

It is possible that phosphorylation inhibits the activity at the very beginning of the catalytic reaction at the stage of the substrate binding. The mechanism by which this is achieved is not understood. In the absence of the crystal structure of the PDC-E1, one can only speculate about the possible localization of the E1 phosphorylation sites. If one hypothesizes E1 structure to be approximately similar to the three-dimensional structure of TK and takes into consideration the sequence alignment of Robinson and Chun (1993), then E1 phosphorylation sites 1 (Ser264), 2 (Ser271), and 3 (Ser203) can be placed at residues: 264, 271, 194 in TK sequence. His263 in E1 is proposed to be in the same position as His263 of TK, which is why residues 264 and 271 can be at the same positions and residue 194 of TK is close to the end of TPP-binding motif and corresponds to residue 203 of E1 by sequence alignment. These sites are close to the TPP binding site. Assuming these positionings are meaningful, it is not surprising that TPP can partly protect the enzyme against phosphorylation and phosphorylation, in its turn affects TPP binding. Three sites also appear to be not far from each other and rather close to the substrate channel proposed for TK, especially Ser 264 next to His263. One cannot exclude the possibility that the substrate channel also exists in E1 and can be closed by phosphorylation of any of these sites. The effect of phosphorylation on substrate binding was proposed for isocitrate dehydrogenase and branched-chain α-keto acid dehydrogenase (Dean and Koshland, 1990; Zhao et al., 1994). Phosphorylation cannot only electrostatically and sterically but also conformationally affect substrate binding. The induction of a conformational change upon phosphorylation is shown for glycogen phosphorylase (Sprang et al., 1988). Interestingly, Ser264 (site 1) is localized close to several essential histidine residues: His84α, His263α and His128β and the possibility of the mechanism similar to the one proposed for 3-hydroxy-3-methylglutaryl-CoA reductase according to which phosphorylation affects the ability of the catalytic histidine to protonate CoA (Omkumar and Rodwell, 1994) cannot be excluded. For example, Ser 264 may be involved in the deprotonation of C2 atom (and hence substrate binding) by interaction with His128β. The mechanism by which phosphorylation causes inactivation of E1 remains to be determined. Also, we do not know yet whether the mechanism of inactivation is the same for each of the three sites.

Concluding remarks

Although the three-dimensional structure of mammalian E1 has not yet been obtained recent studies on the structure-function relationship of E1 provide new insight on the active site structure and localization. Several essential amino acid residues in E1's active site are localized in the α or β subunits of mammalian E1 by chemical modification, namely: Cys-62 in the E1α subunit, Trp-135 and Arg-239 (or Arg-242) in the E1β subunit. The characterization of individually expressed E1α and E1β subunits have shown that both of them are required for catalytic activity and for the coenzyme binding. Comparison to other TPP-requiring enzymes for which there are three-dimensional structures indicates that the structure of the TPP binding site (and the active site itself) of E1 may be similar to the structure of the coenzyme binding site determined by the crystal studies for TK, PD, POX (i.e., the active site consisting of amino acid residues of both subunits). Site-directed mutagenesis of the three phosphorylation sites localized on the E1α subunit near the coenzyme binding site has demonstrated the importance of each of the three sites in regulation of the enzyme activity, as each of the sites can be phosphorylated by the E2-X-kinase resulting in complete inactivation and be dephosphorylated by the phospho-E1-phosphatase with complete reactivation of E1. The three-dimensional structure of E1 will provide new directions for the more detailed investigations of the E1 active site.

Acknowledgments
The work performed in this laboratory and reported in this review was supported by U.S. Public Health Service Grant DK20478.

References

Ali, S.M., Roche, T.E. and Patel, M.S. (1993) Identification of the essential cysteine residue in the active site of bovine pyruvate dehydrogenase. *J. Biol. Chem.* 268: 22353–22356.

Ali, M.S., Shenoy, B.C., Eswaran, D., Andersson, L.A., Roche, T.E. and Patel, M.S (1995) Identification of the tryptophan residue in the thiamin pyrophosphate binding site of bovine pyruvate dehydrogenase. *J. Biol. Chem.* 270: 4570–4574.

Behal, R.H., Browning, K.S. and Reed, L.J. (1989) Nucleotide and deduced amino acid sequence of the alpha subunit of yeast pyruvate dehydrogenase. *Biochem. Biophys. Res. Comm.* 164: 941–946.

Budde, R.J. and Randall, D.D. (1990) Pea leaf mitochondrial pyruvate dehydrogenase complex is inactivated *in vivo* in a light-dependent manner. *Proc. Natl. Acad. Sci. USA* 87: 673–676.

Butler, J.R., Pettit, F.H., Davis, P.F. and Reed, L.J. (1977) Binding of thiamin thiazolone pyrophosphate to mammalian pyruvate dehydrogenase and its effects on kinase and phosphatase activities. *J. Biol. Chem.* 74: 1667–1674.

Candy, J.M. and Duggleby, R.G. (1994) Investigation of the cofactor-binding site of *Zymomonas mobilis* pyruvate decarboxylase by site-directed mutagenesis. *Biochem. J.* 300: 7–13.

Cullingford, T.E., Clark, B. and Phillips, I.R. (1993) Characterization of cDNAs encoding the rat testis-specific E1α subunit of the pyruvate dehydrogenase complex: comparison of expression of the corresponding mRNA with that of the somatic E1α subunit. *Biochim. Biophys. Acta* 1216: 149–153.

Dahl, H.-H.M., Hunt, S.M., Hutcheson, W.M. and Brown, G.K. (1987) The human pyruvate dehydrogenase complex. Isolation of cDNA clones for the E1α subunit, sequence analysis, and characterization. *J. Biol. Chem.* 262: 7398–7403.

Dahl, H.-H.M., Brown, G.K., Brown, R.M., Hansen, L.L., Kerr, D.S., Wexler, I.D., Patel, M.S., De Meirleir, L., Lissens, W., Chun, K., MacKay, N. and Robinson, B.H. (1992) Mutations and polymorphism in the pyruvate dehydrogenase E1α gene. *Human Mutation* 1: 97–102.

Davie, J.R., Wynn, R.M., Cox, R.P. and Chuang, D.T. (1992) Expression and assembly of a functional E1 component ($\alpha_2\beta_2$) of mammalian branched-chain α-ketoacid dehydrogenase complex in *Escherichia coli*. *J. Biol. Chem.* 267: 16601–16606.

Dean, A.M. and Koshland, D.E. (1990) Electrostatic and steric contributions to regulation at the active site of isocitrate dehydrogenase. *Science* 249: 1044–1046.

Diefenbach, R.J., Candy, J.M., Mattick, J.S. and Duggleby, R.G. (1992) Effects of substitution of aspartate-440 and tryptophan-487 in the thiamin diphosphate binding region of pyruvate decarboxylase from *Zymomonas mobilis*. *FEBS Lett.* 296: 95–98.

Dyda, F., Furey, W., Swaminathan, S., Sax, M., Farrenkopf, B. and Jordan, F. (1993) Catalytic centers in the thiamin diphosphate dependent enzyme pyruvate decarboxylase at 2.4 Å resolution. *Biochemistry* 32: 6165–6170.

Eswaran, D., Ali, M.S., Shenoy, B.C., Korotchkina, L.G., Roche, T.E. and Patel, M.S. (1995) Arginine-239 in the beta subunit is at or near the active site of bovine pyruvate dehydrogenase. *Biochim. Biophys. Acta*; *in press*.

Fitzgerald, J., Hutcheson, W.M., Dahl, H.-H.M. (1992) Isolation and characterization of the mouse pyruvate dehydrogenase E1α genes. *Biochim. Biophys. Acta* 1131: 83–90.

Flournoy, D.S. and Frey, P.A. (1989) Inactivation of the pyruvate dehydrogenase complex of *Escherichia coli* by fluoropyruvate. *Biochemistry* 28: 9594–9602.

Hawkins, C.F., Borges, A. and Perham, R.N. (1989) A common structural motif in thiamin pyrophosphate-binding enzymes. *FEBS Lett.* 255: 77–82.

Hawkins, C.F., Borges, A.A. and Perham, R.N. (1990) Cloning and sequence analysis of the genes encoding the α and β subunits of the E1 component of the pyruvate dehydrogenase multienzyme complex of *Bacillus stearothermophilus*. *Eur. J. Biochem.* 191: 337–346.

Ho, L. and Patel, M.S. (1990) Cloning and cDNA sequence of the β-subunit component of human pyruvate dehydrogenase complex. *Gene* 86: 297–302.

Jeng, J., Huh, T.L. and Song, B.J. (1994) Production of an enzymatically active E1 component of human pyruvate dehydrogenase complex in *Escherichia coli*: supporting role of E1 beta subunit in E1 activity. *Biochem. Biophys. Res. Comm.* 203: 225–230.

Johnson, K.R., Komuniecki, R., Sun, Y. and Wheelock, M.J. (1992) Characterization of cDNA clones for the alpha subunit of pyruvate dehydrogenase from *Ascaris suum*. *Molec. Biochem. Parasit.* 51: 37–48.

Kerbey, A.L., Randle, P.J. and Kearns, A. (1981) Dephosphorylation of pig heart pyruvate dehydrogenase phosphate complexes by pig heart pyruvate dehydrogenase phosphate phosphatase. *Biochem. J.* 195: 51–59.

Kerbey, A.L. and Randle P.J. (1985) Pyruvate dehydrogenase kinase activity of pig heart pyruvate dehydrogenase (E1 component of pyruvate dehydrogenase complex). *Biochem. J.* 231: 523–529.

Khailova, L.S. and Korochkina, L.G. (1982) Determination of the number of active centres in the pyruvate dehydrogenase component of the pyruvate dehydrogenase complex from pigeon breast muscle. *Biochem. Int.* 5: 525–532.

Khailova, L.S., Aleksandrovich, O.V. and Severin, S.E. (1983) Study on the role of SH-groups in the activity of muscle pyruvate dehydrogenase. *Biochem. Int.* 7: 223–233.

Khailova, L.S., Aleksandrovich, O.V. and Severin, S.E. (1985) Substrate-dependent inactivation of muscle pyruvate dehydrogenase: identification of the acetyl-substituted enzyme form. *Biochem. Int.* 10: 291–300.

Khailova, L.S. and Gomazkova, V.S. (1986) α-Keto acid dehydrogenases from pigeon breast. *Biokhimiya (Russ.)* 51: 2054–2074.

Khailova, L.S., Korochkina, L.G. and Severin, S.E. (1989) Organization and functioning of muscle pyruvate dehydrogenase active centers. *Ann. N.Y. Acad. Sci.* 573: 36–54.

Khailova, L.S., Korochkina, L.G. and Severin, S.E. (1990) Intersite cooperativity in enzyme action of pyruvate dehydrogenase. *In*: H. Bisswanger and J. Ullrich (eds): *Biochemistry and Physiology of TDP Enzymes*, Blaubeuren, pp 251–265.

Korotchkina, L.G., Khailova, L.S. and Severin, S.E. (1984) Localization of tryptophan residues in thiamine pyrophosphate-binding sites of pyruvate dehydrogenase from pigeon breast muscle. *Biochem. Int.* 9: 491–499.

Korotchkina, L.G., Khailova, L.S. and Severin, S.E. (1991) Investigation of the pyruvate dehydrogenase complex using the circular dichroism method. *Biokhimiya (Russ.)* 56: 1840–1849.

Korotchkina, L.G., Khailova, L.S. and Severin, S.E. (1993) Effect of phosphorylation on the catalytic function of muscular pyruvate dehydrogenase complex. *Biokhimiya (Russ.)* 58: 1503–1512.

Korotchkina, L.G. and Patel, M.S. (1995) Mutagenic studies of the phosphorylation sites of recombinant human pyruvate dehydrogenase. Site-specific regulation. *J. Biol. Chem.* 270: 14297–14304.

Korotchkina, L.G., Tucker, M.M., Thekkumkara, T.J., Madhusudhan, K.T., Pons, G., Kim, H. and Patel, M.S. (1995) Overexpression and characterization of human tetrameric pyruvate dehydrogenase and its individual subunits. *Prot. Express. Purif.* 6: 79–90.

Lessard, I.A.D. and Perham, R.N. (1994) Expression in *Escherichia coli* of genes encoding the E1α and E1β subunits of the pyruvate dehydrogenase complex of Bacillus stearothermophilus and assembly of a functional E1 component (α2β2) *in vitro*. *J. Biol. Chem.* 269: 10378–10383.

Lindqvist, Y., Schneider, G., Ermler, U. and Sundstrom, M. (1992) Three-dimensional structure of transketolase, a thiamine diphosphate dependent enzyme, at 2.5 Å resolution. *EMBO J.* 11: 2373–2379.

Linn, T C., Pettit, F.H. and Reed, L.J. (1969) α-Keto acid dehydrogenase complexes. X. Regulation of the activity of the pyruvate dehydrogenase complex from beef kidney mitochondria by phosphorylation and dephosphorylation. *Proc. Natl. Acad. Sci. USA* 62: 234–241.

Muller, Y.A., Lindqvist, Y., Furey, W., Schulz, G.E., Jordan, F. and Schneider G. (1993) A thiamin diphosphate binding fold revealed by comparison of the crystal structures of transketolase, pyruvate oxidase and pyruvate decarboxylase. *Structure* 1: 95–103.

Muller, Y.A., Schumacher, G., Rudolph, R. and Schulz, G.E. (1994) The refined structures of a stabilized mutant and of wild-type oxidase from *Lactobacillus plantarum*. *J. Mol. Biol.* 237: 315–335.

Nemerya, N.S., Khailova, L.S. and Severin, S.E. (1984) Arginine residues in the active centers of muscle pyruvate dehydrogenase. *Biochem. Int.* 8: 369–376.

Nikkola, M., Lindqvist, Y. and Schneider, G. (1994) Refined structure of transketolase from *Saccharomyces cerevisiae* at 2.0 Å resolution. *J. Mol. Biol.* 238: 387–404.

Omkumar, R.V. and Rodwell, V.W. (1994) Phosphorylaton of Ser[871] impairs the function of His[865] of Syrian hamster 3-hydroxy-3-methylglutaryl-CoA reductase. *J. Biol. Chem.* 269: 16862–16866.

Patel, M.S. and Roche, T.E. (1990) Molecular biology and biochemistry of pyruvate dehydrogenase complexes. *FASEB J.* 4: 3224–3233.

Randall, D.D., Williams, M. and Rapp, B.J. (1981) Phosphorylation-dephosphorylation of pyruvate dehydrogenase complex from pea leaf mitochondria. *Arch. Biochem. Biophys.* 207: 437–444.

Reed, L.J. (1974) Multienzyme complexes. *Acc. Chem. Res.* 7: 40–46.

Robinson B.H. and Chun, K. (1993) The relationship between transketolase, yeast pyruvate decarboxylase and pyruvate dehydrogenase of the pyruvate dehydrogenase complex. *FEBS Lett.* 328: 99–102.

Roche, T.E. and Reed, L.J. (1972) Function of the nonidentical subunits of mammalian pyruvate dehydrogenase. *Biochem. Biophys. Res. Comm.* 48: 840–846.

Russell, G.C., Machado, R.S. and Guest, J.R. (1992) Overproduction of the pyruvate dehydrogenase multienzyme complex of *Escherichia coli* and site-directed substitutions in the E1p and E2p subunits. *Biochem. J.* 287: 611–619.

Sale, G.J. and Randle, P.J. (1982) Role of individual phosphorylation sites in inactivation of pyruvate dehydrogenase complex in rat heart mitochondria. *Biochem. J.* 203: 99–100.

Schwartz, E.R. and Reed, L.J. (1970) α-keto acid dehydrogenase complexes: Reaction of sulfhydryl groups in pyruvate dehydrogenase with organic mercurials. *J. Biol. Chem.* 245: 183–187.

Sprang, S.R., Acharya, K.R., Goldsmith, E.J., Stuart, D.I, Varvill, K., Fletterick, R.J., Madsen, N.B. and Johnson, L.N. (1988) Structural changes in glycogen phosphorylase induced by phosphorylation. *Nature* 336: 215–221.

Stepp, L.R. and Reed, L.J. (1985) Active-site modification of mammalian pyruvate dehydrogenase by pyridoxal 5'-phosphate. *Biochemistry* 24: 7187–7191.

Sugden, P.H. and Randle, P.J. (1978) Regulation of pig heart pyruvate dehydrogenase by phosphorylation. Studies on the subunit and phosphorylation stoichiometries. *Biochem. J.* 173: 659–668.

Sugden, P.H., Kerbey, A.L., Randle, P.J., Waller, C.A. and Reid, K.B.M (1979) Amino acid sequences around the sites of phosphorylation in the pig heart pyruvate dehydrogenase complex. *Biochem. J.* 181: 419–426.

Teague, W.M., Pettit, F.H., Yeaman, S.J. and Reed, L.J. (1979) Function of phosphorylation sites on pyruvate dehydrogenase. *Biochem. Biophys. Res. Comm.* 87: 244–252.

Thissen, J. and Komuniecki, R. (1988) Phosphorylation and inactivation of the pyruvate dehydrogenase from anaerobic parasitic nematode, *Ascaris suum*. *J. Biol. Chem.* 263: 119092–19097.

Uhlinger, D.J., Yang, C.-Y. and Reed, L.J. (1986) Phosphorylation-dephosphorylation of pyruvate dehydrogenase from baker's yeast. *Biochemistry* 25: 5673–5677.

Walsh, D.A., Cooper, R.H., Denton, R.M., Bridges, B.J. and Randle, P.J. (1976) The elementary reactions of the pig heart pyruvate dehydrogenase complex. A study of the inhibition by phosphorylation. *Biochem. J.* 157: 41–67.

Wexler, I.D., Hemalatha, S.G. and Patel, M.S. (1991) Sequence conservation in the α and β subunits of pyruvate dehydrogenase and its similarity to branched-chain α-keto acid dehydrogenase. *FEBS Lett.* 282: 209–213.

Wexler, I.D., Hemalatha, S.G., Liu, T.-C., Berry, S.A., Kerr, D.S. and Patel, M.S. (1992) A mutation in the E1α subunit of pyruvate dehydrogenase associated with variable expression of pyruvate dehydrogenase complex deficiency. *Pediat. Res.* 32: 169–174.

Wieland, O.H., Hartmann, U. and Siess, E.A. (1972) *Neurospora crassa* pyruvate dehydrogenase: interconversion by phosphorylation and dephosphorylation. *FEBS Lett.* 27: 240–244.

Wikner, C., Meshalkina, L., Nilsson, U., Nikkola, M., Lindqvist, Y., Sundstrom, M. and Schneider, G. (1994) Analysis of an invariant cofactor-protein interaction in thiamin diphosphate-dependent enzymes by site-directed mutagenesis. *J. Biol. Chem.* 269: 32144–32150.

Wynn, R.M., Davie, J.R., Cox, R.P. and Chuang, D.T. (1992) Chaperonins GroEL and GroES promote assembly of heterotetramers ($\alpha_2\beta_2$) of mammalian mitochondrial branched-chain α-keto acid decarboxylase in *Escherichia coli. J. Biol. Chem.* 267: 12400–12403.

Yeaman, S.J., Hutcheson, E.T., Roche, T.E., Pettit, F.H., Brown, J.R., Reed, L.J., Watson, D.C. and Dixon, G.H. (1978) Sites of phosphorylation on pyruvate dehydrogenase from bovine kidney and heart. *Biochemistry* 17: 2364–2370.

Zhao, Y., Hawes, J., Popov, K.M., Jaskiewicz, J., Shimomura, Y., Crabb, D.W. and Harris, R.A. (1994) Site-directed mutagenesis of phosphorylation sites of the branched chain α-ketoacid dehydrogenase complex. *J. Biol. Chem.* 269: 18583–18587.

Alpha-Keto Acid Dehydrogenase Complexes
M.S. Patel, T.E. Roche and R.A. Harris (eds)
© 1996 Birkhäuser Verlag Basel/Switzerland

Role of the E2 core in the dominant mechanisms of regulatory control of mammalian pyruvate dehydrogenase complex

T.E. Roche, S. Liu, S. Ravindran, J.C. Baker and L. Wang

Department of Biochemistry, Kansas State University, Manhattan, KS 66506, USA

System and feedback regulation

The mammalian pyruvate dehydrogenase complex (PDC) is a large assembly composed of six components with nine distinct subunits. Four of these components execute the overall reaction through a series of steps linked by cofactor-mediated active site coupling: the pyruvate dehydrogenase component (E1); the dihydrolipoyl acetyltransferase component (E2); the dihydrolipoyl dehydrogenase component (E3); and the E3-binding component (E3BP, previously protein X) (reviewed: Patel and Roche, 1990; Reed and Hackert, 1990; Perham, 1991). The pyruvate dehydrogenase complex has a central and strategic role in metabolism. Control of the PDC reaction is critically important for regulating cellular fuel utilization. PDC activity is controlled primarily by an intricately regulated phosphorylation-dephosphorylation cycle. This regulatory cycle is carried out by dedicated kinase and phosphatase components that operate to decrease and increase the activity of the complex through phosphorylating and dephosphorylating the E1 component, respectively. Here, E1a is used to designate the active form of E1 and E1b to designate the phosphorylated, inactive form.

Figure 1 shows the relationship of PDC reaction and its regulation to other metabolic pathways. In mammalian cells, flux through the PDC reaction controls the oxidative utilization of glucose (Randle, 1986) and results in a net depletion of body carbohydrate reserves by the PDC reaction. The activity of PDC must be reduced when fatty acids or ketone bodies are being preferentially used to provide acetyl-CoA for oxidative energy production by citric acid cycle/oxidative phosphorylation systems – a routine situation in many body organs. Furthermore, under conditions of starvation or diabetes, the activity of PDC is reduced to a minimal level to conserve carbohydrates essential for the operation of the brain and other specialized cells. This critical management role in cellular fuel utilization is primarily achieved through regulatory control of the phosphorylation/dephosphorylation cycle.

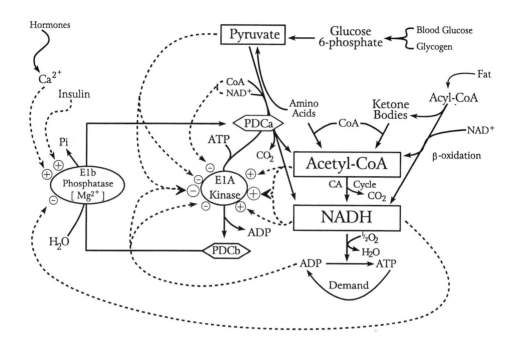

Figure 1. Relationship of the reaction and control of the pyruvate dehydrogenase complex to use of various fuels. The oxidative utilization of glucose-derived pyruvate by the active form of the pyruvate dehydrogenase complex, PDCa, has in common with catabolism of fatty acid and ketone bodies the production of the key metabolites acetyl-CoA and NADH. Acetyl-CoA is the substrate of the citrate acid cycle, CAC, and NADH is produced both by β-oxidation pathway and by CAC. The proportion of PDC in the inactive form, PDCb, is markedly enhanced by NADH and acetyl-CoA stimulating kinase activity while NADH reduces E1b phosphatase activity. These occur by the sensitive signal translation mechanisms described in this review. Regulation in the opposite direction results from direct effects of pyruvate and ADP on E1a kinase activity and Ca^{2+} on E1b phosphatase activity. The operation of these control mechanisms must respond to the needs of specific mammalian tissues that vary from preferential use of glucose (e.g., brain) to preferential use of fat (e.g., liver) to rapidly converting between these uses (e.g., muscle with level of exercise, changes in carbohydrate stores, degree of exercise – short bursts to long endurance, etc.).

Garland et al. (1964) emphasized the importance of feedback suppression of the PDC reaction and presented early evidence for direct product inhibition of PDC. Shortly after Linn et al. (1969) showed PDC was regulated by interconversion of E1 between its a and b forms, the capacity of fatty acids and ketone bodies to promote inactivation of PDC was supported by studies with intact tissues (e.g., Wieland et al., 1971; Taylor et al., 1973) and isolated mitochondria (Wieland and Portenhauser, 1974; Taylor et al., 1975). Pettit et al. (1975), Cooper et al. (1975), and Kerbey et al. (1976) found that the activity of the E1a kinase associated with kidney and heart PDC is greatly enhanced upon elevation of the $NADH:NAD^+$ ratio and the acetyl-CoA:CoA ratio. There is also a small reciprocal reduction in phosphatase activity due to inhibition by NADH (Pettit et al., 1975). Battenburg and Olson (1975, 1976) and Hansford (1976) presented evidence with liver and heart mitochondria that increases in these product to substrate ratios increased the fraction of PDC in the phosporylated (inactive) state. Accordingly, the E1a kinase has a crucial throttle role wherein it more rapidly inactivates PDC in response to increases in the mitochondrial acetylation and reduction potentials. As indicated in Figure 1, kinase stimulation by a combination of NADH and acetyl-CoA is particularly pronounced. Not only are NADH and acetyl-CoA produced as key intermediates in the oxidation of all fuels in mammalian mitochondria, but these are direct products of the PDC reaction (Fig. 1).

Focus

The goal of this review is to summarize recent insights into the elegant molecular mechanisms operative in efficient functioning of the kinase within the confines of the complex and in the steps of a sensitive signal translation system that operates to enhance kinase activity in response to the product buildup described above. These engage unique organizational and operational features of the pyruvate dehydrogenase complex. The E1a kinase binds very tightly to the E2 core; at the same time, a kinase functional unit (considered below) can rapidly phosphorylate a large complement of E1a tetramers which are bound throughout the complex. We have obtained new insights into the dynamic functioning of the E2 core in support maneuvers that facilitate an efficient seek and turn off operation by the kinase within the complex. Understanding this fundamental operation of the kinase is crucial for appreciating the additional roles of the E2 core in mediating the enhancement of kinase activity in response to increased NADH and acetyl-CoA.

We also briefly consider the molecular mechanisms operative in the opposing direction to reduce kinase activity (Fig. 1), particularly under conditions of low energy (ADP↑) and substrate availability (pyruvate↑). Closely linked to the above considerations on the nature of the association of the kinase and its regulation by product buildup, we introduce the finding that the phosphatase binds to the same E2 domain as the kinase and briefly consider the coordinated regulation of the phosphatase (Fig. 1).

Multidomain structure of E2 core

As a context for addressing these questions, the structure of the E2 core must be understood before we consider E2's pivotal role in facilitating PDC regulation. Figure 2 shows the domain structure of E2 and E3BP and the binding interactions of these components. The E2 subunit of mammalian PDC has four domains connected by relatively large (2–3 kDa) and highly mobile linker regions (review articles cited above). In an E2 assembly, 60 inner domains, $E2_I$, associate to form a dodecahedral inner core structure. This precisely organized COOH-terminal inner domain catalyzes the transacetylation reaction. Most of the mass of mammalian E2 subunits is exterior to its inner core structure and consists of three globular domains connected by the linker or hinge regions. The globular domains consist of two ~10 kDa lipoyl domains designated $E2_{L1}$ and $E2_{L2}$ (or for simplicity L1 and L2) and an E1-binding domain, designated $E2_B$, which is located in E2's sequence between the inner core and the N-terminal lipoyl domain region. E3BP has a similar domain structure with an inner domain unrelated to E2's inner domain and distantly related outer domains. The flexibly connected outer domains support implementation of a wide variety of processes associated with catalysis and regulatory control; the role of the outer domains of E2 in kinase function and regulation will be emphasized in this review.

Figure 2 also shows the component associations in mammalian PDC elucidated by Roche and colleagues (cf. references Fig. 2 legend). The inner domains of E2 and E3BP associate with each

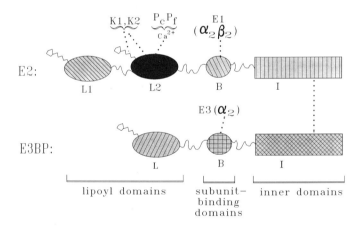

Figure 2. Domain structure of E2 and E3BP and binding interactions of their domains and E2's prosthetic groups with each other and other components of the mammalian PDC. The globular domains of E2 and E3BP are shown with their connecting linker or hinge regions presented as wavy lines. The distinct but related lipoyl domains are designated L1 and L2 in E2 and L in E3BP and are shown with a representation of attached lipoyl groups. The interactions between components and specific domains of E2 are indicated with dashed lines. For the components required for PDC reaction, associations are shown between the inner domain, I, of E2 and of E3BP (Ramatullah et al., 1989a), between β subunit of E1 and the component binding domain, B, of E2 (Rahmatullah et al., 1989b, 1990), between the E3 and E3BP (Rahmatullah et al., 1989b; Powers-Greenwood et al., 1989; Gopalakrishnan et al., 1989) by the B domain of E3BP (Lawson et al., 1991). As detailed in this review, the regulatory E1a kinase (K1 and K2 isoforms) bind to the lipoyl domain region of E2 by and in interaction with the L2 domain that also requires the lipoyl prosthetic group (Rahmatullah et al., 1990; Li et al., 1992; Ono et al., 1993; and Liu et al., 1995a); and the E1b phosphatase (P_c and P_f subunits) also interacts with the L2 domain of E2 (L. Wang, S. Liu and T.E. Roche, unpublished).

other and the outer domains of E2 and E3BP bind the other components. Maeng et al. (1994) found with yeast PDC that 12 of the inner domains of E3BP bind to E2 and Stoops and Reed (unpublished) have determined E3BP binds on the inside of the porous shell of the dodecahedral structure formed by 60 inner domains of E2. E1, an $\alpha_2\beta_2$ tetramer, binds to the B domain of E2 *via* its β subunit and E1 tetramers probably bind to two B domains unless E1 $\alpha\beta$ units are present at levels exceeding E2 subunits, a condition not found in purified complexes. The E3 dimer associates with the B domain of E3BP probably at a ratio of one dimer per E3BP as has been demonstrated by Hipps et al. (1994) for binding of E3 dimers to the B domain of E2 in *B. stearothermophilus* PDC and by Maeng et al. (1994) for B domain of the E3BP component of yeast PDC. The specific binding of the E1a kinase (K1 and K2 are distinct catalytic subunits) to the L2 domain of E2 will be considered in detail below, and a short description of our recent findings on the binding of the phosphatase to the same domain of E2. The phosphatase has a catalytic, Pc, and an FAD-containing subunit, P_f (Reed et al., this volume).

Kinase structure

The K1 and K2 isoforms of E1a kinase of PDC as well as the kinase component of the branched-chain α-keto acid dehydrogenase complex are serine kinases. Nevertheless, based on their cDNA-derived sequences, they have been shown to be structurally related to prokaryotic histidine kinases but not to the extramitochondrial serine and tyrosine protein kinases of eukaryotic cells (Popov et al., 1992, 1993, 1994; Harris and Popov, this volume).

E2-activated kinase function

Rapid phosphorylation of E1 bound to E2 by bound kinase

In typical preparations of bovine PDC, 20–30 $\alpha_2\beta_2$ E1 tetramers and as few as one molecule of E1a kinase are bound to the $E2_{60}$ core. Though very tightly bound to E2, the kinase can efficiently inactivate the complex ($k_{cat} = 0.5$ s^{-1}). Indeed, despite free access by diffusion, the phosphorylation of free E1 by free kinase is several-fold slower than when these components are tightly bound to the E2 core. The dissociation of E1 from the E2 core is slow (Brandt and Roche, 1983; Wu and Reed, 1984), and E1a kinase has an even tighter affinity for the E2 core (Ono et al., 1994). In early efforts to address the question of how an E2-bound kinase molecule rapidly phosphorylates many E2-bound E1 tetramers, Brandt et al. (1983) found, with a very dilute complex, that a major portion of the E1 but little kinase dissociates from the E2 core. Surprisingly, the few bound E1 were rapidly phosphorylated (and PDC activity lost) and phosphorylation of free E1 tetramers occurred at a rate that corresponded to that estimated for E1 reassociating with the E2 core. Only with the knowledge of E2's structure and after characterization of the location and nature of kinase binding to the E2 could the underlying mechanisms begin to be evaluated.

Lipoyl domain-specific, lipoate-dependent kinase binding

The E1a kinase binds to the lipoyl domain region of E2 in bovine PDC (Ramatullah et al., 1990; Li et al., 1992). Radke et al. (1993) demonstrated that selective removal of the lipoyl prosthetic group led to dissociation of the kinase demonstrating an essential role of this cofactor in tight binding of the kinase. Although the kinase binds very tightly to the oligomeric E2 core, Ono et al. (1993) found the kinase efficiently transferred to a fragment of bovine E2 that contained both lipoyl domains. The combination of tight binding and rapid transfer led to the suggestion that transfer involves a direct interchange of the kinase (i.e., without any free kinase) between the exterior lipoyl domain regions of the E2 core. Such movement at the surface of the assembled E2 core would explain how a kinase molecule is able to both bind tightly to E2 core and still phosphorylate 20 to 30 E1 tetramers that are also bound to the $E2_B$ domain at the surface of the E2 core.

To evaluate more precisely the nature of the binding of the kinase to E2 and to gain further insights into the rapid phosphorylation of E1 by the bound kinase, Liu et al. (1995a) conducted studies with recombinant lipoyl domain constructs consisting of the entire lipoyl domain region or the individual lipoyl domains with or without the intervening hinge region that were prepared as described by Liu et al. (1995b). Constructs were made and used both as free lipoyl domains and fused to glutathione S-transferase (GST). Using GSH Sepharose to selectively bind GST-constructs, tightly bound kinase was shown to rapidly transfer from intact E2 core specifically to GST-constructs containing the L2 domain rather than to ones containing only the L1 domain. GST-$E2_{L2}$-kinase complexes could be eluted from GSH Sepharose with glutathione. Figure 3 shows results demonstrating kinase transfer from E2-K1K2 subcomplex (pass through filtrate, hatched bars) to GSH Sepharose-bound constructs containing the L2 domain or both lipoyl domains (solid bars, -LPA samples). Delipoylation by treatment with *Enterococcus faecalis* lipoamidase (+LPA) eliminated binding to the L2 domain-containing constructs establishing that the preferential binding of the kinase to the L2 domain requires its lipoyl prosthetic group.

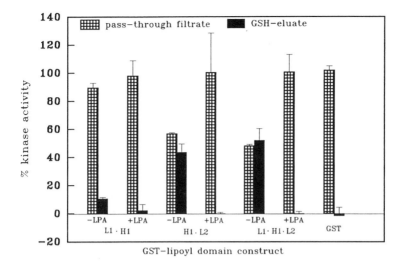

Figure 3. Transfer of the kinase from the E2 core to lipoylated and delipoylated GST-lipoyl domain constructs (25). Following incubation with (+LPA) or without (-LPA) lipoamidase, 750 pmol of the indicated GST-constructs were bound to GSH Sepharose. E2-K1K2 was mixed for 60 s with the indicated gel-bound GST-construct (or GST). Kinase activity was measured in a filtrate that contained all E2 (hatched bars) collected by passing buffer through the gel and then in a eluate containing GST-lipoyl domain constructs collected with a GSH containing buffer. (Reprinted from *Journal of Biological Chemistry* (Liu et al., 1995b) with permission of American Society of Biochemistry and Molecular Biology).

Nature of L2 binding of kinase

Radke et al. (1993) suggested that it is the inner hydrophobic portion of the lipoyl-lysine prosthetic group that directly contributes to binding the kinase because modification of the reactive dithiolane ring portion of the lipoate did not prevent kinase binding. The prospect cannot be eliminated that delipoylation causes a change in the structure of the L2 domain which prevents its binding of the kinase. However, this is not expected due to the role of the lipoyl-lysine as a swinging arm and evidence that lipoates are attached to a Lys in a tight loop at one end of a lipoyl domain (Dardel et al., 1993).

Regardless of how removal of the lipoate prevents kinase binding, unique structural interactions of the kinase with distinctive structural features in the tertiary structure of the L2 domain seem required to explain the effective binding by this domain, but not by the L1 domain which has a highly related sequence with 65% identity in the first 81 amino acids, that can be aligned with bacterial PDC lipoyl domains (Thekkumkara et al., 1988). However, as emphasized by Liu et al. (1995b), the COOH-terminal regions show substantial variation in sequence (35% identities in the next 17 residues) and in size (L2 is a few residues longer). Dardel et al. (1993) first estab-

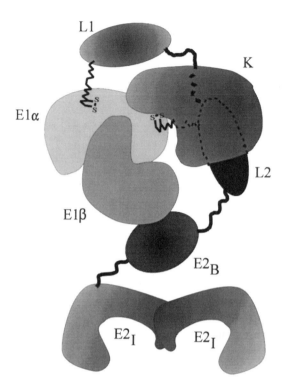

Figure 4. Model of the domains of an E2 subunit binding an αβ unit of the E1 component by its B domain and a kinase subunit by its L2 domain plus the inner part of the lipoyl prosthetic group of L2. In the upper section the model depicts hinge connected L1 and L2 domains with attached lipoyl groups. A major portion of L2 and the inner part of its lipoyl-lysine side chain are positioned under (dashed lines) and interacting with a kinase subunit (K). The β subunit of E1 interacts with the B domain of E2 and the a subunit of E1 is positioned to serve as a substrate for the kinase subunit. To emphasize the oligomer forming role of the inner (I) domain of E2 a second I domain is included but without outer domains. (Reprinted from *Journal of Biological Chemistry* (Liu et al., 1995a) with permission of American Society of Biochemistry and Molecular Biology).

lished the lipoyl domain of the *B. stearothermophilus* PDC-E2 is a flattened β barrel structure. Circular dichroism studies (S. Liu, A.B. Cole, L.A. Andersson, and T.E. Roche, unpublished) indicate that L2 and L1 of mammalian E2 are similar β structures. It is likely that the 22 amino acid region at the C-terminus of L2 (sequence from 208–229: EKEADISAFADYRPTEVTDLK) contributes at least two additional β strands (a proline at residue 217 may participate in a β turn). This 2.3 kDa region of the L2 structure may have a specialized role in interacting with the kinase and/or phosphatase components of mammalian PDC (Liu et al., 1995a, b). This region contains a phenylalanine (residue 217) which has not been found in any of over 35 hinge region sequences of the E2 components of α-keto acid dehydrogenase complexes. However, it is a highly charged region with six acidic and three basic residues. Given this composition, the C-terminal region may be loosely anchored to the rest of L2's polypeptide structure and may even separate upon the binding of the kinase and the phosphatase (cf. further comments below in regard to the phosphatase).

Figure 4 models the interaction of a kinase subunit specifically with the L2 domain and the inner part of its lipoyl-lysine prosthetic group. The model also shows one αβ unit of an E1 tetramer binding *via* its β subunit to the B domain of E2 with its a subunit positioned to interact with the kinase. Liu et al. (1995a) found that free L2 caused a 40% enhancement of the phosphorylation of E1; the much slower rate of phosphorylation of peptide substrate is also increased up to two-fold by the lipoyl domain region of bovine PDC E2 (Ono et al., 1994). Binding to L2 and its prosthetic group probably causes a conformational change in the kinase. The several fold higher activity of the kinase in the intact complex requires additional mechanisms (below) in which an L2-induced conformational change may contribute.

Translocation of the kinase around the surface of the E2 core

"Hand over hand" movement of the kinase

Besides locating the kinase binding site, evidence was obtained that the kinase can rapidly move between L2-containing structures (Ono et al., 1994; Liu et al., 1995). For instance, rapidly passing an E2 core-kinase complex through a column containing GST-anchored L2 domain resulted in rapid transfer of the kinase to the fusion protein-L2 structure. Transfer was too fast to determine its rate (at least 90% complete in 20 s); better approaches are needed to analyze the speed of this translocation. The rapid movement cannot be simply reconciled with the tight binding of the kinase to the E2 core which requires $K_d \leq 3 \times 10^{-9}$ M. Such a high affinity is inconsistent with a half-time of dissociation of a few seconds as would be required for completely dissociated kinase serving as an intermediate in the transfer between the E2 core and other L2-containing structures. Therefore, we have suggested an alternative transfer mechanism that depends on the quaternary structure of the kinase.

While Stepp et al. (1993) suggested the bovine E1a kinase is a heterodimer, there is very limited evidence for that conclusion. The K1 subunit, alone, binds to the E2 oligomer and exhibits all known functions of the kinase (Rahmatullah et al., 1987; Li et al., 1992; Ono et al., 1993). We have recently obtained patterns in native gel electrophoresis that are consistent with the following dimer forms for two kinase subunits – $[K1]_2$, $[K1 \cdot K2]$, $[K2]_2$ (J.C. Baker and T.E. Roche, unpublished). Such a dimer state would help explain the capacity of the kinase to move between L2-containing structures without dissociation since it makes the prospect likely that the kinase has two lipoyl domain binding sites.

Liu et al. (1995a) have proposed that a kinase dimer moves at the surface of the E2 core by a "hand over hand" mechanism involving a continuous process of partial dissociation by one subunit followed by interchange to another lipoyl domain. Figure 5 models such a reversible translocation step for kinase movement across the surface of the E2 core. Movement is proposed to proceed by a repeated series of exchanges *via* the K_{intra} process only with different L2 domains in the same local region at the surface of the E2 core participating in that step. The release of the kinase dimer from one of two L2 domains is rapidly followed by association in the reverse step by the kinase grabbing any of several flexibly held L2 in that region of the surface of the E2 core. That intramolecular association leading to the kinase again hanging onto two lipoyl domains should easily proceed faster than the dissociation step producing completely dissociated kinase (not shown). The model shown also includes the likely, but not absolutely essential, prospect that there is a lesser intrinsic affinity for each L2 domain when two L2 bind the kinase than when only one L2 holds the kinase, i.e., negative cooperativity is invoked which favors the dissociation from being held by two L2 being faster than the dissociation step when being held by one L2. This is indicated in Figure 5 by a conformational change that removes asymmetry when one kinase subunit interacts with an L2 domain of E2. The symmetry of bi-held kinase subunits would allow the dissociation step of the K_{intra} process to occur with either kinase subunit. As indicated above, the effects of L2 on kinase activity are consistent with L2 inducing a conformational change in the kinase.

This mechanism of kinase movement explains how the E1a kinase can patrol the surface of oligomeric E2 core to find and phosphorylate many bound E1. Further studies are underway in our laboratory to establish the quaternary structure of the kinase, to estimate the rate of transfer of the kinase, and to evaluate the influence of changes in the oligomeric state of various $E2_{L2}$-containing structures on kinase binding and activity enhancement.

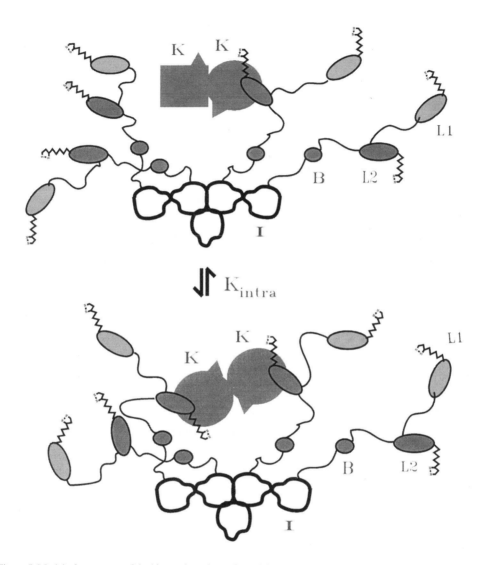

Figure 5. Model of movement of the kinase along the surface of the E2 core. The reversible association of a kinase dimer (labeled K for K1 or K2 for E1a kinase subunits) with one L2 domain and then with a second L2-lipoyl domain (K_{intra} step). "Hand over hand" movement of the kinase is achieved by repeated dissociation and association in the K_{intra} step leading to movement by interchange of L2 domains binding the kinase. The model incorporates a conformational change in the kinase dimer with the binding of lipoyl domains that results in negative cooperativity (i.e., a rapid exchange by a weaker K_{intra} interaction while being held by an interaction that is intrinsically tighter by one L2). The domains of E2 are labeled as shown in Figures 1 and 4. The model shows five $E2_I$ domains out of 60 in a dodecahedron structure and E1 is not included for clarity but would be bound to the B domain as shown in Figure 4 as an $\alpha_2\beta_2$ structure interacting with two B domains. (Reprinted from *Journal of Biological Chemistry* (Liu et al., 1995) with permission of American Society of Biochemistry and Molecular Biology).

The feedback throttle

NADH and acetyl-CoA must react to stimulate kinase activity

We now return to the key question of how the fraction of PDC in the active form is attenuated in response to changes in the ratios of products to substrates for the PDC reaction. This laboratory has presented evidence that the initial steps leading to marked changes in kinase activity involve a translation of those ratios by their competitive utilization in the downstream reactions of the complex leading to adjustment of the fraction of the complex's lipoyl prosthetic groups in the oxidized *versus* reduced *versus* acetylated forms (Roche and Cate, 1976; Cate and Roche, 1978, 1979; Rahmatullah and Roche, 1985, 1987). Thus, the initial steps in our proposed mechanism for how the kinase senses product accumulation and responds in a sensitive fashion to down regulate PDC activity involve NADH reacting in the reverse of the dihydrolipoyl dehydrogenase (E3) reaction to reduce lipoates and acetyl-CoA reacting in the reverse of the dihydrolipoyl acetyltransferase (E2) reaction to acetylate lipoates. Typically a 60–80% enhancement in kinase activity is achieved by just NADH and up to a three-fold enhancement by acetyl-CoA in association with acetylation of only 20% of lipoyl prosthetic groups of the complex. The latter maximal stimulation has been observed under conditions in which a low level of acetyl-CoA is employed and this substrate is completely consumed in the acetylation of lipoyl prosthetic groups of the complex due to the acetylation reaction being driven by CoA removal in a separate reaction. The potential for understanding the molecular basis of this control was greatly improved with new insights into the structure of E2 subunits and the unusual nature of the association of the kinase with the L2 domain of E2 described above.

Minimal requirements

Recently, this mechanism was established and minimal requirements were defined (Ravindran et al., 1996). E2-associated kinase activity was stimulated by NADH and acetyl-CoA using a peptide substrate and with E1 lacking TPP as a substrate. These results ruled out kinase activity being increased due to changes in the reaction state of its E1 substrate (Robertson et al., 1990). E2-free kinase activity was stimulated to a small extent by dihydrolipoamide and by freshly formed acetyl-dihydrolipoamide in the absence of a lipoyl domain source. Since addition of free reduced and acetylated lipoates stimulated kinase not exposed to any oxidized lipoate, stimulation must involve a positive allosteric interaction of the reduced and acetylated forms and not removal of inhibition by the disulfide form of lipoate interacting with the kinase.

Regardless of the source of reduced lipoate, stimulation of the kinase by acetyl-CoA occurred only upon addition of the transacetylase-catalyzing (lipoyl domain free) inner-core portion of E2, firmly establishing a need for acetylation of the prosthetic group of lipoyl domains for acetyl-CoA stimulation. Similarly, addition of NADH to the kinase in the absence of a lipoate source failed to give any stimulation, indicating stimulation of the kinase by NADH requires lipoate reduction by E3 catalysis.

Role of the L2 domain and insights from delipoylation

Ravindran et al. (1996) investigated the relative capacity of fully lipoylated forms of the N-terminal (L1) and the inner lipoyl domain (L2) to mediate kinase stimulation by products. The L1 domain gave a stimulatory effect similar to that achieved with reduced and acetylated forms of free lipoamide. The kinase-binding L2 domain was much more effective than the L1 domain in supporting the enhancement of kinase activity. Reductive acetylation of L2 facilitated > three-fold increase in kinase activity already enhanced by L2, whereas the lower kinase activity in the presence of L1 was stimulated only about 1.8-fold by reductive acetylation of L1. Again stimulation by acetyl-CoA required $E2_I$ to catalyze acetylation along with addition to L1 or L2. Figure 6 models this L2-mediated signal translation regulation for changes in kinase activity in response to changes in the product-to-substrate ratios. The kinase is converted from a non stimulated (K) to partially activated (K^*) to fully stimulated (K^{**}) due to interacting with L2 with oxidized, reduced, or reductively acetylated lipoates, with the latter interaction being favored based on the following findings.

As described above, E2-facilitation of rapid phosphorylation of E1 within the confines of the complex requires lipoyl-dependent binding and the constraint of tightly bound kinase being localized is apparently removed by rapid interchange of a kinase dimer between lipoyl domains. In support of the model shown in Figure 6, full delipoylation of PDC, E2-kinase subcomplex, or L2 removed stimulation as well as the capacity of the E2 oligomer or L2 to enhance kinase activity in the absence of effectors. Upon reintroduction of lipoyl moieties using lipoyl protein ligase, the capacity of the E2 core to greatly increase control activity and typical stimulation by NADH and acetyl-CoA were restored. However, a lipoamidase treatment of PDC, which left lipoyl groups attached to only a few E2 subunits in the complex and caused close to full loss of this E2-enhanced kinase function, still allowed marked stimulation of this low kinase activity upon acetylation of those few domains retaining lipoyl groups (Ravindran et al., 1996). This result forcefully suggests that the kinase has moved to and is bound by the limited L2 domains that were lipo-

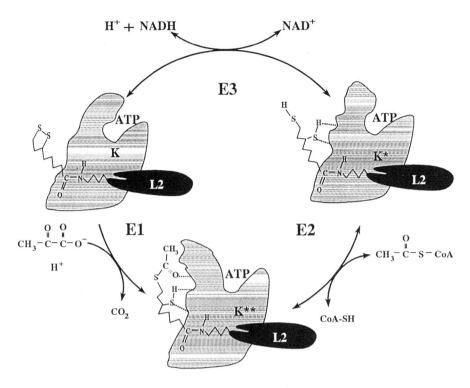

Figure 6. Signal translation regulation mechanism for throttling down PDC activity due to product build-up. Increases in the proportion of the L2 domains of the $E2_{60}$ core that have lipoyl domains in the reduced form due to the reversible E3 reaction responding to increases $NADH:NAD^+$ ratio enhance the rate of kinase inactivation of PDC by up to 80% (indicated by change of the kinase from a K state to a K* state). Due to the E1 reaction or to the reversible E2-catalyzed transacetylation reaction, kinase activity is enhanced by $\geq 300\%$ by acetylation of <25% of the L2 domains. This highly activated form of the kinase is indicated by a change to K** state. Under most metabolic conditions, the E1 reaction is expected to be rate limiting and the level of acetylation of the lipoyl domains of the complex is determined by the acetyl-CoA:CoA ratio and by the $NADH:NAD^+$ ratio since E3 reaction determines the level of reduced lipoates available for the reverse of the E2 reaction.

ylated and, furthermore, that acetylation of just those domains, probably during a short-lived dissociation and reassociation of the kinase, facilitated kinase stimulation. Based on this retained stimulation with few lipoylated domains in PDC and based on the capacity of the L2 domain to give a much larger stimulation than L1, we hypothesize that maximal effector stimulation is mediated by an allosteric effect induced by a reductively acetylated L2 domain that becomes engaged in binding of the kinase (i.e., not by interaction of an acetylated prosthetic group on a neighboring lipoyl domain that is not engaged in binding the kinase). Consideration of the levels of acetylation that are effective in mediating kinase stimulation further supports a highly specific interaction.

The high responsiveness of the kinase to limited reductive acetylation

This enhancement in kinase activity is phenomenally sensitive to the level of acetylation of sites in the intact complex. Half-maximal stimulation is attained following acetylation of only 6% of lipoates and near-maximal stimulation after acetylation of 15% of lipoates in the complex (Cate and Roche, 1979; Rahmatullah and Roche, 1988; Ravindran et al., 1996). Preferential acetylation of L2 is not found. Thus, these results imply that there is a highly specific interaction of the kinase with those very few acetylated L2 domains and as proposed in the mechanism in Figure 6. Indeed, even with free L2, kinase activity is increased to half its maximal level when only ~20% of free L2 are acetylated (Ravindran et al., 1996).

We conclude that fluctuations in the $NADH:NAD^+$ and acetyl-CoA:CoA ratios serve to manipulate kinase activity in a highly sensitive fashion through a substantial activity range as a consequence of translation of those ratios by the rapidly reversible E3 and E2 reactions into changes in the proportion of oxidized, reduced, and acetylated lipoyl moieties of the kinase-binding L2 domain (Fig. 6). None of the above considerations address how the kinase reaction mechanism is specifically changed. Consideration of what is known about other aspects of kinase regulation suggests a possible mechanism.

Inhibition *versus* stimulation of the kinase and phosphatase binding and function

Reciprocal regulation of the kinase and competition among the regulatory effects on the kinase reaction

PDC activity is enhanced by E1b phosphatase (below) catalyzing dephosphorylation and by reducing the activity of the kinase. Acting in the opposite direction to the above mechanisms, kinase stimulation is removed as NAD^+ and CoA accumulate and convert lipoates to the oxidized form since, under these conditions, reductive acetylation catalyzed by E1 is the rate limiting step (Cate et al., 1980; Sumegi and Alkonyi, 1983). As indicated in Figure 1, kinase activity can be markedly reduced by a combination of ADP and pyruvate which act synergistically to inhibit kinase activity (Pratt and Roche, 1979). This stepping down of the kinase reaction is most reasonable since a transient increase in ADP, reflecting a need to meet energy demands, in conjunction with good availability of pyruvate constitute optimal conditions for increasing the PDC reaction. Inhibition of the kinase results from pyruvate binding to the kinase·ADP complex. The affinity of ADP binding to the kinase is greatly increased as K^+ is increased from a low level to 100 mM (i.e., a physiological level) (Roche and Reed, 1974). In the absence of lipoyl domain sources and, therefore, in the absence of reductive acetylation, there is enhanced pyruvate inhibition of the kinase in combination with ADP, particularly at elevated K^+ levels (Rahmatullah and Roche, 1987). The results suggest that at physiological K^+ levels, ADP dissociation is rate limiting favoring pyruvate binding and inhibition of the kinase.

Interestingly, kinase stimulation by NADH and acetyl-CoÅ or by reaction of pyruvate or effective analogs of these substrates is dependent not only on their reaction, but on having similarly elevated K^+ levels (Pettit et al., 1975; Cate and Roche, 1978). If ADP dissociation is the rate limiting step in kinase function under those conditions, it is possible that reduction and acetylation of the L2 domain stimulates kinase activity by enhancing ADP dissociation. For that to be the mechanism, it must operate over a wide range of kinase activities because we have observed stimulation of the very slow phosphorylation of peptide substrate as well as the much faster E2-activated kinase function. Stimulation by enhancing the rate of ADP dissociation would be coordinated with a reduction in pyruvate inhibition *via* the mechanism described above.

Phosphatase binding site and NADH inhibition

The phosphatase component catalyzes a Mg^{2+}-dependent, Ca^{2+}-stimulated reaction that is enhanced by effectors that lower the K_m for Mg^{2+} such as the polyamines or some undefined effector associated with insulin action (Thomas and Denton, 1986), possibly an inositol phosphate glycan (Larner et al., 1989). The regulation of the phosphatase has been reviewed previously by Reed and Damuni (1987) and the properties of the catalytic subunit, P_c, and the regulatory FAD-containing subunit, P_f, of the phosphatase are described in this volume in a review by Reed et al. Here, we summarize our results from a study closely allied to those above – the location and requirements for the binding of the phosphatase to E2.

The binding of the phosphatase to E2 has been known for some time to require μmolar Ca^{2+} and this association with the E2 core causes a large enhancement in phosphatase activity (Pettit et al., 1972). Two Ca^{2+}-binding sites may be involved with one, which requires both E2 and the phosphatase, possibly serving a bridging role (Teague et al., 1982). Now, it has been found that the E1b phosphatase specifically binds to the L2 domain in a Ca^{2+}-requiring process (L. Wang, S. Liu, and T.E. Roche, unpublished). By taking advantage of this specific interaction, the phosphatase has been purified to homogeneity using GST-linked L2 bound to GSH Sepharose. The phosphatase is selectively eluted by chelating Ca^{2+} with the chelator EGTA. As noted above, 6 of the 22 residues of the distinct C-terminal region of L2 are acidic residues. Thus, this seems to be a particularly appropriate region for participating in a Ca^{2+}-bridge between this domain of E2 and the phosphatase.

Since E1b phosphatase binds to the L2 domain, this raises the possibility that NADH inhibition of the phosphatase may involve reduction of L2's lipoyl moiety. Previously, we observed a similar inhibition by free dihydrolipoamide and high sensitivity of phosphatase to arsenite after NADH-dependent reduction of lipoates (Rahmatullah and Roche, 1988).

Conclusion

We have found that the high rates of the kinase and of the phosphatase function, achieved when these regulatory enzymes and the appropriate forms of their E1 substrate are bound to the E2 core, result from both the kinase and the phosphatase binding to the peripheral L2 domain of E2. The enhanced activity of the kinase probably exploits its capacity to rapidly move between L2 domains at the surface of the E2 core. The phosphatase moves between L2 domains by a reversible, Ca^{2+}-requiring association that responds *in vivo* to hormone-induced changes in Ca^{2+} levels. Feedback shutdown of the PDC reaction is mediated by marked stimulation of kinase activity due to an NADH-dependent reduction and acetyl-CoA-dependent acetylation of L2's lipoyl group.

Acknowledgement
We wish to acknowledge the support by National Institutes of Health Grant DK18320, by the Kansas Affiliate of the American Heart Association, and by the Kansas State Agricultural Experiment Station (Contribution 95-471-B). We wish to thank Collin Kilbane and Amy Paulin for help in preparation of figures and Connie Schmidt for help in manuscript preparation.

References

Batenburg, J.J. and Olson, M.S. (1975) The inactivation of pyruvate dehydrogenase by fatty acid in isolated rat liver mitochondria. *Biochem. Biophys. Res. Comm.* 66: 533–540.

Batenburg, J.J. and Olson, M.S. (1976) Regulation of pyruvate dehydrogenase by fatty acid in isolated at liver mitochondria. *J. Biol. Chem.* 251: 1364–1370.

Brandt, D.R. and Roche, T.E. (1983) Specificity of the pyruvate dehydrogenase kinase for pyruvate dehydrogenase component bound to the surface of the kidney pyruvate dehydrogenase complex and evidence for intracore migration of pyruvate dehydrogenase component. *Biochemistry* 22: 2966–2971.

Brandt, D.R., Roche, T.E. and Pratt, M.L. (1983) Heterogeneity of binding sites for the pyruvate dehydrogenase component on the dihydrolipoyl transacetylase component of bovine kidney pyruvate dehydrogenase complex. *Biochemistry* 22: 2958–2966.

Cate, R.L. and Roche, T.E. (1978) A unifying mechanism for stimulation of mammalian pyruvate dehydrogenase kinase activity by NADH, dihydrolipoamide, acetyl-CoA, or pyruvate. *J. Biol. Chem.* 253: 496–503.

Cate, R.L. and Roche, T.E. (1979) Function and egulation of mammalian pyruvate dehydrogenase complex: Acetylation, interlipoyl acetyl transfer, and migration of the pyruvate dehydrogenase component. *J. Biol. Chem.* 254:
  1659–1665.
Cate, R.L., Roche, T.E. and Davis, L.C. (1980) Rapid intersite transfer of acetyl groups and movement of pyruvate
  dehydrogenase component in the kidney pyruvate dehydrogenase complex. *J. Biol. Chem.* 255: 7556–7562.
Cooper, H.C., Randle, P.J. and Denton, R.R. (1975) Stimulation of phosphorylation and inactivation of pyruvate
  dehydrogenase by physiological inhibitors of the pyruvate dehydrogenase reaction. *Nature* 257: 808–809.
Dardel, F., Davis, A.L., Laue, E.D. and Perham, R.N. (1993) Three-dimensional strucutre of the lipoyl domain from
  Bacillus stearothermophilus pyruvate dehydrogenase multienzyme complex. *J. Mol. Biol.* 229: 1037–1048.
Garland, P.B., Newsholme, E.A. and Randle, P.J. (1964) Regulation of glucose uptake by muscle: Effects of fatty
  acids and ketone bodies, and of alloxan-Diabetes and starvation, on pyruvate metabolism and on lactate/Pyruvate
  and 1-Glycerol 3-Phosphate/Dihydroxyacetone phosphate concentration ratios in rat heart and rat diaphragm
  muscles. *Biochem. J.* 93: 665–678.
Gopalakrishnan, S., Rahmatullah, M., Radke, G.A., Powers-Greenwood, S.L. and Roche, T.E. (1989) Role of protein x in the function of mammalian pyruvate dehydrogenase complex. *Biochem. Biophys. Res. Comm.* 160:
  715–721.
Hansford, R.G. (1976) Studies on the effects of coenzyme A–SH: acetyl coenzyme A, nicotinamide adenine dinucleotide: reduced nicotinamide adenine dinucleotide, and adenosine diphosphate: Adenosine triphosphate ratios on
  the interconversion of active and inactive pyruvate dehydrogenase in isolated rat heart mitochondria. *J. Biol.
  Chem.* 251: 5483–5489.
Hipps, D.S., Packman, L.C., Allen, M.D., Fuller, C., Sakaguchi, K., Appella, E. and Perham, R.N. (1994) The peripheral subunit-binding domain of the dihydrolipoyl acetyltransferase component of the pyruvate dehydrogenase
  complex of *Bacillus stearothermophilus*: preparation and characterization of its binding to the dihydrolipoyl
  dehydrogenase component. *Biochem. J.* 297: 137–143.
Kerbey, A.L., Randle, P.J., Cooper, R.H., Whitehouse, S., Pask, H.T. and Denton, R.M. (1976) Regulation of pyruvate dehydrogenase in rat heart: mechanism of regulation of proportions of dephosphorylated and phosphorylated
  enzyme by oxidation of fatty acids and ketone bodies and of effects of diabetes: Role of coenzyme A, acetyl-
  Coenzyme A and reduced and oxidized nicotinamide-Adenine dinucleotide. *Biochem. J.* 154: 327–348.
Larner, J., Huang, L.C., Suzuki, S., Tong, G., Zheng, C., Schwartz, C.F.W., Romero, G., Luttrell, L. and
  Kennington, A.S. (1989) insulin mediators and the control of pyruvate dehydrogenase complex. *N.Y. Acad. Sci.*
  573: 297–305.
Lawson, J.E., Behal, R.H. and Reed, L.J. (1991) Disruption and mutagenesis of the *Saccharomyces cerevisiae*
  PDX1 gene encoding the protein x component of the pyruvate dehydrogenase complex. *Biochemistry* 30: 2834–
  2839.
Li, L., Radke, G.A., Ono, K. and Roche, T.E. (1992) Additional binding sites for the pyruvate dehydrogenase
  kinase but not for protein x in the assembled core of the mammalian pyruvate dehydrogenase complex. binding
  region for the kinase. *Arch. Biochem. Biophys.* 296: 497–504.
Linn, T.C., Pettit, F.H. and Reed, L.J. (1969) α-keto acid dehydrogenase complexes, X. regulation of the activity
  of pyruvate dehydrogenase complex from beef kidney mitochondria by phosphorylation and dephosphorylation.
  Proc. Nat. Acad. Sci. USA 62: 234–241.
Liu, S., Baker, J.C. and Roche, T.E. (1995a) Binding of the pyruvate dehydrogenase kinase to recombinant constructs containing the inner lipoyl domain of the dihydrolipoyl acetyltransferase component. *J. Biol. Chem.* 270:
  793–800.
Liu, S., Baker, J.C., Andrews, P.C. and Roche, T.E. (1995b) Recombinant expression and evaluation of the lipoyl
  domains of the dihydrolipoyl acetyltransferase component of human pyruvate dehydrogenase complex. *Arch.
  Biochem. Biophys.* 316: 926–940.
Maeng, C.-Y., Yazdi, M.A., Niu, X.-D., Lee, H.Y. and Reed, L.J. (1994) expression, purification, and characteriza-
  tion of the dihydrolipoamide dehydrogenase-Binding protein of the pyruvate dehydrogenase complex from
  Saccharomyces cerevisiae. *Biochemistry* 33: 13801–13807.
Ono, K., Radke, G.A., Roche, T.E. and Rahmatullah, M. (1993) Partial activation of the pyruvate dehydrogenase
  kinase by the lipoyl domain region of E2 and interchange of the kinase between lipoyl domain regions. *J. Biol.
  Chem.* 268: 26135–26143.
Patel, M.S. and Roche, T.E. (1990) Molecular biology and biochemistry of pyruvate dehydrogenase complexes.
  FASEB J. 4: 3224–3233.
Perham, R.N. (1991) Domains, motifs, and linkers in 2-Oxo acid dehydrogenase multienzyme complexes: A para-
  digm in the design of a multifunctional protein. *Biochemistry* 30: 8501–8512.
Pettit, F.H., Roche, T.E. and Reed, L.J. (1972) Function of calcium ions in pyruvate dehydrogenase phosphatase
  activity. *Biochem. Biophys. Res. Comm.* 49: 563–571.
Pettit, F.H., Pelley, J.W. and Reed, L.J. (1975) Regulation of pyruvate dehydrogenase kinase and phosphatase by
  acetyl-CoA/CoA and NADH/NAD ratios. *Biochem. Biophys. Res. Comm.* 65: 575–582.
Popov, K.M., Zhao, Y., Shimomura, Y., Kuntz, M.J. and Harris, R.A. (1992) Branched-chain α-Ketoacid dehydro-
  genase kinase: Molecular cloning, expression, and sequence similarity with histidine protein kinases. *J. Biol.
  Chem.* 267: 13127–13130.

Popov, K.M., Kedishvili, N.Y., Zhao, Y., Shimomura, Y., Crabb, D.W. and Harris, R.A. (1993) primary structure of pyruvate dehydrogenase kinase establishes a new family of eukaryotic protein kinases. *J. Biol. Chem.* 268: 26602–26606.

Popov, K.M., Kedishvili, N.Y., Zhao, Y., Guidi, R. and Harris, R.A. (1994) Molecular cloning of the p45 subunit of pyruvate dehydrogenase kinase. *J. Biol. Chem.* 269: 29720–29724.

Powers-Greenwood, S.L., Rahmatullah, M., Radke, G.A. and Roche, T.E. (1989) Separation of protein x from the dihydrolipoyl transacetylase component of the mammalian pyruvate dehydrogenase complex and role of protein x. *J. Biol. Chem.* 264: 3655–3657.

Pratt, M.L. and Roche, T.E. (1979) Mechanism of pyruvate inhibition of kidney pyruvate dehydrogenase$_a$ kinase and synergistic inhibition by pyruvate and ADP. *J. Biol. Chem.* 254: 7191–7196.

Radke, G.A., Ono, K., Ravindran, S. and Roche, T.E. (1993) Critical role of a lipoyl cofactor of the dihydrolipoyl acetyltransferase in the binding and enhanced function of the pyruvate dehydrogenase kinase. *Biochem. Biophys. Res. Comm.* 190: 982–991.

Rahmatullah, M. and Roche, T.E. (1985) Modification of bovine kidney pyruvate dehydrogenase kinase activity by CoA esters and their mechanism of action. *J. Biol. Chem.* 260: 10146–10152.

Rahmatullah, M., Jilka, J.M., Radke, G.A. and Roche, T.E. (1986) Properties of the pyruvate dehydrogenase kinase bound to and separated from the dihydrolipoyl transacetylase-Protein x subcomplex and evidence for binding of the kinase to protein x. *J. Biol. Chem.* 261: 6515–6523.

Rahmatullah, M. and Roche, T.E. (1987) The catalytic requirements for reduction and acetylation of protein x and the related regulation of various forms of resolved pyruvate dehydrogenase kinase. *J. Biol. Chem.* 262: 10265–10271.

Rahmatullah, M. and Roche, T.E. (1988) Component requirements for NADH inhibition and spermine stimulation of pyruvate dehydrogenase$_b$ phosphatase activity. *J. Biol. Chem.* 263: 18106–8110.

Rahmatullah, M., Gopalakrishnan, S., Radke, G.A. and Roche, T.E. (1989a) Domain structures of the dehydrolipoyl transacetylase and the protein x components of mammalian pyruvate dehydrogenase complex – Selective cleavage by protease arg C. *J. Biol. Chem.* 264: 11245–1251.

Rahmatullah, M., Gopalakrishnan, S., Andrews, P.C., Chang, C.L., Radke, G.A. and Roche, T.E. (1989b) Subunit associations in the mammalian pyruvate dehydrogenase complex: structure and role of protein x and the pyruvate dehydrogenase component binding domain of the dihydrolipoyl transacetylase component. *J. Biol. Chem.* 264: 12221–2227.

Rahmatullah, M., Radke, G.A., Andrews, P.C., Roche, T.E. (1990) Changes in the core of the mammalian-Pyruvate dehydrogenase complex upon selective removal of the lipoyl domain from the transacetylase component but not from the protein x component. *J. Biol. Chem.* 265: 114512–14517.

Randle, P.J. (1986) Fuel selection in animals. *Biochem. Soc. Trans.* 14: 1799–806.

Ravindran, S., Radke, G.A., Guest, J.R. and Roche, T.E. (1996) Lipoyl domain-Based mechanism for the integrated feedback control of the pyruvate dehydrogenase complex by enhancement of pyruvate dehydrogenase kinase activity. *J. Biol. Chem.* 271: 653–662.

Reed, L.J. and Damuni, Z. (1987) Mitochondrial protein phosphatases. *Adv. Prot. Phosphatases* 4: 159–76.

Reed, L.J. and Hackert, M.L. (1990) Structure-Function relationships in dihydrolipoamide acyltransferases. *J. Biol. Chem.* 265: 18971–8974.

Robertson, J.G., Barron, L.L. and Olson, M.S. (1990) Bovine heart pyruvate dehydrogenase kinase stimulation by α-Ketoisovalerate. *J. Biol. Chem.* 265: 116814–16820.

Roche, T.E. and Reed, L.J. (1974) Monovalent Cation requirement for ADP inhibition of pyruvate dehydrogenase kinase. *Biochem. Res. Comm.* 59: 11341–1348.

Roche, T.E. and Cate, R.L. (1976) Evidence for lipoic acid mediated NADH and acetyl-CoA stimulation of liver and kidney pyruvate dehydrogenase kinase. *Biochem. Biophys. Res. Comm.* 72: 11375–1383.

Roche, T.E. and Cate, R.L. (1977) Purification of porcine liver pyruvate dehydrogenase complex and characterization of its catalytic and regulatory properties. *Arch. Biochem. Biophys.* 183: 1664–677.

Stepp, L.R., Pettit, F.H., Yeaman, S.J. and Reed, L.J. (1983) Purification and properties of pyruvate dehydrogenase kinase from bovine kidney. *J. Biol. Chem.* 258: 19454–9458.

Sumegi, B. and Alkonyi, I. (1983) Elementary steps in the reaction of the pyruvate dehydrogenase complex from pig heart: kinetics of thiamin diphosphate binding to the complex. *Eur. J. Biochem.* 136: 1347–353.

Taylor, S.I., Mukherjee, C. and Jungas, R.L. (1973) Studies on the mechanism of activation of adipose tissue pyruvate dehydrogenase by insulin. *J. Biol. Chem.* 248: 173–81.

Taylor, S.I., Mukherjee, C. and Jungas, R.L. (1975) Regulation of pyruvate dehydrogenase in isolated rat liver mitochondria. *J. Biol. Chem.* 250: 12028–2035.

Teague, W.M., Pettit, F.H., Wu, T.-L., Silberman, S.R. and Reed, L.J. (1982) Purification and properties of pyruvate dehydrogenase phosphatase from bovine heart and kidney. *Biochemistry* 21: 15585–5592.

Thekkumkara, T.J., Ho, L., Wexler, I.D., Pons, G., Liu, T.-C. and Patel, M.S. (1988) nucleotide sequence of a cDNA for the dihydrolipoamide acetyltransferase component of human pyruvate dehydrogenase complex. *FEBS Lett.* 240: 145–48.

Thomas, A.P. and Denton, R.M. (1986) Use of toluene-permeabilized mitochondria to study the regulation of adipose tissue pyruvate dehydrogenase *in situ*. *Biochem. J.* 238: 193–101.

Wieland, O., von Funcke, H. and Laffler, G. (1971) *FEBS Lett.* 15: 1295–298.

Wieland, O.H. and Portenhauser, R.L. (1974) Regulation of pyruvate-dehydrogenase interconversion in rat-Liver mitochondria as related to the phosphorylation state of intramitochondrial adenine nucleotides. *Eur. J. Biochem.* 45: 1577–588.

Wu, T.-L. and Reed, L.J. (1984) Subunit binding in the pyruvate dehydrogenase complex from bovine kidney and heart. *Biochemistry* 23: 1221–226.

Alpha-Keto Acid Dehydrogenase Complexes
M.S. Patel, T.E. Roche and R.A. Harris (eds)
© 1996 Birkhäuser Verlag Basel/Switzerland

Lipoamide dehydrogenase

A. de Kok and W.J.H. van Berkel

Department of Biochemistry, Agricultural University, Dreijenlaan 3, NL-6703 HA Wageningen, The Netherlands

Summary. In this review the structure and function of the lipoamide dehydrogenase component of the keto acid dehydrogenase complexes is discussed. Three structural models are available and in recent years many new genes have been sequenced and expressed. Unsolved is the role that many non-complex-bound lipoamide dehydrogenases play in metabolism and the physiological function of free lipoic acid. Central in catalysis is the problem of stabilization of the 2-electron reduced enzyme. Mutagenesis studies and the use of a modified flavin cofactor revealed the various factors involved in this stabilization in which also the core component of the complex has a function.

Introduction

Dihydrolipoamide dehydrogenase or lipoamide dehydrogenase belongs to the family of the homodimeric FAD-dependent disulfide oxidoreductases. Well known other members of this family are glutathione reductase, mercuric ion reductase, thioredoxin reductase and trypanothione reductase. Lipoamide dehydrogenase is the only member that belongs to a multienzyme complex, however an increasing number of non-complex bound functions of this enzyme appears in the literature. An excellent review on the enzymology of this group of enzymes has been written by Williams (1992). A review on the structural characteristics of the pyruvate dehydrogenase complex appeared in 1992 (Mattevi et al.).

New enzymes and genes

Dihydrolipoamide dehydrogenase is usually described as the common component (Enzyme 3 or E3) of the 2-oxoacid dehydrogenase complexes. This is only partially true. In some organisms, thus far only in microorganisms, up to three different *lpd* genes are observed, each assigned to a specific complex. In *P. putida* three genes have been found, one for the pyruvate and oxoglutarate dehydrogenase complexes, one for the branched-chain complex and one of unknown function, termed *lpd*-glc, *lpd*-val and *lpd*-3 respectively (Burns et al., 1989; Palmer et al., 1991a, b). It has been suggested that *lpd*-3 might be part of the newly discovered acetoin dehydrogenase complex

(Oppermann and Steinbüchel, 1994). In *Enterococcus faecalis* two *lpd* genes are found: one is part of PDHC (Allen and Perham, 1991), this organism does not have an OGDHC, and one of unknown origin (Claiborne et al., 1994). This latter enzyme is part of an operon that probably encodes the components of the branched-chain complex (A. Claiborne, personal communication). Of interest is that *Enterococcus* is the only known organism that expresses PDHC under anaerobic conditions. The complex can function under anaerobic conditions because the lipoamide dehydrogenase component is insensitive to inhibition by NADH (Snoep et al., 1992).

Very exciting are the new gene clusters found for the acetoin dehydrogenase complex in *Clostridium magnum* and for the pyruvate dehydrogenase complexes in *Alcaligenes eutrophus* and *Neisseria meningitidis*. These clusters possess a lipoamide dehydrogenase component containing a lipoyl domain, connected with a linker of about 30 amino acids to the N-terminus (Hein and Steinbüchel, 1994; Krüger et al., 1994; Silva et al., 1994; Ala'Aldeen et al., 1995). Two of the three lipoyl domains usually found on the transacetylase (Enzyme 2, PDHC or E_2P) component from organisms of Gram-negative origin remain present. As the function of more than one lipoyl domain per E2-chain remains unclear, it would be interesting to know whether a lipoyl domain attached to the E3 component can take over the function of the domains connected to the E2 component. The protein from pathogenic *Neisseria meningitidis*, the causative agent of meningitis, has been patented by a Cuban group (Silva et al., 1994) for use as a vaccine candidate as they claim it to be an outer membrane protein. In collaboration with Dr Ala'Aldeen (Nottingham, UK) we have expressed this gene in *E. coli* and purified the enzyme. Its general features are very similar to other lipoamide dehydrogenases, without properties characteristic for a membrane protein. An E2E3 subcomplex could be reconstituted. Antibodies against E2 were cross reactive with E3 and with anti-outer membrane preparations, indicating that these antibodies were directed against the lipoyl domain. In view of the high antigenicity of the lipoyl domain, the antibodies may have been caused by a non-outer membrane impurity in the preparation. Immunogold studies are in progress to localize the protein.

Several reports indicate a plasma-membrane or even a membrane-bound ribosomal localization of the pyruvate dehydrogenase complex (Hamilä et al., 1990; Hamilä, 1991; Berks et al., 1993). A physiological explanation for this localization cannot be given at present.

Lipoamide dehydrogenases not associated with the multienzyme complexes

Several lipoamide dehydrogenases which are not part of the multienzyme complexes have been reported, but none of them have been well characterized. Most of these proteins have been identified as disulfide reductases, based on the reaction of the reduced disulfide bridge with arsenite. In some cases this identification is confirmed by sequence information. Without knowledge of a physiological function for free lipoamide or lipoic acid it is not possible to classify them as true lipoamide dehydrogenases. Sequence information is not of much help as there is no sequence motif for lipoamide binding. Furthermore dihydrolipoamide may not be a natural substrate, while the reaction of the known complex-bound lipoamide dehydrogenases with dihydrolipoic acid is very poor.

In the blood-stream form of *Trypanosoma brucei* a plasma membrane bound lipoamide dehydrogenase is found (Else et al., 1993). The gene has been cloned and sequenced and the derived amino acid sequence shows homology with other lipoamide dehydrogenases. In this organism, the substrate dihydrolipoate may be required for stability of the plasma membrane as arsenic-based drugs cause cell lysis.

In *E. coli* lpd⁻ mutants a second lipoamide dehydrogenase is found that is involved in sugar transport (Richarme, 1989). Unfortunately, the gene is not cloned and no information is available on the role of this enzyme in the transport process.

Two lipoamide dehydrogenases from archae have been purified and characterized (Danson, 1986; Vettakkorumakankav, 1992). These enzymes are not associated with any of the ketoacid dehydrogenase complexes, but could be associated with the glycine cleavage complex (Danson, 1986). Disruption of the gene in *Haloferax volcanii* had no effect on growth (Vettakkorumakankav, 1994), so the function of this enzyme is still a mystery.

Several unusual NAD(P)H dependent "diaphorases" have been described. Lipoamide dehydrogenase activity has been observed when a thioredoxin-like protein of 14 kDa was combined with a larger 70 kDa flavoprotein, termed electron-transfer flavoprotein, in *Clostridium litoralis* and *Eubacterium acidaminophilum* (Meyer et al., 1991). A similar system might operate, with a different disulfide substrate, in *Streptomyces clavuligerus* (Aharonowitz et al., 1993). Another role for lipoamide dehydrogenase was found in the vitamin K cycle (Thijssen et al., 1994). Here lipoamide dehydrogenase is thought to provide the epoxide reductase with reducing equivalents. A comparable system might operate in *Xanthobacter* where alkane epoxide oxidation requires the combination of an epoxide reductase and a disulfide reductase (Swaving et al., 1995). The FAD and NADP-binding domains, including the disulfide-bridge, are homologous with known E3 sequences, but no Glu-His diad (see below) is found.

Structures

Three X-ray structures have been described thus far, all from lipoamide dehydrogenases of prokaryotic origin. The enzyme from *Azotobacter vinelandii* at 2.2 Å resolution (Mattevi et al., 1991),
from *Pseudomonas putida* (LPD-val) at 2.45 Å with NAD⁺ bound (Mattevi et al., 1992) and
from *P. fluorescens* at 2.8 Å resolution (Mattevi et al., 1993). The three structures have very
similar characteristics, which resemble those of other disulfide reductases (Fig. 1). Each of its two
identical subunits comprises three domains: the FAD-binding domain, the NAD-binding domain
and the interface domain. The central domain, defined in analogy with the glutathione reductase
structure, is considered here as part of the FAD-domain. The catalytic center is located in an 11 Å
deep cleft at the interface between the two chains. The isoalloxazine separates the nicotinamide
ring at the *re*-side of the flavin from the redox active disulfide bridge at the *si*-side (Fig. 2). No
structures with bound lipoamide or of reduced forms are available. Crystals crack when they are

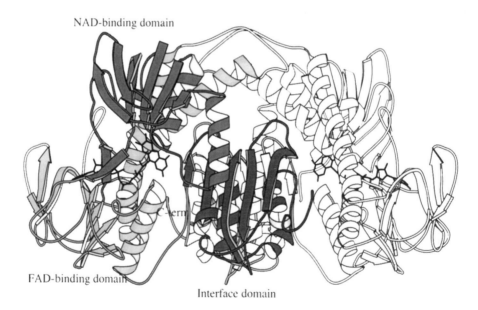

Figure 1. Ribbon (Kraulis, 1991) representation of the structure of *P. fluorescens* lipoamide dehydrogenase
viewed perpendicular to the twofold axis. Subunit 1 is indicated in grey shading, the other in white. FAD is a line
drawing. The FAD-binding domain (residues 1–150 and 281–350), the NAD-binding domain (151–280) and the
interface domain (351–472) are indicated by different intensities. The position of the C-terminal tail from subunit
2 is also indicated.

soaked with NADH or dihydrolipoamide, indicating extensive structural changes upon reduction. R-dihydrolipoamide has been modelled in the active site (De Kok et al., 1994) which indicates that S-8 is close to the active site residues Cys48 and Nε2(His450) and thus in a good position for proton abstraction and mixed disulfide formation. S-8 is also involved in the acyltransfer reactions (Mattevi et al., 1992).

In the original *A. vinelandii* structure the last 10 residues were not observed, indicating multi-conformational behavior. Deletion mutagenesis indicated the importance of these residues in subunit interaction and catalysis (Benen et al., 1992a). This was confirmed by the other two structures which show that the C-terminal alpha helix (residues 451 to 465, *A. vinelandii* numbering) aligns the active site and that the C-terminal tail (residues 466 to 473) folds back towards the active site and by doing so forms several hydrogen bridges with the other subunit as well as a contribution to van der Waals interaction. In *P. fluorescens* E3, the last five residues are not visible. They are probably on the outside of the protein and disordered.

Of interest is that the C-terminal alpha helix aligns a 20 Å-long, narrow, solvent-accessible channel towards the active site (De Kok et al., 1994). This channel runs perpendicular to the substrate binding cavity at the subunit interface. The function of this channel is unclear: is it required for the escape of bound water molecules when dihydrolipoamide enters the hydrophobic substrate binding channel or is it involved in the rapid release of protons produced in the reaction, or does it have other functions? The function has been probed by mutagenesis, with the aim to block this channel (see below).

No structures are yet available of E3-E2 subcomplexes. Direct binding studies with the wild type components of *A. vinelandii* PDHC have shown a 1:1 stoichiometry between E2 and the E3 or E1 dimer, while binding is mutually exclusive (Bosma et al., 1984). This stoichiometry could be determined by ultracentrifuge analysis, because the large E2 core dissociates into trimers upon binding. These results were confirmed more recently by binding studies of a lipoyl domain-binding domain construct to E3 and E1 from *Bacillus stearothermophilus* (Lessard and Perham, 1995) and by the effect of binding of E2p and E2o on the restoration of catalytic activity of an E3 mutant with weakened subunit interaction, Y16F (Fig. 3) (De Kok et al., 1994). These results indicate a binding mode where the asymmetric E2 binding domain binds to the interface domain near the twofold symmetry axis, thereby excluding the binding of a second domain.

The increase in activity of Y16F was used as an assay in the purification of a binding domain-E3 complex. An E2p binding domain-catalytic domain complex with E3 was treated with chymotrypsin, which cleaves the E2 presumably at Tyr373. E3 protected the binding domain against internal cleavage by chymotrypsin. This indicates that Phe349 in helix 1 of the homologous binding domain of *B. stearothermophilus* (Kalia et al., 1993), is part of the E2p-E3 interface.

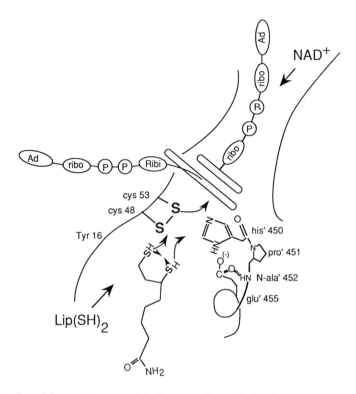

Figure 2. Detailed view of the catalytic center of *P. fluorescens* lipoamide dehydrogenase. The drawing outlines the two substrate binding channels separated by the flavin ring. The position of the redox-active disulfide bridge, the His-Glu diad and some important adjacent residues (see text) are indicated.

Similarly, residues of helix 2 might be involved in E1 binding (Schulze et al., 1991). The purified complex was crystallized with a resolution limit of 2.8 Å, but the structure could not yet be solved due to unexpected symmetry problems (W.G.J. Hol, personal communication).

Catalytic properties

The database presently contains 16 sequences of E3 enzymes. All important residues indicated by structural comparison with other disulfide reductases and mutagenesis are conserved. These are residues of the adenine binding folds of FAD and NAD, Lys57 (*A. vinelandii* numbering) interacting with O-4 of FAD, the CLXXGC consensus sequence of the redox active disulfide bridge, the ribitol binding motif TXXXXIYAIGD (Eggink et al., 1990), Ala326 forming a H-

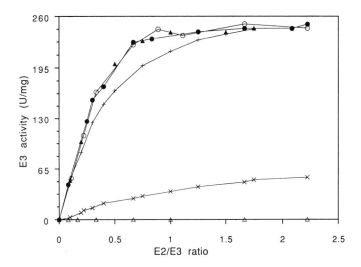

Figure 3. Titration of *A. vinelandii* E2p, E2o binding and some E2p mutants to E3(Y16F). All experiments were performed in 50 mM potassium phosphate buffer, pH 7.0. (○) varying amounts of E2p wild type were incubated with 96 nM E3; (●) E2o wild type with 960 nM E3; (+) E2o wild type with 96 nM E3; (×) E2o wild type with 19 nM E3; (△) E2p binding domain-catalytic domain with N-terminus at Pro340 of E2p sequence with 96 nM E3; (▲) E2p binding domain-catalytic domain with N-terminus at Lys335 of E2p sequence with 96 nM E3. From de Kok et al. (1994) with permission.

bridge with O-2 of FAD and the HPXXXE motif of the catalytic site, containing the *cis*-proline, interacting with the N-3 of FAD, and the His-Glu diad in the other subunit.

The flavin ring separates the NAD^+ binding site from the redox-active disulfide (Fig. 2). In the reductive half reaction, electrons are transferred from dihydrolipoamide to the disulfide bridge, forming the two-electron reduced enzyme. The proximal thiolate (Cys53) forms a charge-transfer complex with the oxidized flavin. Depending on the species and conditions several enzyme forms are in equilibrium at this oxidation state (Benen et al., 1991; Williams, 1992). In the oxidative half reaction the flavin is reduced and the electrons are transferred to NAD^+. The role of the active site His-Glu diad was demonstrated by site directed mutagenesis (Benen et al., 1991, 1992b). It is not only involved in the disulfide interchange reaction but also in the subsequent electron transfer to and from the flavin. The pKa of the histidine is tuned by the interaction with Glu455 and possibly also by the dipole moments of four alpha helices that point with their positive dipoles towards the active site (helices involving residues 11–24, 324–341, 47–64 and 451–465). These dipoles may also assist in regulating the redox potential of the FAD and the redox active disulfide (F. van den Akker, personal communication).

The weak binding of NAD$^+$ in the oxidized enzyme, a consequence of the ping-pong mechanism, is revealed by the *P. putida* structure with bound NAD$^+$. A tyrosine residue (Tyr181) is positioned between the flavin and the nicotinamide ring and, as a consequence, the nicotinamide ring points away from the isoalloxazine moiety. This situation is similar to that in glutathione reductase, where the phenyl ring of Tyr197 covers the NADPH binding site in the oxidized enzyme (Karplus and Schulz, 1989). Presumably this residue, or a Val or Ile at this position in the other known enzymes, moves away when the two-electron reduced enzyme is formed. Replacement of Ile184 by Tyr in the *E. coli* enzyme inhibited the transfer of electrons from the flavin to the disulfide bridge (Maeda-Yorita et al., 1991). Probably the stacking interaction between tyrosine and flavin alters the redox potential and the interaction of NAD$^+$ with the flavin. These results show the importance of residues in this position in the regulation of the catalytic activity. Analysis of flavin fluorescence in lipoamide dehydrogenase has shown that the flavin environment is dynamic (Bastiaens et al., 1992). The results were interpreted in terms of equilibria between conformational substates. A minor population of an "open" substate, which might be ascribed to a conformation in which the residue at position 181 (*P. putida* numbering) has moved away from the flavin and binds the pyridine nucleotide substrate, is in equilibrium with a major population of a "closed" substate that does not bind the substrate. Equilibria between flavin substates are also present in *p*-hydroxybenzoate hydroxylase where the isoalloxazine ring can move between productive and unproductive substates (van Berkel et al., 1994).

The redox potential of the FAD plays a crucial role in catalysis: if it is too high the electrons are passed over from the disulfide to the flavin, facilitating the formation of the inactive four-electron reduced enzyme. Mutagenesis and other studies indicate that several factors are involved. The redox potential of the flavin is directly influenced by the electrostatic properties of residues in the vicinity of the isoalloxazine. The most important residue seems to be the conserved Lys57 mentioned above. By changing this residue into Arg, the effect on the redox potential of the FAD was demonstrated in the *E. coli* enzyme (Maeda-Yorita et al., 1994). Rapid reduction to the inactive four-electron reduced enzyme indicates a rise in redox potential of the FAD, which was confirmed by direct measurements of this potential.

Lipoamide dehydrogenase from *A. vinelandii* is moderately stable against over-reduction. Nevertheless, steady state kinetics has to be studied by stopped-flow like the *E. coli* enzyme (Sahlman and Williams, 1989) to prevent the build up of NADH. In the first 2 s after mixing, the turnover in the physiological direction is 830 s^{-1}, compared to 330 s^{-1} measured under the usual steady-state conditions. The enzyme dissociates upon reduction to the four-electron reduced state. For this enzyme it has therefore been concluded that the subunit-subunit interaction plays an important role in over-reduction. The C-terminal tail contributes significantly to this interaction as indicated by the 275 Å^2 of the total surface of 3510 Å^2 that is buried upon dimerization. Four hydrogen bonds, Tyr16-His470' and Gln24-Ala468' are present in this region. C-terminal deletion mutants in which these interactions are disrupted form dimers with decreased conformational stability (Benen et al., 1992a). They are almost inactive due to rapid transfer of electrons from the disulfide to the flavin. The more subtle mutations Y16F or Y16S, preventing the formation of two intersubunit H-bonds, gave similar results. Spectral and kinetic details of these mutant enzymes are described in Benen et al. (1992a). The turnover rate of the mutant in the stopped flow is still Y16F 560 s^{-1}, but the inhibition by NADH is severe and rapid inactivation takes place (Westphal et al., 1995). Inhibition is purely competitive with respect to NAD$^+$, with a Ki of 10 μM (Fig. 4), a

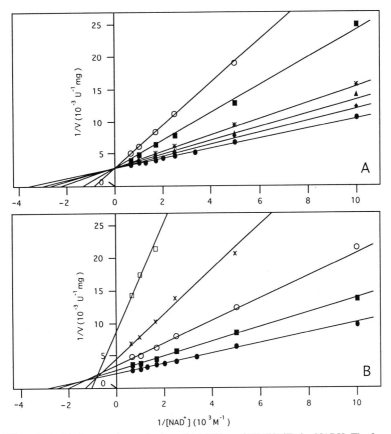

Figure 4. Effect of E2p binding on the mechanism of inhibition of E3 (Y16F) by NADH. The forward reaction was measured in 50 mM sodium pyrophosphate, 0.5 mM EDTA, pH 8.0 at 25°C. 1 mM dihydrolipoamide was used and the NAD$^+$ concentration was varied. Rates were measured 2 s after mixing in the stopped-flow apparatus. The NADH concentration was (●) 0 μM; (◆) 5 μM; (▲) 10 μM; (★) 25 μM; (■) 50 μM, (○) 100 μM; (×) 167 μM and (□) 250 μM. (A) Free Y16F mutated lipoamide dehydrogenase. (B) Y16F mutated lipoamide dehydrogenase in the presence of a stoichiometric amount of E2p.

clear example of dead-end inhibition in a Ping-Pong Bi Bi system (Reed, 1973). The wild-type enzyme and the complex between Y16F and E2p (see below) show mixed inhibition.

So far no mutations have been made that strengthen the subunit interaction to show the opposite effect. However, the structure of the enzyme from *P. fluorescens* provides a direct clue (Mattevi et al., 1993). The enzyme has 84% sequence identity with the *A. vinelandii* enzyme, but is much more resistant against over-reduction. In this enzyme all interface residues are conserved with one exception. In *P. fluorescens* E3 the oxygen atom of Thr452 forms a hydrogen bond with the carbonyl oxygen of Ser449. In *A. vinelandii* E3 Thr452 is replaced by Ala, so the carbonyl oxygen of Ser449 is left as an unsatisfied H-bond acceptor. The higher resistance of the *P. fluorescens* enzyme towards over-reduction may be caused by this substitution (Mattevi et al., 1993). Another clue is provided by the interaction with the E2 component of the complex. In *A. vinelandii* E3, the interaction of wild-type E3 with E2 leads to a somewhat decreased susceptibility for NADH. However, with enzyme Y16F the effect is dramatic. The activity is almost completely restored. This is probably due to tightening of the subunit-subunit interaction and demonstrates clearly that E2 binds to the subunit interface (see above) (Westphal et al., 1995).

Despite the importance of the subunit interface for substrate binding and catalysis, the percentage change in subunit residues compares very well with the overall residue changes (Grishin et al., 1994). Replacement of a few residues, causing small changes in subunit interaction discussed above for *A. vinelandii* and *P. fluorescens* enzymes, may be a general mechanism for adaptation to different intracellular NADH/NAD ratio's. One extreme is the *E. coli* enzyme which is very easily reduced to the inactive four-electron reduced form and causes PDHC to be functional only under aerobic conditions. The other extreme is the *E. faecalis* enzyme that is very stable and allows the complex to be functional under both aerobic and anaerobic conditions (Snoep et al., 1993).

Table 1. Unfolding of (mutant) lipoamide dehydrogenase from *A. vinelandii* by Gdn/HCl

Wild-type	C_m (Gdn/HCl)	Mutant	C_m (Gdn/HCl)
	M		M
E_{ox}	2.4	Y16F	2.1
EH_2	1.5	Y16S	2.1
EH_4	0.5	Δ5	2.4
apoenzyme	0.7	Δ9	2.0
arabino-FAD	2.2	Δ14	0.8
		L462F	0.9

The experiments were performed as described in the legend to Figure 6 and in Van Berkel et al. (1991b).

To probe the function of the long solvent accessible channel mentioned above, we have replaced Leu462 by Phe. Modelling shows that the larger side chain could effectively block this channel. However, the mutation resulted probably in local distortion of the subunit interface, causing similar catalytic effects as the Y16F mutation. The effect on the conformational stability is dramatic: the midpoint concentration of unfolding by guanidine. HCl (C_m Gdn/HCl) shifted from 2.4 M to 0.9 M, close to the corresponding value of the monomeric apoenzyme (Tab. 1). This indicates large effects on the strength of subunit interaction.

Another approach to study the redox properties of lipoamide dehydrogenase is to replace the natural cofactor by flavin analogs. To probe the effect on the stabilization of the two-electron reduced enzyme, we reconstituted the *A. vinelandii* enzyme with arabino-FAD. This optical isomer is spontaneously formed from natural FAD in alcohol oxidase from methylotrophic yeasts (Kellog et al., 1992). The structure of the arabityl side chain was recently confirmed by crystallographic analysis of *p*-hydroxybenzoate hydroxylase reconstituted with the modified FAD (van Berkel et al., 1994). Arabino-FAD is expected to be a good reporter group in lipoamide dehydrogenase because the C2 carbon of the sugar side chain is located in the vicinity of the redox active cysteine pair.

Reconstitution of dimeric lipoamide dehydrogenase from the monomeric apoprotein and arabino-FAD is much slower than with natural FAD, indicating that the interaction between the sugar side chain and the apoprotein plays an important role in the dimerization process. The spectral properties of the arabino-FAD complexed enzyme indicate that in the oxidized state the microenvironment of the isoalloxazine is highly conserved. Unfolding experiments revealed a C_m Gdn/HCl = 2.2 M (cf. Tab. 1). These results suggest that binding of arabino-FAD influences the kinetic rather than the thermodynamic stabilization of the dimer.

Reduction of lipoamide dehydrogenase is accompanied by characteristic changes in the absorption spectrum of protein-bound FAD (Williams, 1992). Reduction of the native enzyme by dihydrolipoamide results in formation of charge-transfer absorption around 530 nm. With the arabino-FAD complexed enzyme no stabilization of 530-nm absorption is observed and the addition of a small molar excess of dihydrolipoamide results in a rapid and complete bleaching of the 450 nm absorption band (Fig. 5). When the arabino-FAD complexed enzyme is anaerobically titrated with NADH, again no thiolate-flavin charge transfer absorbance is observed. Instead, the flavin is reduced as evidenced from the decrease of absorbance at 450 nm. Flavin reduction is accompanied with the appearance of long wavelength absorbance indicative for the formation of a charge transfer complex between reduced flavin and NAD^+ (Williams, 1992). These results indicate that in the arabino-FAD complexed enzyme the thiolate anion fails to stabilize the charge transfer interaction, thus prohibiting the lowering of the reduction potential of the flavin.

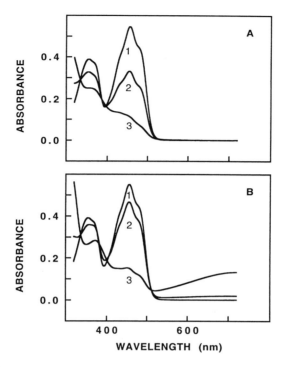

Figure 5. Visible absorption spectra of lipoamide dehydrogenase from *A. vinelandii*, reconstituted with arabino-FAD. Spectra were recorded under anaerobic conditions at 25°C in 50 mM sodium pyrophosphate, 0.5 mM EDTA, pH 8.0. (A) 50 µM arabino-FAD complexed lipoamide dehydrogenase in the absence (1) and presence of (2) 0.6 mol/mol or (3) 1.2 mol/mol dihydrolipoamide. (B) 50 µM arabino-FAD complexed lipoamide dehydrogenase in the absence (1) and presence of (2) 0.6 mol/mol or (3) 1.2 mol/mol NADH.

The catalytic properties of lipoamide dehydrogenase reconstituted with arabino-FAD are comparable to those of the Y16F mutant (Benen et al., 1992a). Strong dead-end inhibition is observed, which can be overcome by using AcPyAde+ as the electron acceptor at pH 7. This again indicates that replacement of normal FAD with arabino-FAD leads to an increase in the redoxpotential of the flavin. In contrast to the Y16F mutant, the addition of E2 does not improve the turnover rate.

The structure of the EH$_2$-NADPH complex of glutathione reductase may provide a clue to the observed effects. In the reduced state the charge transfer thiolate (Cys63, glutathione reductase numbering) is in H-bond distance of the 2OH of the ribitol which may help to stabilize the thiolate (Karplus and Schulz, 1989). A comparable geometry is also seen in the EH$_2$-NADP+ form of mercuric reductase (Schiering et al., 1991). From the conserved active-site geometry around the disulfide (Mattevi et al., 1991) it is likely that reduction of native lipoamide dehydrogenase results in a comparable structural stabilization of the thiolate. Changing the stereochemistry of the C2 group of the sugar side chain clearly affects this geometry, probably by changing the optimal orientation and/or the pKa of the proximal cysteine (Cys53, *A. vinelandii* numbering).

As discussed above, several lines of evidence have indicated that a tight subunit interaction prevents over-reduction. Understanding the factors influencing dimer stabilization has been facili-

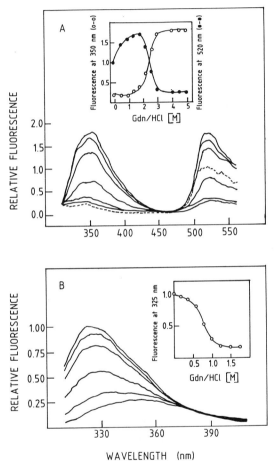

Figure 6. Tryptophan emission spectra of holo-lipoamide dehydrogenase and apo-lipoamide dehydrogenase from *A. vinelandii* in Gdn/HCl. 2 μM enzyme was incubated for 30 min in 100 mM potassium phosphate buffer pH 7.0 and various concentrations of Gdn/HCl at 25 °C. The excitation wavelength was 295 nm. (A) Holoenzyme in the presence of 0.5, 1.0, 1.8, 2.6, 3.0 and 5.0 M Gdn/HCl. The inset shows the relative fluorescence at 350 nm and 520 nm as a function of the unfolding agent, assigning the relative fluorescence of the native holoenzyme at 520 nm a value of 1.0 (dashed line). (B) Apoenzyme. For clarity only some of the recorded spectra are shown. In the inset, the relative fluorescence at 325 nm is plotted as a function of the unfolding agent. (From van Berkel et al. (1991) *Eur. J. Biochem.* 202: 1049–1055 (with permission)).

tated by the development of a mild and convenient method of preparing the monomeric apoenzyme (van Berkel et al., 1988). Briefly, the enzyme is bound in high salt to phenyl Sepharose and the flavin prosthetic group is quantitatively stripped off at pH 4. Through subsequent elution with ethylene glycol, fully monomeric apoenzyme with negligible residual activity is obtained in high yield.

Reconstitution experiments showed that FAD-induced dimerization of the *A. vinelandii* apoenzyme is minimally a two-step sequential process (van Berkel et al., 1991a). Initial flavin binding results in regaining of DCIP activity and quenching of tryptophan fluorescence. In the second step, dimerization occurs as reflected by the regain of lipoamide activity, strongly increased FAD fluorescence and increased hyperchroism of the visible absorption spectrum. Furthermore, from stopped flow reconstitution experiments it was demonstrated that the rate of dimerization is temperature dependent and that the dissociation constant of the dimer is less than 1 nM.

Lipoamide dehydrogenase from *A. vinelandii* contains a single tryptophan residue (Trp199). In the apoenzyme, this tryptophan is highly fluorescent which served as a probe to monitor the conformational stability (van Berkel et al., 1991b). Unfolding of the apoenzyme is a rapid fully reversible process, following a simple two-state mechanism. Loss of energy transfer from Trp199 to flavin was used as a reliable method to monitor the unfolding of the holoenzyme (van Berkel et al., 1991b). As can be seen from Figure 6 and Table 1, the holodimer is far more resistant towards unfolding than the monomeric apoenzyme. From Table 1 it is also clear that the conformational stability is strongly dependent on the degree of reduction, supporting the idea that over-reduction is accompanied by conformational changes that promote subunit dissociation.

Medical implications

Although α-lipoic acid is known to be a poor substrate for lipoamide dehydrogenase, racemic α-lipoic acid is therapeutically applied in pathologies in which free radicals are involved (Biewenga et al., 1994). This has drawn attention to its anti-oxidant profile. It appears that dihydrolipoic acid in many cases is a better anti-oxidant than lipoic acid. This is partly due to direct effects and in part by regeneration of oxidized glutathione and vitamin C. Mammalian cells within minutes convert α-lipoic acid to the dithiol form and excrete dihydrolipoic acid in the medium (Handelman et al., 1994).

In the complex, the R-enantiomer is the natural substrate (Gunsalus et al., 1956). Biphasic kinetics in the reduction of lipoamide dehydrogenase by R, S-dihydrolipoamide was explained by a rapid reaction with the R-enantiomer, while the slow phase could be due to a slow reduction of the enzyme by the S-enantiomer or by a slow nonenzymatic reduction of the R-enantiomer. Arscott and Williams (1994) have recently shown that the methyl esters of both enantiomers are substrates of lipoamide dehydrogenase, although the S-enantiomer is a much poorer substrate. On the other hand, S-lipoic acid is a better substrate than R-lipoic acid for mammalian glutathione reductase (Pick et al., 1995).

In humans a single gene codes for lipoamide dehydrogenase (Otulakowski et al., 1988). Therefore E3 deficiency causes a reduction in the activities in all three alpha keto acid dehydrogenase complexes involved and in the glycine cleavage system as well. The availability of a full-length human cDNA (Pons et al., 1988) has made it possible to characterize the molecular characterization of E3 deficiencies (Liu et al., 1993). Two point mutations, K37E and P453L, were found in a single patient. K37 is in contact with the solvent and the effect of the mutation would probably be negligible. The effects of the P453L substitution however would be very severe. P453 is in *cis*-conformation and positions the backbone with the his-glu diad in the active site and orients the backbone carbonyl of the adjacent His452 for interaction with N-3 of the isoalloxazine. We have mutated P451 to alanine in *A. vinelandii* E3 (Benen et al., 1992a). Although the holo enzyme could be expressed and isolated in high yield, the enzyme is over-reduced in a few turnovers and therefore inactive.

Acknowledgement
This research was supported by the Netherlands Foundation for Chemical Research (SON) with financial aid from the Netherlands Organization for Scientific Research (NWO).

References

Aharonowitz, Y.A., Gay, Y., Schreiber, R. and Cohen, G. (1993) Characterization of a broad-range disulfide reductase from *Streptomyces clavuligerus* and its possible role in beta-lactam biosynthesis. *J. Bacteriology* 175: 623–629.

Ala'Aldeen, D.A.A., Weston, V, Baldwin, T.J. and Boriello, S.P. (1995) The gene cluster encoding components of the pyruvate dehydrogenase complex of *Neisseria meningitidis*: detection and sequence analysis. *J. Med. Microbiol.* 42: 148.

Allen, A.G. and Perham, R.N. (1991) Two lipoyl domains in the dihydrolipoamide acetyltransferase chain of the pyruvate dehydrogenase complex of *Streptococcus faecalis*. *FEBS Lett.* 287: 206–210.

Arscott, L.D. and Williams, C.H., Jr. (1994) R- and S-dihydrolipoic acid derivatives as substrates of lipoamide dehydrogenase. *In*: K. Yagi (ed.): *Flavins and Flavoproteins 1993*, Walter de Gruyter, Berlin, pp 527–530.

Bastiaens, P.I.H., van Hoek, A., Wolkers, W.F., Brochon, J.C. and Visser, A.J.W.G. (1992) Comparison of the dynamical structures of lipoamide dehydrogenase and glutathione reductase by time resolved polarized flavin fluorescence. *Biochemistry* 31: 7050–7060.

Benen, J., van Berkel, W., Zak, Z., Visser, T., Veeger, C. and De Kok, A. (1991) Lipoamide dehydrogenase from *Azotobacter vinelandii*: site-directed mutagenesis of the His450-Glu455 diad. Spectral properties of wild type and mutated enzymes. *Eur. J. Biochem.* 202: 863–872.

Benen, J., van Berkel, W., Veeger, C. and de Kok, A. (1992a) Lipoamide dehydrogenase from *Azotobacter vinelandii*. The role of the C-terminus in catalysis and dimer stabilization. *Eur. J. Biochem.* 207: 499–505.

Benen, J., van Berkel, W., Dieteren, N., Arscott, D., Williams, C.H., Jr., Veeger, C. and De Kok, A. (1992b) Lipo-amide dehydrogenase from *Azotobacter vinelandii*: site-directed mutagenesis of the His 450-Glu455 diad. Kinetics of wild type and mutated enzymes. *Eur. J. Biochem.* 207: 487–497.

Berks, B.C., McEwan, A.G. and Ferguson, S.J. (1993) Membrane-associated NADH-dehydrogenase activities in *Rhodobacter capsulatus*: purification of a dihydrolipoyl dehydrogenase. *J. Gen. Microbiol.* 139: 1841–1851.

Biewenga, G.P., de Jong, J. and Bast, A. (1994) Lipoic acid favors thiolsulfinate formation after hypochlorous acid scavenging: a study with lipoic acid derivatives. *Arch. Biochem. Biophys.* 312: 114–120.

Bosma, H.J., de Kok A., Westphal, A.H. and Veeger, C. (1984) The composition of the pyruvate dehydrogenase complex from *Azotobacter vinelandii*. Does a unifying model exist for the complexes from gram-negative bacte-ria? *Eur. J. Biochem.* 142: 541–549.

Burns, G., Brown, T., Hatter, K. and Sokatch, J.R. (1989) Sequence analysis of the lpdV gene for lipoamide dehy-drogenase of branched-chain-oxoacid dehydrogenase of *Pseudomonas putida*, *Eur. J. Biochem.* 179: 61–69.

Claiborne, A., Ross, R.P., Ward, D., Parsonage, D. and Crane III, E.J. (1994) Flavoprotein peroxide and disulfide reductases and their roles in Streptococcal oxidative metabolism. *In:* K. Yagi (ed.): *Flavins and Flavoproteins 1993*, Walter de Gruyter, Berlin, pp 587–596.

De Kok, A., Berg, A., van Berkel, W., Fabish-Kijowska, A., van den Akker, F., Mattevi, A. and Hol, W.G.J. (1994) The pyruvate dehydrogenase complex from *Azotobacter vinelandii*. *In:* K. Yagi (ed.): *Flavins and Flavoproteins 1993*, Walter de Gruyter, Berlin, pp 535–544.

Danson, M.J., McQuattie, A. and Stevenson, K.J. (1986) Dihydrolipoamide dehydrogenase from halophilic archae-bacteria: purification and properties of the enzyme from *Halobacterium halobium*. *Biochemistry* 25: 3880–3884.

Eggink, G., Engel, H., Vriend, G., Terpstra, P. and Witholt, B. (1990) Rubredoxin reductase of *Pseudomonas oleo-vorans*. Structural relationship to other flavoprotein oxidoreductases based on one NAD and two FAD finger-prints. *J. Mol. Biol.* 212: 135–142.

Else, A.J., Hough, D.W. and Danson, M.J. (1993) Cloning, sequencing, and expression of *Trypanosoma brucei* dihydrolipoamide dehydrogenase. *Eur. J. Biochem.* 212: 423–429.

Grishin, N.V. and Phillips, M.A. (1994) The subunit interfaces of oligomeric enzymes are conserved to a similar extent to the overall protein sequence. *Protein Sci.* 3: 2455–2458.

Gunsalus, I.C., Barton, L.S. and Gruber, W. (1956) Biosynthesis and structure of lipoic acid derivatives. *J. Am. Chem. Soc.* 78: 1763–1766.

Hamilä, H., Palva, A., Paulin, L., Arvidson, S. and Palva, I. (1990) Secretory S complex of *Bacillus subtilis*: sequence analysis and identity to pyruvate dehydrogenase. *J. Bacteriology* 172: 5052–5063.

Hamilä, H. (1991) Lipoamide dehydrogenase of *Staphylococcus aureus*: nucleotide sequence and sequence analy-sis. *Biochim. Biophys. Acta* 1129: 119–123.

Handelman, G.J., Han, D., Tritschler, H. and Packer, L. (1994) α-lipoic acid reduction by mammalian cells to the dithiol form, and release into the culture medium. *Biochem. Pharmacol.* 47: 1725–1730.

Hein, S. and Steinbüchel, A. (1994) Biochemical and molecular characterization of the *Alcaligenes eutrophus* pyruvate dehydrogenase complex and identification of a new type of dihydrolipoamide dehydrogenase. *J. Bacteriology* 176: 4394–4408.

Kalia, Y.N., Brocklehurst, S.M., Hipps, D.S., Appela, E., Sakaguchi, K. and Perham, R.N. (1993) The high resolu-tion structure of the peripheral subunit binding domain of dihydrolipoamide acetyltransferase from the pyruvate dehydrogenase complex of *Bacillus stearothermophilus*. *J. Mol. Biol.* 230: 323–341.

Karplus, P.A. and Schulz, G. (1989) Substrate binding and catalysis by glutathione reductase as derived from refined enzyme: substrate crystal structures at 2 Å resolution. *J. Mol. Biol.* 210: 163–180.

Kellog, R.M., Kruizinga, W., Bystrykh, L.V., Dijkhuizen, L. and Harder, W. (1992) Structural analysis of a stereo-chemical modification of flavin adenine dinucleotide in alcohol oxidase from methylotrophic yeasts. *Tetrahedron* 48: 4147–4162.

Kraulis, P.J. (1991) MOLSCRIPT: a program to produce both detailed and schematic plots of protein structures. *J. Appl. Cryst.* 24: 946–950.

Krüger, N., Oppermann, F.B., Lorenzl, H. and Steinbüchel, A. (1994) Biochemical and molecular characterization of the *Clostridium magnum* acetoin dehydrogenase enzyme system. *J. Bacteriology* 176: 3614–3630.

Lessard, I.A.D. and Perham, R.N. (1995) Interaction of component enzymes with the peripheral subunit-binding domain of the pyruvate dehydrogenase multienzyme complex of *Bacillus stearothermophilus*: stoichiometry and specificity in self assembly. *Biochem. J.* 306: 727–733.

Liu, T.-C., Kim, H., Arizmendi, C., Kitano, A. and Patel, M.S. (1993) Identification of two missense mutations in a dihydrolipoamide dehydrogenase-deficient patient. *Proc. Natl. Acad. Sci. USA* 90: 5186–5190.

Maeda-Yorita, K., Russell, G.C., Guest, J.R., Massey, V. and Williams, C.H., Jr. (1991) Properties of lipoamide dehydrogenase altered by site-directed mutagenesis at a key residue (I184Y) in the pyridine nucleotide binding domain. *Biochemistry* 30: 11788–11795.

Maeda-Yorita, K., Russell, G.C., Guest, J.R., Massey, V. and Williams, C.H., Jr. (1994) Modulation of the oxida-tion-reduction potential of the flavin in lipoamide dehydrogenase from *Escherichia coli* by alteration of a nearby charged residue, K53R. *Biochemistry* 33: 6213–6220.

Mattevi, A., Schierbeek, A.J. and Hol, W.G.J. (1991) Refined crystal structure of lipoamide dehydrogenase from *Azotobacter vinelandii* at 2.2 Å resolution. A comparison with the structure of glutathione reductase. *J. Mol. Biol.* 220: 975–994.

Mattevi, A., de Kok, A. and Perham, R.N. (1992) The pyruvate dehydrogenase multienzyme complex. *Curr. Opin. Struct. Biol.* 2: 877–887.

Mattevi, A., Obmolova, G., Kalk, K.H., Sokatch, J., Betzel, C.H. and Hol, W.G.J. (1992) The refined crystal structure of *Pseudomonas putida* lipoamide dehydrogenase complexed with NAD+ at 2.45 Å resolution. *Proteins* 13: 336–351.

Mattevi, A., Obmolova, G., Kalk, K.H., Van Berkel, W.J.H. and Hol, W.G.J (1993) The refined crystal structure of *Pseudomonas fluorescens* lipoamide dehydrogenase at 2.8 Å resolution: Analysis of redox and thermostability properties. *J. Mol. Biol.* 230: 1200–1215.

Meyer, M., Dietrichs, D., Schmidt, B. and Andreesen, J.R. (1991) Thioredoxin elicits a new dihydrolipoamide dehydrogenase activity by interaction with the electron-transferring flavoprotein in *Clostridium litoralis* and *Eubacterium acidaminophilum*. *J. Bacteriology* 173: 1509–13.

Oppermann, F.B. and Steinbüchel, A. (1994) Identification and molecular characterization of the *aco* genes encoding the *Pelobacter carbinolicus* acetoin dehydrogenase enzyme system. *J. Bacteriology* 176: 469–485.

Otulakowski, G., Robinson, B.H. and Willard, H.F. (1988) Gene for lipoamide dehydrogenase maps to human chromosome 7. *Som. Cell. Mol. Gen.* 14: 411–414.

Palmer, J.A., Hatter, K. and Sokatch, J.R. (1991a) Cloning and sequence analysis of the LPD-glc structural gene of *Pseudomonas putida*. *J. Bacteriology* 173: 3109–3116.

Palmer, J.A., Madhusudhan, K.T., Hatter, K. and Sokatch, J.R. (1991b) Cloning, sequencing and transcriptional analysis of the structural gene of LPD-3, the third lipoamide dehydrogenase of *Pseudomonas putida*. *Eur. J. Biochem.* 202: 231–240.

Pick, U., Haramaki, N., Constantinescu, A., Handelman, G.J., Tritschler, H.J. and Packer, L. (1995) Glutathione reductase and lipoamide dehydrogenase have opposite stereospecificities for α-lipoic acid enantiomers. *Biochem. Biophys. Res. Comm.* 206: 724–730.

Pons, G., Raefsky-Estrin, C., Carothers, D.J., Pepin, R.A., Javed, A.A., Jesse, B.W., Ganapathi, M.K., Samols, D. and Patel, M.S. (1988) Cloning and cDNA sequence of the dihydrolipoamide dehydrogenase component of human alpha ketoacid dehydrogenase complexes. *Proc. Natl. Acad. Sci. USA* 85: 1422–1426.

Reed, J.K. (1973) Studies on the kinetic mechanism of lipoamide dehydrogenase from rat liver mitochondria. *J. Biol. Chem.* 248: 4834–4839.

Richarme, G. (1989) Purification of a new dihydrolipoamide dehydrogenase from *Escherichia coli*. *J. Bacteriology* 171: 6580–6585.

Sahlman, L. and Williams, C.H., Jr. (1989) Lipoamide dehydrogenase from *Escherichia coli*. Steady state kinetics of the physiological reaction. *J. Biol. Chem.* 264: 8039–8045.

Schiering, N., Kabsch, W., Moore, M.J., Distefano, M.D., Walsh, C.T. and Pai, E.F. (1991) Structure of the detoxification catalyst mercuric ion reductase from *Bacillus* sp. strain RC607. *Nature* 352: 168–172.

Schulze, E., Westphal, A.H., Boumans, H. and De Kok, A. (1991) Site-directed mutagenesis of the dihydrolipoyl transacetylase component (E2p) of the pyruvate dehydrogenase complex from *Azotobacter vinelandii*. Binding of the peripheral components E1p and E3. *Eur. J. Biochem.* 202: 841–848.

Silva, R., Selman, M., Guillen, G., Herrera, L., Fernandez, J.R., Novoa, L.L., Morales, J., Morrera, V., Gonzalez, S., Tamargo, B., del Valle, J.A., Caballero, E., Alvarez, A., Coizeau, E., Cruz, S. and Mussachio, A. (1991) Nucleotide sequence coding for an outer membrane protein from *Neisseria meningitidis* and use of said protein in vaccine preparation. *Eur. Patent Application* 0 474 313 A2.

Snoep, J.L., Westphal, A.H., Benen, J.A.E., Teixeira de Mattos, M.J., Neijssel, O.M. and de Kok, A. (1992) Isolation and characterization of the pyruvate dehydrogenase complex of anaerobically grown *Enterococcus faecalis* NCTC 775. *Eur. J. Biochem.* 203: 245–250.

Snoep, J.L., De Graef, M.R., Westphal, A.H., De Kok, A., Teixeira de Mattos, M.J. and Neijssel, O.M. (1993) Differences in sensitivity to NADH of purified pyruvate dehydrogenase complexes of *Enterococcus faecalis*, *Lactococcus lactis*, *Azotobacter vinelandii* and *Escherichia coli*: Implications for their activity *in vivo*. *FEMS Microbiol. Lett.* 114: 279–284.

Swaving, J., Weijers, C.A.G.M., van Ooyen, A.J.J. and de Bont, J.A.M. (1995) Complementation of *Xanthobacter* Py2 mutants defective in epoxyalkane degradation, and expression and nucleotide sequence of the complementing DNA fragment. *Microbiology* 141: 477–484.

Thijssen, H.H.W., Janssen, Y.P.G. and Vervoort, L.T.M. (1994) Microsomal lipoamide reductase provides vitamin K equivalents with reducing equivalents. *Biochem. J.* 297: 277–280.

Van Berkel, W.J.H., van den Berg, W.A.M. and Müller, F. (1988) Large-scale preparation and reconstitution of apo-flavoproteins with special reference to butyryl-CoA dehydrogenase from *Megasphaera elsdenii*. *Eur. J. Biochem.* 178: 197–207.

Van Berkel, W.J.H., Benen, J.A.E. and Snoek, M.C. (1991a) On the FAD-induced dimerization of apo-lipoamide dehydrogenase from *Azotobacter vinelandii* and *Pseudomonas fluorescens*. Kinetics of reconstitution. *Eur. J. Biochem.* 197: 769–779.

Van Berkel, W.J.H., Regelink, A.G., Beintema, J.J. and de Kok, A. (1991b) The conformational stability of the redox states of lipoamide dehydrogenase from *Azotobacter vinelandii*. *Eur. J. Biochem.* 202: 863–872.

Van Berkel, W.J.H., Eppink, M.H.M. and Schreuder, H.A. (1994) Crystal structure of *p*-hydroxybenzoate hydroxylase reconstituted with the modified FAD present in alcohol oxidase from methylotrophic yeasts: Evidence for an arabinoflavin. *Protein Science* 3: 2245–2253.

Vettakkorumakankav, N.N. Danson, M.J., Hough, D.W., Stevenson, K.J., Davison, M. and Young, J. (1992) Dihydrolipoamide dehydrogenase from the halophilic archaebacterium *Haloferax volcanii*: characterization and N-terminal sequence. *Biochem. Cell. Biol.* 70: 70–75.

Vettakkorumakankav, N.N., Stevenson, K.J., Schalkwyk, L.C. and Doolittle, W.F. (1994) Disruption of the gene coding for dihydrolipoamide dehydrogenase in *Haloferax volcanii* by homologous recombination. *In*: K. Yagi (ed.): *Flavins and Flavoproteins 1993*, Walter de Gruyter, Berlin, pp 519–522.

Westphal, A.H., Fabisz-Kijovska, A., Kester, H., Obels, P.P. and DeKok, A. (1995) The interaction between lipoamide dehydrogenase and the peripheral-component-binding domain from *Azotobacter vinelandii* pyruvate dehydrogenase complex. *Eur. J. Biochem.*; *in press.*

Williams, C.H., Jr. (1992) Lipoamide dehydrogenase, Glutathione reductase, Thioredoxin reductase and Mercuric reductase – Family of flavoprotein transhydrogenases. *In*: F. Müller (ed.): *Chemistry and Biochemistry of Flavoenzymes*, Vol. 3, CRC Press, Boca Raton, pp 121–211.

Alpha-Keto Acid Dehydrogenase Complexes
M.S. Patel, T.E. Roche and R.A. Harris (eds)
© 1996 Birkhäuser Verlag Basel/Switzerland

Plant pyruvate dehydrogenase complexes

M.H. Luethy, J.A. Miernyk, N.R. David and D.D. Randall

Department of Biochemistry, University of Missouri, 117 Schweitzer Hall, Columbia, MO 65211, USA

Introduction

While there is considerable overall metabolic similarity between plant and animal cells, drastically different anatomy, physiology, and organismal requirements have led to increasing diversity between these two classes of eukaryotes. Analyses of the pyruvate dehydrogenase complex (PDC) in plant cells serve to illustrate both the similarities inherent in pyruvate metabolism and differences dictated by the need to respond to diverse external stimuli. Plants contain two distinct, spatially separated PDCs, one within the mitochondrial matrix and the other in the plastid stroma. Each PDC isoform has characteristic structural, catalytic, and regulatory properties (Miernyk et al., 1985; Randall et al., 1989). The mitochondrial location of PDC is typical of the eukaryotic cell, where it serves as a primary entry point for carbon into the citric acid cycle. The plastid PDC provides the acetyl-CoA and NADH required for fatty acid and isoprenoid biosynthesis. Thus the first and most important mechanism for regulation of plant PDCs is compartmentalization of each of the enzymes.

In contrast to mammalian cells and microbes, all *de novo* fatty acid biosynthesis by plant cells occurs within the plastids (Ohlrogge et al., 1979). Plant cells also contain two glycolytic pathways, the cytosolic pathway and a second metabolic complement within the plastids (Dennis and Miernyk, 1982). The search to connect the pyruvate produced by plastid glycolysis with the acetyl-CoA necessary for fatty acid synthesis led Dennis and his associates (Reid et al., 1977) to investigate the PDC within the leucoplasts of developing castor oil seeds. Observations concerning the occurrence of plastid PDC were subsequently extended to photosynthetic tissues (Williams and Randall, 1979; Elias and Givan, 1979). To date, PDC has been found in plastids isolated from every plant source investigated with the exception of the leucoplasts of endosperm from germinating castor oil seeds (Rapp and Randall, 1980). Chloroplast PDC not only produces acetyl-CoA for fatty acid biosynthesis, but it is also the only known source of NADH in chloroplasts (Camp and Randall, 1985).

Another feature unique to green plant cells is the existence of two separate systems for electron transport-coupled synthesis of ATP: oxidative phosphorylation in mitochondria and photophosphorylation in chloroplasts. In the dark, only oxidative phosphorylation is operative. Conversely, it has been long believed that only photophosphorylation occurs in the light. However, Krömer et al. (1992) have shown that inhibition of mitochondrial oxidative phosphorylation resulted in inhibition of photosynthesis, indicating that both pathways function in the light. It is not clear if oxidative phosphorylation is supported by the citric acid cycle, by an alternative substrate such as glycine from photorespiration, or a combination of these two. The mitochondrial PDC occupies an ideal position for regulation of carbon flow into the citric acid cycle during photosynthesis.

Herein we will describe our current understanding of the regulation of the activities of plant PDCs and present the results from recent molecular characterization of plant mitochondrial PDC. An emphasis has been placed on the differences between plant and other eukaryotic PDC systems.

General properties of plant mitochondrial pyruvate dehydrogenase complexes

Mammalian mitochondrial PDC has an $\underline{S}_{20, w}$ of 70–90 (Linn et al., 1972), corresponding to a native M_r of $7-8 \times 10^6$. The broccoli mitochondrial PDC is somewhat smaller at 59.3 $\underline{S}$. Analysis of pea mitochondrial PDC by SDS-polyacrylamide gel electrophoresis (SDS-PAGE) in combination with immunoblotting has shown subunits of M_r 76 000, 57 000, 53 000, 43 000, and 38 500 (Fig. 1). The M_r 57 000 subunit is dihydrolipoamide dehydrogenase (E3), while the 43 000 and 38 500 polypeptides are the α and β subunits, respectively, of the pyruvate dehydrogenase (PDH, E1). The polypeptide at M_r 76 000 is assigned to E2 based on the calculated M_r of the E2 deduced amino acid sequence (Guan et al., 1995). The identity of the 53 000 polypeptide is not known but it is always present in immunoprecipitations and with PDC which has been immunopurified using a monoclonal antibody raised against E1α. This polypeptide could be the E3-binding protein described for mammalian complexes. Plant PDH kinase is proposed also to be of this M_r (Miernyk and Randall, 1987c). Polyclonal antibodies raised against purified L-protein of the pea mitochondrial glycine decarboxylase complex (GDC) (Turner et al., 1992), which also serves as the E3 subunit for PDC, specifically recognizes the M_r 57 000 subunit. Polyclonal antibodies raised against pig heart lipoamide dehydrogenase also recognize the protein at M_r 57 000 (J.A. Miernyk, unpublished observations). Polyclonal antibodies raised against recombinant *Arabidopsis thaliana* E1β subunit decorate the M_r 38 500 subunit (Fig. 1). A monoclonal antibody raised against maize mitochondrial E1α subunit (Luethy et al., 1995b)

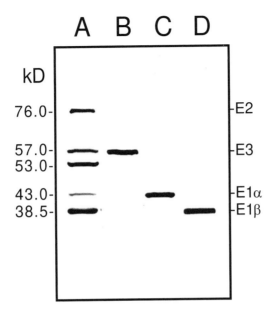

Figure 1. Immunoblot analyses of pea mitochondrial matrix proteins. Pea mitochondrial matrix proteins were separated by SDS-PAGE and transferred to a nitrocellulose membrane (Luethy et al., 1995b). Individual lanes were excised from the nitrocellulose and probed separately with: (A) polyclonal antibodies raised against purified broccoli PDC (Camp and Randall, 1985); (B) polyclonal antibodies raised against the pea mitochondrial L-protein of GDC (Turner et al., 1992); (C) a monoclonal antibody raised against maize E1α (Luethy et al., 1995b); (D) polyclonal antibodies raised against recombinant *Arabidopsis thaliana* E1β (M.H. Luethy, unpublished results). Immunoblotting and color development were carried out according to Luethy et al. (1995b).

specifically recognizes the M_r 43 000 subunit. This monoclonal recognizes the mitochondrial E1α subunit from all plants tested, but does not recognize the E1 subunit of plastid PDC, *E. coli*, of *S. cerevisiea* (Luethy et al., 1995b)

Regulation

Product inhibition is observed with both NADH and acetyl-CoA which are linearly competitive with NAD^+ and CoASH, respectively. The complex was more sensitive to the $NADH/NAD^+$ ratio than to the acetyl-CoA/CoASH ratio, i.e., the K_i value for NADH is 5–10 fold lower than the K_m for NAD^+, whereas the K_i for acetyl-CoA is twice the K_m for CoASH (Rubin et al., 1978; Randall et al., 1977). All plant mitochondrial PDCs examined undergo reversible phosphorylation. While numerous other potential modulators of PDC activity have been examined, none have been reported to affect PDC activity (Randall et al., 1989; Rubin and Randall, 1977b).

Plant mitochondrial pyruvate dehydrogenase kinase

The PDH kinase has been studied with the mitochondrial PDC from pea seedlings and from the endosperm of *Ricinus communis* seeds. The kinase is lost from the complex or is inactivated during purification, thus, all studies to date with the plant enzyme have used either partially purified PDC that retains kinase activity or mitochondrial extracts (100 000 g supernatant). Activity of plant PDH-kinases is optimal at pH 7.5, with Mg-ATP as the preferred phosphoryl donor but the kinase shows broad nucleotide specificity. The K_m for Mg-ATP is 2.5–5 μM, certainly supporting the conclusion that if the PDH kinase is not regulated, the PDC would always be inactivated (phosphorylated) (Miernyk and Randall, 1987c).

Since the low abundance of the plant PDH kinase has thus far precluded its isolation and purification, we estimated its M_r using 8-azido[^{32}P-α]ATP. This strategy indicated that the M_r is about 53 000, which is considerably larger than the M_r 48 000 subunit of the mammalian kinase (Miernyk and Randall, 1987c).

In vitro effectors of the plant PDH kinase activity include ADP, pyruvate, acetyl-CoA, NADH, citrate, 2-oxoglutarate, and monovalent cations (Miernyk and Randall, 1987c; Schuller and Randall, 1989). Inhibition by ADP is competitive with ATP, as it is for the mammalian PDH-kinase, however, in contrast to the mammalian PDH kinase, K$^+$ has no effect on this inhibition (Miernyk and Randall, 1987b). Similarly, *in contrast to their stimulatory effect on the mammalian kinase, acetyl-CoA and NADH inhibit the plant enzyme*. Inhibition by acetyl-CoA is competitive with Mg-ATP, and NADH is non-competitive with respect to Mg-ATP.

Initially, we reported that pyruvate was a competitive inhibitor of kinase with respect to ATP in the presence of TPP and that TPP was a competitive inhibitor with respect to ATP in the presence of pyruvate (Budde and Randall, 1988a), suggesting the pyruvate-TPP effect was due to formation of a reaction intermediate on PDH. However, reexamination using dialyzed PDC and kinase established that pyruvate was an uncompetitive inhibitor of kinase with respect to ATP, but TPP or TPP plus pyruvate exhibited non-linear Lineweaver-Burke plots with respect to ATP (Schuller and Randall, 1989).

In vitro, pea leaf PDH kinase is stimulated by 10–80 μM NH$_4^+$ and 10–100 mM K$^+$, but inhibited by 100 mM Na^{2+} (Schuller and Randall, 1989). Since NH$_4^+$ is a product of photorespiratory carbon metabolism that occurs within the mitochondria of illuminated leaves (Fig. 2), this NH_4^+ stimulation has significant regulatory potential.

Plant mitochondrial phospho-pyruvate dehydrogenase phosphatase

The plant phospho-PDH (P-PDH) phosphatase requires divalent cations for activity, with activation by $Mg^{2+} > Mn^{2+} > Co^{2+}$ and *in vitro* K_m values of 3.8, 1.7, and 1.4 mM, respectively. In contrast to the mammalian phosphatase, Ca^{2+} does not activate the pea leaf P-PDH phosphatase, but $10-100\ \mu M\ Ca^{2+}$ antagonizes Mg^{2+}-dependent dephosphorylation of the plant PDC. Monovalent cations, and polyamines do not activate the plant phosphatase. The only metabolite that has any effect on plant P-PDH phosphatase is orthophosphate, which inhibits slightly at

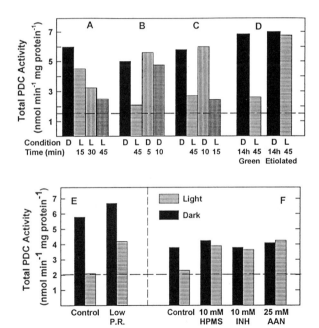

Figure 2. Light-dependent PDC inactivation. Pea seedlings were dark adapted overnight and then illuminated ($250\ \mu E \cdot m^{-2} \cdot s^{-1}$, 25°C) for the indicated time (A). Plants were also returned to the dark for various time intervals after an initial illumination (B). PDC activity was assessed after cycling through dark and light conditions (C) and with etiolated tissue (D). Conditions which reduce photorespiration (Low P.R.; 7% O_2, 1 100 ppm CO_2) (E) or inhibit photosynthesis (DCMU: 3-(3, 4-dichlorophenyl)-1, 1 dimethylurea) and photorespiratory metabolism (HPMS: 2-pyridylhydroxymethane sulfonate; INH: isonicotinic acid hydrozide; AAN: aminoacetonitrile) were also examined (F). The dashed horizontal line indicates the activity due to the plastid PDC. Activity measurements were made according to Gemel and Randall (1992).

physiological concentrations (16% at 5 mM). The *in vitro* rates of P-PDH phosphatase are usually 10–20% of the PDH-kinase rates (Miernyk and Randall, 1987a).

Results of *in vitro* studies indicate that plant mitochondrial PDC is capable of being regulated by product inhibition and reversible phosphorylation. The activity rates of the regulatory enzymes, *in vitro*, are such that in the absence of effectors, the *in vivo* complex will be inactive at all times, i.e., our data indicate that PDH kinase rates are about six times the P-PDH phosphatase rates. Pyruvate appears to be the metabolite with the greatest potential for controlling PDH kinase activity and thus adjusting the steady-state level of PDC *in vivo*.

In organello regulation of mitochondrial pyruvate dehydrogenase complex

When intact mitochondria purified from green pea seedlings (Fang et al., 1987) were used to examine respiration rates under various conditions as measured with an O_2 electrode, with concomitant determination of steady-state PDC activity, most *in vitro* results were generally verified. There were differences in the degree of regulation and potentially in the order of importance of the various effectors. We have established that the *in organello* PDC inactivation or phosphorylation is not the result of the overall respiratory rate or the redox poise of the quinone pool but that inactivation reflects the production of ATP within the matrix (Moore et al., 1993).

In contrast to mammalian PDC, changes in the ADP:ATP ratio over a 20-fold range (0.25–5) have little or no effect on the activation state of the plant mitochondrial PDC *in organello* (Budde and Randall, 1988a). PDC in intact mitochondria can be inactivated by exogenously added ATP and remains inactivated until the ATP is removed or exhausted (Budde and Randall, 1988b). Once the ATP is depleted, PDC is rapidly reactivated, and this reactivation occurred in the presence of only 0.5 mM Mg^{2+}, which is 20 to 40-fold less Mg^{2+} than was needed for optimal *in vitro* reactivation (Budde and Randall, 1988b). Ca^{2+}, which inhibits mammalian PDH kinase and stimulates mammalian P-PDH phosphatase activity, has no effect on the *in organello* activity or reactivation of the plant PDC (Budde and Randall, 1988b).

Pulse chase experiments have shown that reversible phosphorylation of PDC is a steady-state phenomenon with both the kinase and the phosphatase active at the same time (Budde and Randall, 1988b). Our results indicate that the steady-state PDC activity reflects kinase regulation or inhibition, particularly by pyruvate (Budde et al., 1988).

When mitochondria are oxidizing other substrates, including the photorespiratory metabolite glycine, the PDC is rapidly phosphorylated and results in a new steady-state activity level (Budde et al., 1988). The inactivation of PDC when the mitochondria are oxidizing glycine reflects: ATP generation from oxidative phosphorylation; increases in NADH and NH$_4^+$ levels from rapid glycine oxidation; and NH$_4^+$stimulation of the PDH-kinase (Schuller and Randall, 1989). This, in turn, provides a mechanism for potentially regulating citric acid cycle activity during photosynthesis.

In organello studies have shown that the mtPDC activity is seemingly more sensitive to the acetyl-CoA:CoA-SH ratio than the *in vitro* studies suggest (Budde et al., 1991; Fang et al., 1988).

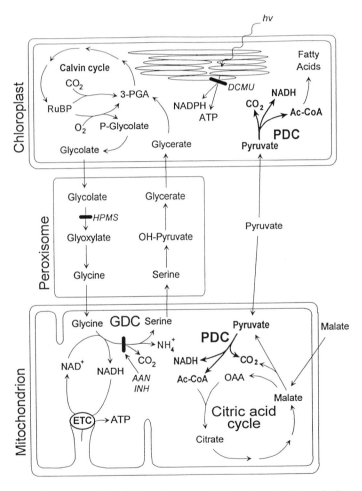

Figure 3. Interactions of the photosynthetic, photorespiratory, and citric acid cycles. Schematic diagram of the photorespiratory cycle, including portions of the Calvin and citric acid cycles. The sites of inhibitor action (DCMU, HPMS, INH, and AAN) are indicated. Not all of the intermediates are shown and stoichiometry has not been maintained. (GDC) glycine decarboxylase complex. (PDC) pyruvate dehydrogenase complex.

The fact that CoA has a small pool size suggests that a slow turnover of acetyl-CoA would have significant effects on PDC *in planta*.

The effect of light and photosynthesis on PDC activity and PDH phosphorylation status

To understand how the mitochondrial PDC may be regulated during photosynthesis, we determined the steady-state PDC activity *in planta*. Upon illumination of the pea leaves, steady-state mtPDC activity decreased to 10 to 20% of the level found in dark-adapted leaves (Fig. 2). PDC activity remained high if the normal, dark conditions were extended. After mtPDC has reached its final steady-state activity in the light, the inactivation can be reversed by returning the tissue to the dark. This reversible inactivation can be cycled rapidly by turning the light off and on. Etiolated tissue did not exhibit light-dependent inactivation of mtPDC, nor did green pea seedlings that had been treated overnight with an inhibitor of photosynthetic electron transport, indicating that a functional photosynthetic apparatus is essential for light-dependent inactivation of the mitochondrial complex (Fig. 2). We also determined that no small molecular weight effectors were produced in the light which inhibited PDC activity (Budde and Randall, 1990; Gemel and Randall, 1992).

Photorespiration (PR) describes the metabolic pathways used to recycle carbon lost from the Calvin cycle due to the condensation of O_2 rather than CO_2 with ribulose-bisphosphate creating a 2 carbon molecule which cannot be metabolized by the Calvin cycle enzymes (Fig. 3) (Tolbert, 1983). The PR pathway also spans three organelles as depicted in Figure 3. Experiments designed to reduce or inhibit PR in the plants showed a decrease in the extent of light-dependent inactivation of mtPDC (Fig. 2). Similarly, inhibitors of photosynthesis and photorespiratory metabolism fed through the transpiration stream prevented the light-dependent inactivation as well (Figs 2 and 3). Glycine is the only photorespiratory intermediate that is metabolized in the mitochondria during photosynthesis and feeding glycine to leaf tissue in the dark leads to inactivation of PDC. However, inhibitors of mitochondrial glycine oxidation prevented this inactivation (Budde and Randall, 1990; Gemel and Randall, 1992).

In planta results corroborate the *in organello* and *in vitro* data, which have shown the mtPDC is regulated by products of photorespiratory metabolism, i.e., the mitochondrial oxidation of glycine and the presence of NH_4^+. The light-dependent and PR-dependent inactivation of mtPDC likely results from ATP production during the photorespiratory metabolism of glycine. Light-dependent glycine production and oxidation occurs at two to three times the rate of flux through the citric acid cycle. The fact that both photosynthesis and photorespiration are required for inactivation of

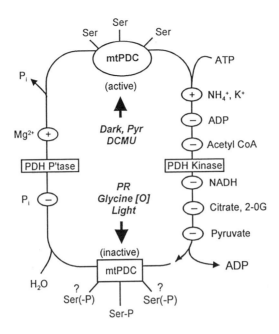

Figure 4. Model illustrating effectors of the reversible phosphorylation of pea leaf mitochondrial PDC.

PDC *in planta* reflects the fact that mitochondrial glycine oxidation will be significant only when there is a functional photosynthetic apparatus to produce the precursor of glycine (Tolbert, 1983). This inactivation would be further enhanced by the stimulation of PDH-kinase by NH_4^+, also produced by glycine oxidation during photorespiration. Figure 4 provides a model summarizing the information we have about the reversible phosphorylation of the pea leaf mitochondrial PDC.

General properties of the plastid pyruvate dehydrogenase complex

Plastid PDC catalyzes the same overall reaction and has the same kinetic mechanism as the mito-chondrial PDC. A marked instability has hampered a definitive estimation of the native M_r of the pea and maize mesophyll chloroplast PDCs (Camp and Randall, 1985; Camp et al., 1988; Treede and Heise, 1986), but it is intermediate in size between that of *Escherichia coli* (M_r 4.3×10^6) and bovine mitochondrial (M_r $6-8 \times 10^6$) PDCs (Camp and Randall, 1985). Subunit analyses of plastid and mitochondrial PDCs by SDS-PAGE plus immunoblotting showed significant differ-ences. At this time, the only subunit of the plastid PDCs which has been identified is that of dihydrolipoamide dehydrogenase, with an M_r of 57 000–58 000 (Camp and Randall, 1985; N. David, unpublished observations; Taylor et al., 1992).

The *in vitro* catalytic characteristics of plastid PDCs are generally similar to those of mitochondrial PDCs. A possible exception is the Michaelis constant for NAD^+ (49 μM [four species]), which is approximately half of the value for the mitochondrial complex (94 μM [six species]). The plastid complex is most active at pH 8.0 and requires a higher Mg^{2+} concentration (10 mM) for maximal activity than does the mitochondrial PDC (Camp and Randall, 1985).

Product inhibition

As is the case with bacterial and mitochondrial PDCs, the plastid complex is sensitive to product inhibition by NADH and acetyl-CoA. NADH is linearly competitive (K_i, 10–20 μM) with NAD^+ (K_m, 49 μM). Inhibition by acetyl-CoA is competitive with respect to CoA-SH (K_m, 20 μM), and K_i values are also around 20 μM. The plastid complex, like the mitochondrial complex, was much more sensitive *in vitro* to the NADH:NAD ratio, with a ratio of 0.2 giving 50% inhibition, than to that of acetyl-CoA to CoA, which required a ratio of 2.2 for 50% inhibition (Camp et al., 1988).

Phosphorylation and metabolite effects

Plastid PDCs resemble the bacterial complexes rather than mitochondrial PDCs in that they are not regulated by reversible phosphorylation (Randall et al., 1989). The *in vitro* activity of pea chloroplast PDC is insensitive to intermediates of glycolysis, the reductive pentose-phosphate pathway, or amino acid metabolism. Only inorganic phosphate (24% inhibition at 10 mM), oleic acid (57% inhibition by 50 μM), and palmitic acid (36% stimulation by 50 μM) affected pea chloroplast PDC (Camp et al., 1988).

Light regulation

During photosynthesis the chloroplast stroma becomes more alkaline (pH 8.0–8.3 in the light as opposed to pH 7.3–7.5 in the dark) and this pH shift is accompanied by an increase in free Mg^{2+} levels. The pH optimum (pH 8.0) of the plastid PDCs is more alkaline than those of mitochondrial PDCs. The K_m of the pea chloroplast PDC for pyruvate is lowest at the pH optimum for the complex and increases sharply at pH values more acidic or alkaline than 8.0. The values for V_{max} are largely unaffected by changes in pH between 7.0 and 9.0. Plastid PDCs require higher divalent cation concentrations (K_m, 1 mM) for maximum *in vitro* activity than do mitochondrial PDCs (K_m, 0.4 mM). Thus, the increase in pH and free Mg^{2+} in the plastid stroma in the light would increase the relative activity of PDC (Camp and Randall, 1985; Miernyk et al., 1985; Treede and Heise, 1986; Williams and Randall, 1979). Finally, plastid PDCs are relatively insensitive to inhibition by ATP and NADPH (Reid et al., 1977; Thompson et al., 1977; Camp and Randall, 1985) metabolites which would increase in concentration in illuminated chloroplasts. Overall, then, there is substantial evidence supporting the proposal that chloroplast PDC activity is regulated by the changes in the biochemical environment of the chloroplast stroma (i.e., increasing pH and Mg^{2+}) which accompany the dark-light transition and not by reversible phosphorylation.

Role of plastid PDC in fatty acid biosynthesis

There are at least two pathways for acetyl-CoA production within the plastids. One pathway employs the plastid PDC, generating acetyl-CoA *via* the decarboxylation of pyruvate. The second proposed pathway converts acetate to acetyl-CoA through the action of acetyl-CoA synthetase (Zeiher and Randall, 1991). Acetate for this latter pathway is proposed to originate in the mitochondria by action of acetyl-CoA hydrolase (Zeiher and Randall, 1990) on acetyl-CoA produced by the mitochondrial PDC. Plastids isolated from developing pea cotyledons contained all of the enzymes required to synthesize acetyl-CoA from triose phosphates (Deyner and Smith, 1988), however, activity of acetyl-CoA synthetase was not assessed.

Kang and Rawsthorne (1994) observed that plastids isolated from the developing embryos of oilseed rape were capable of synthesizing fatty acids from ^{14}C-labeled glucose-6-phosphate (G-6-P), dihydroxyacetone phosphate, pyruvate, malate, and acetate. Pyruvate was the most effective substrate for fatty acid synthesis, i.e., rates of ^{14}C incorporation from pyruvate were approximately five times greater than the rates observed for incorporation of label from acetate. Similar results were observed for the leucoplasts of the developing endosperm of castor oil seeds (Miernyk and Dennis, 1983; Smith et al., 1992) and the chloroplasts of mustard cotyledons (Liedvogel and Bäuerle, 1986). However, acetate has been shown to be a better substrate for fatty acid synthesis in other cases (Roughan et al., 1979; Springer and Heise, 1989; Masterson et al., 1990). Smith et al. (1992) reported that incorporation of label from malate into fatty acids occurred at a higher rate than from either pyruvate or acetate, however, malate conversion to fatty acids follows a pathway that includes pyruvate and PDC. It would appear that rates of substrate incorporation into fatty acids is highly dependent upon the species and tissue type investigated.

The species-to-species variations in substrate incorporation into fatty acids is probably dependent on the levels of various enzymes present in the plastids of these tissues. For instance, castor oil seed endosperm leukoplasts have higher levels of NADP-malic enzyme relative to other plastids (Smith et al., 1992), allowing for higher rates of malate incorporation into fatty acids *via* PDC. The levels of acetyl-CoA synthetase in barley chloroplasts are lower than those found in the chloroplasts of spinach leaves (Preiss et al., 1993). It has also been observed that the levels of plastid PDC relative to the mtPDC are highly dependent upon the plant species. Mitochondrial PDC levels in pea leaves represent approximately 75–90% of the total cellular PDC (Camp and Randall, 1985; Budde and Randall, 1990). On the other hand, the chloroplast PDC of barley leaves represents approximately 60% of the total PDC activity (Krömer et al., 1994). In this case, the specific activity of the plastid PDC is comparable to the levels measured in pea chloroplasts but the specific activity of the mitochondrial PDC is approximately 15 times lower. Lernmark and Gardeström (1994) reported that the plastid PDC activity in protoplasts of barley, wheat, and spinach represents approximately 85% of the total PDC activity. In the same study, the mitochondrial PDC in pea protoplasts was found to represent 85% of the total activity.

We conclude that plastid PDC is a primary source of acetyl-CoA for fatty acid and isoprenoid biosynthesis, and in addition, plastid demands for acetyl-CoA could be supplied by a combination of mtPDC and the carnitine:acetyl-carnitine cycle to move mitochondrially produced acetyl CoA to the plastids (Wood et al., 1992).

Recent advances in the study of plant mitochondrial PDC

Molecular characterization of the plant pyruvate dehydrogenase complex subunits

We have cloned and sequenced a cDNA encoding the E1α subunit of mtPDC from *Arabidopsis thaliana* (AtE1α; GenBank accession number U21214; Luethy et al., 1995a). This cDNA is 1436 bp in length, with a 1176 bp open reading frame, encoding a polypeptide of 389 amino acids. The N-terminal residues have properties consistent with a mitochondrial targeting signal. The overall homology of the deduced amino acid sequence of AtE1α with other eukaryotic E1α sequences falls within the range of 47–51% identity and 63–69% sequence similarity (Tab. 1). Several regions of high homology are noteworthy. The region surrounding phosphorylation sites 1 and 2 is highly conserved among the eukaryotic sequences (Fig. 5). Over a stretch of 19 residues, from Tyr285 to Arg303 of the *A. thaliana* sequence, the yeast, rat and human subunits differ by only two amino acids from *Arabidopsis*. Of the two residues which differ from the

```
                                                       Targeting signal
                                                             ◇
Arabidopsis thaliana...............MAL SRLSSRSNII TRPFSA-AFS RLISTDTTPI  32
Saccharomyces cerevisiae..........MLA ASFKRQPSQL V*GLG*VLRT PTRIGHVRTM
Homo sapiens....................MRKMLA A------VSR V-LSG*SQKP ASRVLVASRN  -1
Ascaris suum.........................MIFVF ANIFKVPTVS PSVMAISVRL
Bacillus stearothermophilus..........  .......... .......... ..........


A.t.  ---------- ---------- --TIETSLPF TAHLCDPPSR SVESSSQELL DFFRTMALMR  70
S.c.  ATLKTTDKKA PEDIEGSDTV QIELPE*-S* ESYMLE**DL *Y***KAT** QMYKD*VII*
H.s.  ---------- ---FANDATF EIKKCKKCDL HRLEEG**VT T*LTREDG*K Y-Y*M*QTV*  47
A.s.  ---------- ------ASTE ATFQTKPFKL HKLDSG*DIN VHVTKEDAVH Y-YTQ*LTI*
B.s.  ........MG VKTFQFPFAE QLEKVAEQFP TFQILNEEGE VVNEEAMPEL SDEQLKE***

                     ◆
A.t.  RMEIAAD--- ---SLYKANV IRGFCHLYDG QEAVAIGMEA AITKKDAIIT AYRDHCIFLG  124
S.c.  ***M*C*--- ---A****KK ******SV* ***I*V*I*N ****L*S*** S**C*GFTFM
H.s.  ***------- ---Q***QKI ******C** ***CCV*L** G*NPT*HL** ***A*GFTFT  97
A.s.  ***S*AG--- ---N***EKK V*******S* ***C*V*TK* *MDAG**AV* ***C*GWTYL
B.s.  **VYTRILDQ RSIS*NRQGR L-**YAPTA* ***SQ*ASHF *LE*E*F*LP G***VPQIIW


A.t.  RGGSLHEVFS ELMGRQAGCS KGKGGSMHFY KKESSFYGGH GIVGAQVPLG CGIAFAQKYN  184
S.c.  **A*VKA*LA *****R**V* Y*******L* --APG****N ********* A*L***HQ*K
H.s.  **L*VR*ILA **T**KG**A ********M* --AKN****N ********* A***L*C***  155
A.s.  S*S*VAK*LC **T**IT*NV Y*******M* --GEN***N *****Q*** T*****M**R
B.s.  H*LP*YQA*L FSR*HFH*NQ IPE*V----- ----NVLPPQ I*I***YIQA A*V*LGL*MR


                                               3▾
A.t.  KEEAVTFALY GDGAANQGQL FEALNISALW DLPAILVCEN NHYGMGTAEW RAAKSPSYYK  244
S.c.  N*D*CSFT** ****S****V **SF*MAK** N**VVFC*** *K******AS *SSAMTE*F*
H.s.  GKDE*CLT** *********I ***Y*MA*** K**C*FI*** *R*****SVE ***A*TD***  215
A.s.  **KN*CITMF ****T***** **SM*MAK** ***VLY**** *G******AA *SSA*TD**T
B.s.  GKK**AITYT ***GTS**DF Y*GI*FAGAF KA***F*VQ* *RFAIS*PVE KQTVAKTLAQ


                                    1▾        2▾
A.t.  R--GDYVPGL KVDGMDAFAV KQACKFAKQH AL-EKGPIIL EMDTYRYHGH SMSDPGSTYR  301
S.c.  *--***I*** **N***IL** Y**S****DW C*SG***LV* *YE****G** ******T***
H.s.  *--**FI*** R*****ILC* RE*TR**AAY CRSG****LM *LQ******* ******VS**  273
A.s.  *--******I W*****VL** R**VRW**EW CNAG***LMI *MA****S** ******TS**
B.s.  KAVAAGI**I Q*****PL** YA*V*A*RER *INGE**TLI *TLCF**GP* T**GDDP*RY


A.t.  TRDEISGVRQ ERDPIERIKK LVLSHDLATE KELKDMEKEI RKEVDDAIAK AKDCPMPEP-  360
S.c.  ****QHM*S KN***AGL*M HLIDLGI*** A*V*AYD*SA **Y**EQVEL ADAA*P**AK
H.s.  **E**QE**S KS***MLL*D RMVNSN**SV E***EIDV*V ***IE**AQF *TAD--***P  331
A.s.  **E*VQE**K T****TGF*D KIVTAG*VT* D*I*EID*QV ***I*A*VKQ *HTD--K*SP
B.s.  RSK*LENEWA KK**LV*FR* FLEAKG*WS* E*ENNVIEQA KE*IKE**K* *DET--*KQK


A.t.  -SELFTNVYV KGFGTESF-- --GPDRKEVK ASLP                           389
S.c.  L*I**ED*** **TE*PTLRG RIPE*TWDF* KQGFASRD
H.s.  LE**GYHI*- --SSDPP*EV RGANQWIKF* SVS                            361
A.s.  VELML*DI*- --YN*PAQYV RCTT*EVLQ* YVTSEEALKALAK
B.s.  VTD*ISIMF- ---EELP*NL KEQYEIYKE* E*K
```

Figure 5. Alignment of the deduced *Arabidopsis thaliana* E1α amino acid sequence with those of *Saccharomyces cerevisiae* (Swiss Protein Database (SW):P16387), *Homo sapiens* (SW:P08559), *Ascaris suum* (SW:P26267), and *Bacillus stearothermophilus* (SW:P21873). Asterisks have been placed to indicate sequence identity with the *A. thaliana* sequence. The cleavage point is indicated by ◇, designating the processing site of the targeting sequence. Spaces (−) have been inserted to maximize homology. The three residues that are phosphorylated on the mammalian E1α subunit have been designated by ▼. The cysteine residue implicated in the active site of the bovine subunit has been designated by ◆. Alignment of the amino acid sequences was performed using the GCG software package from the University of Wisconsin.

Table 1. Amino acid identities and similarities between *Arabidopsis thaliana* and *Pisum sativum* PDC subunits and other sequenced subunits

	AtE1α	AtE1β	AtE2	PsE3
S. cerevisiea	51 (69)	62 (73)	43 (62)	56 (74)
H. sapiens	50 (67)	58 (75)	41 (60)	55 (73)
A. suum	47 (63)	57 (74)		
B. stearothermophilus	27 (54)	40 (63)	31 (52)	44 (64)
E. coli			34 (54)	40 (62)

Numbers represent percent identity (and percent similarity in parentheses) and were determined using the GCG software, University of Wisconsin.

mammalian subunits, the serine residue at phosphorylation site 2 is a threonine residue in the plant subunit (previous studies have shown that only Ser residues are phosphorylated).

The putative subunit interaction domain (residues 246–289 of AtE1α), as depicted by Wexler et al. (1991) has several regions of high sequence identity, however, the overall sequence identity is weak, especially in the middle of that region. Furthermore, as that region abuts the phosphorylation site 1, the sequence homology at the C-terminal portion of the putative subunit interaction region may, in fact, be related to the phosphorylation event rather than subunit interactions. Portions of the putative thiamin pyrophosphate binding motif (residues 182–232 of AtE1α) show high sequence identity, particularly among residues in the N-terminal region, where the homology also appears to extend to several deduced amino acid sequences of the branched chain keto acid dehydrogenase complex (BCKDC) E1α subunit (Wexler et al., 1991).

Additionally, a region (residues 85–97) of high sequence identity surrounds and includes an essential cysteine found in the active site of the bovine E1α (Ali et al., 1993). This residue, Cys62 of the bovine subunit (corresponding to Cys89 of AtE1α), is thought to play a role in the oxidation of hydroxyethylidene-TPP (Ali et al., 1993). Interestingly, this region is highly conserved among the PDC E1α subunits and not at all conserved among the BCKDC counterparts, indicating that this region may play a role in either substrate specificity or specificity towards binding exclusively with the PDC E1β subunit.

A cDNA encoding the mitochondrial E1β subunit was identified from the *A. thaliana* EST database. This cDNA is a full-length clone of 1230 bp, containing an open reading frame of 1089 bp (Luethy et al., 1994). The open reading frame encoded 363 amino acids, predicting a polypeptide of 39, 190 daltons. The amino acid composition of the N-terminal 30–35 amino acids is consistent with a mitochondrial targeting sequence rather than a chloroplast transit peptide (von Heijne et al., 1989).

```
                                                              Targeting signal
                                                                     ◇
Arabidopsis thaliana.....................  .MLGILRQRA IDGASTLRRT RFALVSARSY  29
Saccharomyces cerevisiae.............MFS RLPTSLARNV ARR*PTSFVR PS*AAA*LRF
Homo sapiens..........................  .MAAVSGLVR RPLREVSGLL KRRFHWTAPA  -1
Ascaris suum..........................  ...MAVNGCM RLLRNG*TSA CALEQ*V*RL
Bacillus stearothermophilus............. .......... .......... ..........
```

```
                                                    Region 1
A.t.    AAGAKEMTVR DALNSAIDEE MSADPKVFVM GEEVGQYQGA YKITKGLLEK YGPERVYDTP  89
S.c.    -SST*T**** E*****MA** LDR*DD**LI ****A**N** **VS****DR F*ER**V***
H.s.    ---*VQV*** **I*QGM*** LER*E***LL *******D** **VSR**WK* **DK*II***  57
A.s.    *SGTLNV*** ****A*L*** IKR*DR**LI *******D** ***S***WK* **DG*IW***
B.s.    ---MAQ**MV Q*ITD*LRI* LKN**N*LIF **D**VNG*V FRA*E**QAE F*ED**F***
```

```
A.t.    ITEAGFTGIG VGAAYAGLKP VVEFMTFNFS MQAIDHIINS AAKSNYMSAG QINVPIVFRG 149
S.c.    ***Y****LA ****LK**** *****S**** ******VV** ***TH***G* TQKCQM****
H.s.    *S*M**A**A ****M***R* IC******** *****QV*** ***TY***G* LQP*******  117
A.s.    ***MAIA*LS ****MN***R* IC***SM*** **G******* ***AH***** RFH*******
B.s.    LA*S*IG*LA I*L*LQ*FR* *P*IQF*G*V YEVM*S*CGQ M*RIR*RTG* RYHM**TI*S
```

```
                          *                         Region 2
A.t.    PNGAAAGVGA QHSQCYAAWY ASVPGLKVLA PYSAEDARGL LKAAIRDPDP VVFLENELLY 209
S.c.    *****V*L** ****DFSP** G*I******V ********** ********N* **********
H.s.    ****S***A* *****F**** GHC*****VS *WNS***K** I*S****NN* **V*****M*  177
A.s.    A****V**AQ ****DFTA*F MHC**V**VV **DC****** ****V**DN* *IC***I**
B.s.    *F*GGVHTPE L**DSLEGLV *QQ*****VI *STPY**K** *IS****N** *I***HLK**
```

```
A.t.    GESFPISEEA LDSSFCLPIG KAKIEREGKD VTIVTFSKMV GFALKAAEKL AEE-GISAEV 268
S.c.    ****E***** *SPE*T**Y- ********T* IS***YTRN* Q*S*E***I* QKKY*V****
H.s.    *VP*EFPP** QSKD*LI*** *****Q*TH I*V*SH*RP* *HC*E**AV* SK*-*VEC**  236
A.s.    *MK**V*P** QSPD*V**F* Q***Q*P*** I***SL*IG* DVS*H**DE* *KS-**DC**
B.s.    -R**--RQ*V PEGEYTI*** **D*K***** I**IAYGA** HES****AE* EK*-******
```

```
             *          Region 3                                Region 4
A.t.    INLRSIRPLD RATINASVRK TSRLVTVEEG FPQHGVCAEI CASVVE-ESF SYLDAPVERI 327
S.c.    ********** TEA*IKT*K* *NH*I***ST **SF**G*** V*Q*M*S*A* D*****IQ*V
H.s.    **M*T***M* ME**E***M* *NH*****G* W**F**G*** **RIMEGPA* NF****AV*V  296
A.s.    ****CV**** FQ*VKD**I* *KH*****S* W*NC**G*** S*R*T*SDA* G***G*IL*V
B.s.    VD**TVQ*** IE**IG**E* *G*AIV*Q*A QR*A*IA*NV V*EIN*-RAI LS*E***L*V
```

```
A.t.    AGADVPIPYT ANLERLALPQ IEDIVRASKR ACYRSK                          363
S.c.    T*****T**A KE**DF*F*D TPT**K*V*E -VL-*IE
H.s.    T*****M**A KI**DNSI** VK**IF*I*K -TL-NI                          330
A.s.    T*V***M**A QP**TA**** PA*V*KMV*K -CL-NVQ
B.s.    *AP*TVY*F- *QA*SVW**N FK*VIETA*K -VM-NF
```

Figure 6. Alignment of the deduced *Arabidopsis thaliana* E1β sequence with those of *Saccharomyces cerevisiae* (SW:P32473), *Homo sapiens* (SW:P11177), *Ascaris suum* (SW:P26269), and *Bacillus stearothermophilus* (SW:P21874). Asterisks have been placed to indicate sequence identity with the *A. thaliana* sequence. Spaces (-) have been inserted to maximize homology. Homology regions are based on a review by Wexler et al. (1991). Essential residues have been designated by *. Alignment of the amino acid sequences was performed using the GCG software package from the University of Wisconsin.

Four conserved E1β regions, outlined by Wexler et al. (1991) are indicated in Figure 6. These regions have been designated as regions of high homology between the deduced amino acid sequences of E1β subunits from three PDC and three BCKDC. It was postulated that region 1 (residues 58–92 of AtE1β) is a site at which E1β binds to E2, since this is the only other region (along with the TPP binding domain) which shares homology with the *E. coli* E1 (Stephens et al., 1983). Regions 2 (residues 173–209 of AtE1β) and 3 (residues 263–298 of AtE1β) share homology with other oxidative decarboxylating enzymes, suggesting that these regions may be involved in the oxidative decarboxylation of pyruvate (Wexler et al., 1991). A role has not been assigned to region 4 (residues 303–336 of AtE1β), at the C-terminus.

Comparison of the deduced amino acid sequences of the mature *A. thaliana* E1β subunit with previously reported sequences revealed high levels of sequence identity. The *A. thaliana* sequence has the highest amino acid identity with the *S. cerevisiae* subunit (65%) (Tab. 1). A high degree of identity was also observed between the sequences for the *R. norvegicus* (60%), *H. sapiens* (59%), and *A. suum* (58%) subunits. In contrast, the *B. stearothermophilus* sequence has only 38% identity with the *A. thaliana* sequence. A region of high sequence identity among all of the E1β sequences is located between regions 1 and 2 (residue 100–169). This region was not previously noted by Wexler et al. (1991), possibly because it lacks homology with the BCKDCs. This highly conserved PDC domain could be related to specificity of PDC for pyruvate.

Essential residues within the bovine E1β subunit have been identified through chemical modification followed by proteolysis and amino acid sequencing (Ali et al., 1995). Trp135 and Arg239 are implicated in the active centers and are conserved among all of the eukaryotic sequences presented in Figure 6, including the *A. thaliana* sequence (Trp168 and Arg272 of AtE1β). The Arg239 residue is also conserved in the *B. stearothermophilus* sequence.

Figure 7 is an alignment of the deduced amino acid sequence of an *Arabidopsis* cDNA encoding the E2 subunit (Guan et al., 1995; GenBank accession number Z46230) with other E2 sequences. The *Arabidopsis* E2 sequence contains regions which are highly homologous to the human and yeast E2 subunits. Two regions within the *A. thaliana* sequence (AtE2 residues 83–120 and 210–247) represent a repeating unit and show homology with the lipoyl binding sites of the human subunit. It appears, therefore, that the *A. thaliana* subunit contains two lipoyl binding domains, similar to the human and rat E2 subunits (Perham, 1991; Matuda et al., 1992). Alignment within the putative E3 and/or E1 binding sites (residues 306–354 of AtE2) reveals a relatively low degree of sequence identity. The sequence motif G-X-G-X-X-G, a proposed nucleotide-binding site (Soderling, 1990) is only partially conserved in the *Arabidopsis* sequence (residues 324–329 of AtE2), with the initial glycine residue replaced by an alanine. The *Arabidopsis* sequence also contains the conserved CoA-binding site D-H-R-X-X-D-G within the inner catalytic core at the C-terminus (AtE2 residues 582–588).

A cDNA encoding lipoamide dehydrogenase (E3) has been cloned from a pea cDNA library (Turner et al., 1992). This subunit was originally cloned using antibodies directed against the L-protein of the pea mitochondrial glycine decarboxylase complex (GDC). It has been postulated

```
A. thaliana          EFHSRFSNGL YHLDDKISSS NGVRSASIDL ITRMDDSSPK PILRFGVQNF  50
S. cerevisiae        .......... .......... .......... .......... ..........
H. sapiens           ....*VTSRS GPAPARRN*V TTGYGGVRA* CGWTPS*GAT *RN*LLL*LL
E.coli.              .......... .......... .......... .......... ..........
B.stearothermophilus .......... .......... .......... .......... ..........

A.t.    SSTGPI---- -SQTVLAMPA LSPTMSHGNV VKWMKKEGDK VEVGDVLCEI ETDKATVEFE 105
S.c.    .......... .......... .......... .......... .......... ..........
H.s.    G*P*RRYYSL PPHQKVPL*S ****QA*TI AR*K****** INE**LIA*V ******G**
E.c.    .......... .......... .......... .......... .......... ..........
B.s.    .......... .......... .......... .......... .......... ..........

A.t.    SQEEGFLAKI LVTEGSKDIP VNEPIAIMVE EEDDIKNVPA TIEGGRDGKE ETSAHQVMKP 165
S.c.    .......... .......... .......... .......... ....MSAFV RVVPR--ISR
H.s.    *L**CYM*** **A**TR*V* IGAI*C*T*G KPE**E-AFK NYTLDSSAAP TPQ*APAPT*
E.c.    .......... .......... .......... .......... .......... ..........
B.s.    .......... .......... .......... .......... .......... ..........

A.t.    -DESTQQKSS IQPDASDLPP HVVLEMPALS PTMNQGNIAK WWKKEGDKIE VGDVIGEIET 224
S.c.    SSVLTRSLRL QLRCYASY*E *TIIG***** ***T***L*A *T*****QLS P*E**A****
H.s.    AATASPPTP* A*APG*SY** *MQVLL**** ***TM*TVQR *E**V*E*LS E**LLA****
E.c.    .......... ........MS S*DILV*D*P ESVADATV*T *H**P**AVV RDE*LV****
B.s.    .......... .......... AFEFKL*DIG EGIHE*E*V* *FV*P**EVN ED**LC*VQN

A.t.    DKATLEFESL E-EGYLAKIL IPEGSKDVAV GKPIALIVED AESIEAIK-- ---------- 271
S.c.    ***QMD**FQ *-D******* V***T**IP* N****VY*** KADVP*F*DF KLEDSGSDSK
H.s.    ****IG**VQ *-******** V***TR**PL *T*LCI***K EAD*S*FADY RPTEVTD---
E.c.    **VV**VP-A SAD*I*DAV* ED**T-T*TS RQILGRLR*G N------SAG KETSAKSEEK
B.s.    ***VV*IP-S PVK*KVLE** V***T-VAT* *QTLITLDAP GYENMTF*GQ EQEEEAKKEEK

A.t.    -SSSAGSSEV DTVKEVPDSV VDKPTERKAG ---------- -----FTKIS PAAKLLILEH 315
S.c.    T*TK*QPA*P QAE*KQEAPA EETK*SAPEA KKSDVAAPQG RIF-----A* *L**TIA**K
H.s.    LKPQVPPPTP PPVAA**PTP QPLAPTPS*P CPATPAGPKG RVF-----V* *L**K*AV*K
E.c.    A*TP*QRQQA SLEEQNN*AL S--------- ---------- ---------- **IRR*LA**
B.s.    TETVSKEEK* *A*APNAPAA EAEAGPNRRV ---------- -------IAM *SVRKYAR*K

                GX  GXXG
A.t.    GLEASSIEAS GPYGTLLKSD VVAAIASGKA SKSSASTKKK QPSKETPSKS SSTSKPSVTQ 375
S.c.    *ISLKDVHGT **R*RIT*A* IESYLEKSSK QS*QT*GAAA ATPAAAT*ST TAG*A**PS.
H.s.    *IDLTQVKGT **D*RIT*K* IDSFV----- ----P*KVAP A*AAVV*PTG P---GMAPV-
E.c.    N*D**A*KGT *VG*R*TRE* *EKHL*KAP* KE*APAAAAP AAQPALAARS EKRVPM----
B.s.    *VDIRLVQGT *KN*RV**E* ID*FL*G*AK PAPA*AEE*A A*AAAK*ATT EGEFPE----

A.t.    SDNNYEDFPN SQIRKIIAKR LLESKQKIPH LYLQSDVVLD PLLAFRKELQ E--NHG--VK 431
S.c.    *TAS***V*I *TM*S**GE* **Q*T*G**S YIVS*KISIS K**KL*QS*N ATA*DK--Y*
H.s.    PTGVFT*I*I *N**RV**Q* *MQ***T*** Y**SI**NMG EV*LV****N KILEGR--S*
E.c.    .....TR--- --L**RV*E* ***A*NSTAM *TTFNE*NMK *IMDL**QYG *AFEKRHGIR
B.s.    .....TREKM *G**RA***A MVH**HTA** VT*MDEADVT K*V*H**KF- KAIAAEKGI*

A.t.    VSVNDIVIKA VAVALRNVRQ ANAFWDAEKG DIVMCDSVDI SIAVATEKGL MTPIIKNADQ 491
S.c.    L*I**LLV** IT**AKR*PD ***Y*LPNEN V*RKFKN**V *V****PT** L***V*NCEA
H.s.    I****FI*** S*L*CLK*PE **SS*M--DT V*RQNHV**V *V**S*PA** I***VF**HI
E.c.    LGFMSFYV** *VE**KRYPE V**SI*G--D DV*YHNYF*V *M**S*PR** V**VLRDV*T
B.s.    LTFLPYVV** LVS***EYPV L*TSI*D*TE E*IQKHYYN* G**AD*DR** LV*V**H**R

A.t.    KSISAISLEV KELAQKARSG KLAPHEFQGG TFSISNLGMY -PVDNFCAII NPPQAGILAV 550
S.c.    *GL*Q**N*I ***VKR**IN ****E***** *IC***M**N NA*NM*TS** ****ST***I
H.s.    *GVET*AND* VS**T***E* **T*G*MK*A SCT*T*I*SA *E**K**KT VI*NPLEM**.
E.c.    LGMAD*EKKI ****V*G*D* **TVEDLT** N*T*T*G*VF GSL-MSTP** ****SA**GM
B.s.    *PIF*LAQ*I N***E***D* **T*G*MK*A SCT*T*I*SA GGQ-W*TPV* *H*EVA**GI

              DHRXXDG
A.t.    GRGNKEVEPV IGLDGIEKPC VVTKMNVTLS ADHRIFDGQV GASFMSELRS NFEDVRRLLL.. 610
S.c.    ATVERVA--* EDAAAENGFS FDNQVTI*GT F***TI**AK **E**K**KT VI*NPLEM**..
H.s.    *ASE-DK--L VPA*NEKGFD *ASM*S**** C***VV**A* **QWLA*F*K YL*KPITM**..
E.c.    HAIKDR---- -PMAVNGQVE ILPM*YLA** Y***LI**RE SVG*LVTIKE LL**PT****DV
B.s.    **IAEK---- -PIVRDGEIV AAPMLALS** F***MI**AT AQKALNHIKR LLS*PEL**MEA
```

Figure 7. Alignment of the deduced amino acid sequences of the *Arabidopsis thaliana* E2 subunit with those of *Saccharomyces cerevisiae* (SW:P12696), *Homo sapiens* (SW:P10515), *Escherichia coli* (SW:P07016), and *Bacillus stearothermophilus* (SW:P11961). The amino acid numbers indicate the residues of the *A. thaliana* sequence. Asterisks indicate identity with the *A. thaliana* sequence and spaces (–) have been inserted to maximize homology. The lipoyl-binding sites have been indicated by overstriking with a solid line. The proposed nucleotide binding site and a conserved CoA binding site have also been indicated. Alignment of the amino acid sequences was performed using the GCG software package from the University of Wisconsin.

```
P.sativum  ......MAMA NLARRKGYSL LSSETLRYSF SLRSRAFASG SDENDVVIIG GGPGGYVAAI  54
S.cerevis  .......... ......MLRI R*LLNNKRA* *STV*TLTIN KSH.****** ***A*******
H.sapiens  MQSWSRVYCS LAK*GHFNRI SHGLQGLSAV P--L*TY*DQ PIDA**TV** S*********
B.stearot  .......... .......... .......... ....MVVGDF AI*TETLVV* A*********
E.coli     .......... .......... .......... ........*T EIKTQ**VL* A**A**S**F


P.s.       KAAQLGFKTT CIEKRGALGG TCLNVGCIPS KALLHSSHMY HEAKH-SFAN HGVKVS-NVE  114
S.c.       *******N*A ******K*** ********** ****NN**LF *QMH-TEAQK R*ID*NGDIK
H.s.       *********V ****NET*** ********** ****NN**Y* *M*HGTD**S R*IEM-SE*R
B.s.       R*****Q*V* IV**GN-*** V********* ***ISA**R* EQ***--SEE M*I-KAE**T
E.c.       RC*D**LE*V IV*RYNT*** V********* *****VAKVI E***A--L*E **I-*FGEPK


P.s.       IDLAAMMGQK DKAVSNLTRG IEGLFKKNKV TYVKGYGKFV SPSEISVDTI EG------EN  169
S.c.       *NV*NFQKA* *D**KQ**G* **L******* **Y**N*S*E DETK*R*TPV D*LEGTVK*D
H.s.       LN*DK**E** ST**KA**G* *AH***Q*** VH*N****IT GKNQVTATKA D*------GT
B.s.       **F*KVQEW* ASV*KK**G* V***L*G*** EI***EAY** DANTVR*VNG DS-------A
E.c.       T*IDKIRTW* E*VINQ**G* LA*MA*GR** KV*N*L***T GANTLE*EGE N*-------K


P.s.       TVVKGKHIII ATGSDVKSLP GVTIDEKKIV SSTGALALSE IPKKLVVIGA GYIGLEMGSV  229
S.c.       HILDV*N**V ****E*TPF* *IE**E*** ******S*KE ***R*TI**G *I********
H.s.       Q*IDT*N*L* ****E*TPF* *I****DT*V ******S*KK V*E*M***** *V**V*L***
B.s.       QTYTF*NA** ****RPIE** NFKFSN.R*L D*****N*G* V**S*****G ****I*L*TA
E.c.       **INFDNA** *A**RPIQ** FIPHEDPR*W D**D**E*K* V*ER*L*M*G *I******T*


P.s.       WGRIGSEVTV VEFASEIVP- TMDAEIRKQF QRSLEKQGMK FKLKTKVVGV -DTSGDGVKL  278
S.c.       YS*L**K*** ***QPQ*GA- S**G*VA*AT *KF*K***LD ***S***ISA KRNDDKN*VE
H.s.       *Q*L*AD**A ***LGHVGGV GI*M**S*N* **I*Q***F* ***N***T*A TKK*DGK-ID
B.s.       YANF*TK**I L*G*G**LS- GFEKQMAAII KKR*K*K*VE VVTNALAK*A EEREDG-*TV
E.c.       YHAL**QID* **MFDQVI*- AA*KD*V*V* TKRIS*K-FN LM*E***TA* EAKEDG-IYV


P.s.       TVEPS-AGGE QTIIEADVVL VSAGRTPFTS GLNLDKIGVE TDKLGRILVN ERFSTNVSGV  337
S.c.       I*VEDTKTNK *ENL**E*L* *AV**R*YIA *LGAEKI*L* V**R**LVID DQ*NSKFPHI
H.s.       VSIEAAS**K AEV*TC**L* *CI**R***K N*G*EEL*I* L*PR***P** T**Q*KIPNI
B.s.       *Y*---*N** TKT*D**Y** *TV**R*N*D E*G*EQ**IK MTNR*L*E*D QQCR*S*PNI
E.c.       *M*GKK*PA* PQRYD*--** *AI**V*NGK N*DAG*A*** V*DR*F*R*D KQLR***PHI


P.s.       YAIGDVI-PG PMLAHKAEED GVACVEYLAG KVGHVDYDKV PGVVYTNPEV ASVGKTEEQV  396
S.c.       KVV***T-F* *********E *I*A**M*KT GH***N*NNI *S*M*SH*** *W*******L
H.s.       ******V-A* ********DE *II***GM** GAV*I**NC* *S*I**H*** *W***S***L
B.s.       F****IV-** *A*****SYE *KVAA*AI** HPSA***VAI *A**FSD**C ****YF*Q*A
E.c.       F****IV-GQ ******GVHE *HVAA*VI** *KHYF*PKVI *SIA**E*** *WV*L**KEA


P.s.       KETGVEYRVG KFPFMANSRA KAIDNAEGLV KIIAEKETDK ILGVHIMAPN AGELIHEAAI  456
S.c.       **A*ID*KI* ****A***** *TNQDT**F* **LIDSK*ER ***A**IG** ***M*A**GL
H.s.       **E*I**K** ****A***** *TNADTD*M* **LGQ*S**R V**A**LG*G ***MVN***L
B.s.       *DE*IDVIAA ****A**G** L*LNDTD*FL *LVVR**DGV *I*AQ*IG** *SDM*A*LGL
E.c.       **K*IS*ETA T**WA*SG** I*S*C*D*MT *L*FD**SHR VI*GA*VGT* G***LG*IGL


P.s.       ALQYDASSED IARVCHAHPT MSEAIKEAAM AT-YDKPHSH LKSWLLLSSL VFIFVQGFTL  515
S.c.       **E*G**A** V********* L***F***N* *A-***AIHC .......... ..........
H.s.       **E*G**C** ********** L***FR**NL *ASFG*SINF .......... ..........
B.s.       *IEAGMTA** **LTI***** LG*IAM***- EVALGT*IHI ITK....... ..........
E.c.       *IEMGCDA** **LTI***** LH*SVGL**- EVFEGSITDL PNPKAKKK.. ..........


P.s.       TWRRYFVC                                                          523
S.c.       ........
H.s.       ........
B.s.       ........
E.c.       ........
```

Figure 8. Alignment of the *Pisum sativum* L-protein (E3) deduced amino acid sequence with *Saccharomyces cerevisiae* (SW:P09624), *Homo sapiens* (SW:P09622), *Escherichia coli* (SW:P00391), and *Bacillus stearothermophilus* (SW:P11959). The amino acid numbers indicate the residues of the *P. sativum* sequence. Asterisks indicate identity with the *P. sativum* sequence and spaces (-) have been inserted to maximize homology. Alignment of the amino acid sequences was performed using the GCG software package from the University of Wisconsin.

that the lipoamide dehydrogenase is shared among the mitochondrial and plastid PDCs, GDC, and 2-oxoglutarate dehydrogenase complex. The deduced amino acid sequence shows high homology with the sequences from yeast and humans (Fig. 8, Tab. 1), particularly within the functional domains (e.g., the FAD and NAD binding sites). Southern blot analysis of pea genomic DNA with a L-protein probe from within the coding region indicated that potentially two copies of the gene might be present in the genome (Turner et al., 1992), leaving open the possibility that the products of more than one L-protein gene are present in the plant mitochondria. Alternatively, this second copy of the L-protein gene could represent the plastid E3 component.

Conclusions

Plants are unique in having two distinct types of PDCs localized within different subcellular compartments. Many of the kinetic and regulatory properties of plant mitochondrial PDCs are similar to those of the mammalian complexes. In common with mammalian PDCs, plant mitochondrial PDC exhibits typical product inhibition and undergoes reversible phosphorylation. Primary differences due exist, primarily with regard to regulation of the reversible phosphorylation, e.g., Ca^{2+} has no apparent role with either the plant kinase or phosphatase, NADH and acetyl CoA inhibit the kinase rather than stimulate kinase, the ATP/ADP ratio is relatively unimportant in the plant system, and NH_4^+ stimulates the plant kinase. The PDCs within plastids of plant cells are in many ways more similar to the bacterial complexes than to mitochondrial PDCs. Both types of plant PDCs have evolved unique regulatory features in response to particular metabolic requirements and environmental factors. Plastid PDC lacks the reversible phosphorylation mechanism of control and requires a higher pH and Mg^{2+} environment in comparison to either mitochondrial or prokaryotic PDCs.

While in many cases the results of *in organello* PDC activity measurements have confirmed and extended results obtained *in vitro* with the partially purified or purified complex, there have also been several instances where the results of experiments with isolated intact mitochondrial have clarified interpretations of *in vitro* results. Examples include observations on the divalent cation concentrations required for activation of P-PDH phosphatase, and the regulatory importance of acetyl-CoA/CoA-SH. By using a combination of both methods, a more thorough understanding of this important potential regulatory system has resulted.

Molecular characterization of plant mitochondrial PDCs has revealed important similarities between the plant components and their eukaryotic and prokaryotic counterparts, as well as unique properties. Regions of high homology among the deduced amino acid sequences include but are not limited to regions previously determined to play important roles in the structural and catalytic properties of the PDC from both eukaryotic and prokaryotic sources. Analyses of the primary amino acid sequences have and will continue to play an important role in the furthering of our understanding of the plant as well as non-plant PDCs.

Acknowledgement
Authors' research supported by NSF-IBN-9419489.

References

Ali, M.S., Roche, T.E. and Patel, M.S. (1993) Identification of the essential cysteine residue in the active site of bovine pyruvate dehydrogenase. *J. Biol. Chem.* 268: 22353–22356.

Ali, M.S., Shenoy, B.C., Eswaran, D., Andersson, L.A., Roche, T.E. and Patel, M.S. (1995) Identification of the tryptophan residue in the thiamin pyrophosphate binding site of bovine pyruvate dehydrogenase. *J. Biol. Chem.*; *in press.*

Behal, R.H., Browning, K.S. and Reed, L.J. (1989) Nucleotide and deduced amino acid sequence of the alpha subunit of yeast pyruvate dehydrogenase. *Biochem. Biophys. Res. Comm.* 164: 941–946.

Buchanan, B.B. (1980) Role of light in regulation of chloroplast enzymes. *Ann. Rev. Plant Physiol.* 31: 341–374.

Budde, R.J.A. and Randall, D.D. (1988a) Regulation of pea mitochondrial pyruvate dehydrogenase complex activity: Inhibition of ATP-dependent inactivation. *Arch. Biochem. Biophys.* 258: 600–606.

Budde, R.J.A. and Randall, D.D. (1988b) Regulation of steady state pyruvate dehydrogenase complex activity in plant mitochondria: Reactivation constraints. *Plant Physiol.* 88: 1026–1030.

Budde, R.J.A., Fang, T.K. and Randall, D.D. (1988) Regulation of the phosphorylation of mitochondrial pyruvate dehydrogenase comples *in situ:* Effects of respiratory substrates and calcium. *Plant Physiol.* 88: 1031–1036.

Budde, R.J.A. and Randall, D.D. (1990) Pea leaf mitochondrial pyruvate dehydrogenase complex is inactivated *in vivo* in a light-dependent manner. *Proc. Natl. Acad. Sci. USA* 87: 673–676.

Budde, R.J.A., Fang, T.K., Randall, D.D. and Miernyk, J.A. (1991) Acetyl-Coenzyme A can regulate activity of the mitochondrial pyruvate dehydrogenase *in situ. Plant Physiol.* 95: 131–136.

Camp, P.J. and Randall, D.D. (1985) Purification and characterization of the pea chloroplast pyruvate dehydrogenase complex. A source of acetyl-CoA and NADH for fatty acid biosynthesis. *Plant Physiol.* 77: 571–577.

Camp, P.J., Miernyk, J.A. and Randall, D.D. (1988) Some kinetic and regulatory properties of the pea chloroplast pyruvate dehydrogenase complex. *Biochim. Biophys. Acta* 993: 269–275.

Dennis, D.T. and Miernyk, J.A. (1982) Compartmentation of non-photosynthetic carbohydrate metabolism. *Ann. Rev. Plant Physiol.* 33: 27–50.

Deyner, K. and Smith, A.M. (1988) The capacity of plastids from developing pea cotyledons to synthesize acetyl-CoA. *Planta* 173: 172–182.

Elias, B.A. and Givan, C.V. (1979) Localization of pyruvate dehydrogenase complex in *Pisum sativum* chloroplasts. *Plant Sci. Lett.* 17: 115–122.

Fang, T.K., David, N.R., Miernyk, J.A. and Randall, D.D. (1987) Isolation and purification of functional pea leaf mitochondria free of chlorophyll contamination. *Curr. Top. Plant Biochem. Physiol.* 6: 175.

Fang, T.K., Budde, R.J.A., Randall, D.D. and Miernyk, J.A. (1988) *In situ* regulation of the plant mitochondrial pyruvate dehydrogenase complex by product inhibition. *Plant Physiol.* 86S: 61.

Gemel, J. and Randall, D.D. (1992) Light regulation of leaf mitochondrial pyruvate dehydrogenase complex. *Plant Physiol.* 100: 908–914.

Guan, Y., Rawsthorne, S., Scofield, G., Shaw, P. and Doonan, J. (1995) Cloning and characterization of a dihydrolipoamide acetyltransferase (E2) subunit of the pyruvate dehydrogenase complex from *Arabidopsis thaliana. J. Biol. Chem.* 270: 5412–5417.

Hawkins, C.F., Borges, A. and Perham, R.N. (1990) Cloning and sequence analysis of the genes encoding the α and β subunits of the E1 component of the pyruvate dehydrogenase multienzyme complex of *Bacillus stearothermophilus. Eur. J. Biochem.* 191: 337–346.

von Heijne, G., Steppuhn, J. and Herrmann, R.G. (1989) Domain structure of mitochondrial and chloroplast targeting peptides. *Eur. J. Biochem.* 180: 535–545.

Ho, L., Wexler, I.D., Liu, T.-C., Thekkumkara, T.J. and Patel, M.S. (1989) Characterization of cDNAs encoding human pyruvate dehydrogenase α subunit. *Proc. Natl. Acad. Sci. USA* 86: 5330–5334.

Hoppe, P., Heintze, A., Riedel, A., Creuzer, C. and Schultz, G. (1993) The plastidic 3-phosphoglycerate → acetyl CoA pathway in barley leaves and its involvement in the synthesis of amino acids, plastidic isoprenoids, and fatty acids during chloroplast development. *Planta* 190: 253–262.

Kang, F. and Rawsthorne, S. (1994) Starch and fatty acid synthesis in plastids from developing embryos of oil-seed rape (*Brassica napus* L.). *Plant J.* 6: 795–805.

Korotchkina, L.G., Tucker, M.M., Thekkumkara, T.J., Madhusudhan, K.T. Pons, G., Kim, H. and Patel, M.S. (1995) Overexpression and purification of human tetrameric pyruvate dehydrogenase and its individual subunits. *Prot. Express. Purif.* 6: 79–90.

Krömer, S., Stitt, M. and Heldt, H.W. (1988) Mitochondrial oxidative phosphorylation participating in photosynthetic metabolism of a leaf cell. *FEBS Lett.* 226: 352–356.

Krömer, S., Lernmark, U. and Gardeström, P. (1994) *In vivo* mitochondrial pyruvate dehydrogenase activity, studied by rapid fractionation of barley leaf protoplasts. *J. Plant Physiol.* 144: 485–490.

Lernmark, U. and Gardeström, P. (1994) Distribution of pyruvate dehydrogenase complex activities between chloroplasts and mitochondria from leaves of different species. *Plant Physiol.* 106: 1633–1638.

Liedvogel, B. and Bäuerle, R. (1986) Fatty acid synthesis in chloroplasts from mustard (*Sinapis alba* L.) cotyledons: formation of acetyl coenzymeA by intraplastid glycolytic enzymes and a pyruvate dehydrogenase complex. *Planta* 169: 481–489.

Linn, T.C., Pelleg, J.W., Pettit, F.H., Hucho, F., Randall, D.D. and Reed, L.J. (1972) α-Keto acid dehydrogenase complexes. XV. Purification and properties of the component enzymes of the pyruvate dehydrogenase complexes from bovine kidney and heart. *Arch. Biochem. Biophys.* 148: 327–342.

Luethy, M.H., Miernyk, J.A. and Randall, D.D. (1994) The nucleotide and deduced amino acid sequences of a cDNA encoding the E1β-subunit of the *Arabidopsis thaliana* mitochondrial pyruvate dehydrogenase complex. *Biochim. Biophys. Acta* 1187: 95–98.

Luethy, M.H., Miernyk, J.A. and Randall, D.D. (1995a) Molecular analyses of the pyruvate dehydrogenase complex. The nucleotide and deduced amino acid sequences of a cDNA encoding the E1α-subunit of the *Arabidopsis thaliana* mitochondrial pyruvate dehydrogenase. *Gene; in press.*

Luethy, M.H., David, N.R., Elthon, T.E., Miernyk, J.A. and Randall, D.D. (1995b) Characterization of a monoclonal antibody recognizing the E1α subunit of the plant mitochondrial pyruvate dehydrogenase. *J. Plant Physiol.* 145: 443–449.

Maeng, C.Y., Yazdi, M.A., Niu, X.D., Lee, H.Y. and Reed, L.J. (1994) Expression, purification, and characterization of the dihydrolipoamide-binding protein of the pyruvate dehydrogenase complex from *Saccharomyces cerevisiae.* *Biochemistry* 33: 13801–13807.

Masterson, C., Wood, C. and Thomas, D.R. (1990) L-acetylcarnitine, a substrate for chloroplast fatty acid synthesis. *Plant, Cell, and Environ.* 13: 755–765.

Matuda, S., Nakano, K., Ohta, S., Saheki, T., Kawanishi, Y. and Miyata, T. (1991) The α-ketoacid dehydrogenase complexes. Sequence similarity of rat pyruvate dehydrogenase with *Eschericia coli* and *Azotobacter vinelandii* α-ketoglutarate dehydrogenase. *Biochim. Biophys. Acta* 1089: 1–7.

Matuda, S., Nakano, K., Ohta, S., Shimura, M., Yamanaka, T., Nakagawa, S., Titani, K. and Miyata, T. (1992) Molecular cloning of dihydrolipoamide acetyltransferase of the rat pyruvate dehydrogenase complex: sequence comparison and evolutionary relationship to other dihydrolipoamide acyltransferases. *Biochim. Biophys. Acta* 1131: 114–118.

Miernyk, J.A. and Dennis, D.T. (1983) The incorporation of glycolytic intermediates into lipids by plastids isolated from the developing endosperm of castor oil seeds (*Ricinus communis* L.). *J. Exp. Bot.* 34: 712–718.

Miernyk, J.A., Camp, P.J. and Randall, D.D. (1985) Regulation of plant pyruvate dehydrogenase complexes. *Curr. Top. Plant Biochem. Physiol.* 4: 175–190.

Miernyk, J.A. and Randall, D.D. (1987a) Some properties of pea mitochondrial phosphopyruvate dehydrogenase phosphatase. *Plant Physiol.* 83: 311–315.

Miernyk, J.A. and Randall, D.D. (1987b) Some kinetic and regulatory properties of the pea mitochondrial pyruvate dehydrogenase complex. *Plant Physiol.* 83: 306–310.

Miernyk, J.A. and Randall, D.D. (1987c) Some properties of plant mitochondrial pyruvate dehydrogenase kinase. *In:* A.L. Moore and R.B. Beachey (eds): *Plant Mitochondria.* Plenum Publishing Corp., New York, pp 223–226.

Miernyk, J.A. and Randall, D.D. (1989) A synthetic peptide-directed antibody as a probe of the phosphorylation site of pyruvate dehydrogenase. *J. Biol. Chem.* 264: 9141–9144.

Moore, A.L., Gemel, J. and Randall, D.D. (1993) The regulation of pyruvate dehydrogenase activity in pea leaf mitochondria. *Plant Physiol.* 103: 1431–1435.

Ohlrogge, J.B., Kuhn, D.N. and Stumpf, P.K. (1979) Subcellular localization of acyl carrier protein in leaf protoplasts of *Spinacia oleracia.* *Proc. Natl. Acad. Sci. USA* 76: 1194–1198.

Perham, R.N. (1991) Domains, motifs and linkers in 2-oxo acid dehydrogenase multienzyme complexes: a paradigm in the design of a multifunctional protein. *Biochemistry* 30: 8501–8512.

Pratt, M.L. and Roche, T.E. (1979) Mechanism of pyruvate inhibition of kidney pyruvate dehydrogenase kinase and synergistic inhibition of pyruvate and ADP. *J. Biol. Chem.* 254: 7191–7196.

Preiss, M., Rosidi, B., Hoppe, P. and Schultz, G. (1993) Competition of CO_2 and acetate as substrates for fatty acid synthesis in immature chloroplasts of barley seedlings. *J. Plant Physiol.* 142: 525–530.

Randall, D.D. and Rubin, P.M. (1977) Plant pyruvate dehydrogenase complex purification, characterization and regulation by metabolites and phosphorylation. *Biochim. Biophys. Acta* 485: 336–349.

Randall, D.D., Williams, M. and Rapp, B.J. (1981) Phosphorylation-dephosphorylation of pyruvate dehydrogenase complex from pea leaf mitochondria. *Arch. Biochem. Biophys.* 207: 437–444.

Randall, D.D., Miernyk, J.A., Fang, T.K., Budde, R.J.A. and Schuller, K.A. (1989) Regulation of the pyruvate dehydrogenase complex in plants. *Ann. N.Y. Acad. Sci.* 573: 192–205.

Rapp, B.J. and Randall, D.D. (1980) Pyruvate dehydrogenase complex from germinating castor bean endosperm. *Plant Physiol.* 65: 314–318.

Rapp, B.J., Miernyk, J.A. and Randall, D.D. (1987) Pyruvate dehydrogenase complexes from *Ricinus communis* endosperm. *J. Plant Physiol.* 127: 293–306.

Reid, E.E., Thompson, P., Lyttle, C.R. and Dennis, D.T. (1977) Pyruvate dehydrogenase complex from higher plant mitochondria and proplastids. *Plant Physiol.* 59: 842–848.

Roughan, P.G., Holland, R., Slack, C.R. and Mudd, J.B. (1979) Acetate is the preferred substrate for long-chain fatty acid synthesis in isolated spinach chloroplasts. *Biochem. J.* 184: 565–569.

Roughan, P.G., Kagawa, T. and Beevers, H. (1980) On the light dependency of fatty acid synthesis by isolated spinach chloroplasts. *Plant Sci. Lett.* 18: 221–228.

Rubin, P.M. and Randall, D.D. (1977a) Purification and characterization of pyruvate dehydrogenase complex. *Arch. Biochem. Biophys.* 178: 342–349.

Rubin, P.M. and Randall, D.D. (1977b) Regulation of plant pyruvate dehydrogenase complex by phosphorylation. *Plant Physiol.* 60: 34–39.

Rubin, P.M., Zahler, W.L. and Randall, D.D. (1978) Plant pyruvate dehydrogenase complex purification, characterization and regulation by metabolites and phosphorylation. *Biochim. Biophys. Acta* 485: 70–77.

Schuller, K.A. and Randall, D.D. (1989) Regulation of pea mitochondrial pyruvate dehydrogenase complex: Does photorespiratory ammonium influence mitochondrial carbon metabolism? *Plant Physiol.* 89: 1207–1212.

Schuller, K.A. and Randall, D.D. (1990) Mechanism of pyruvate inhibition of plant pyruvate dehydrogenase kinase and synergism with ADP. *Arch. Biochem. Biophys.* 278: 211–216.

Smith, R.G., Gauthier, D.A., Dennis, D.T. and Turpin, D.H. (1992) Malate- and pyruvate-dependent fatty acid synthesis in leucoplasts from developing castor endosperm. *Plant Physiol.* 98: 1233–1238.

Soderling, T.R. (1990) Protein kinases. *J. Biol. Chem.* 265: 1823–1826.

Springer, J. and Heise, K.-P. (1989) Comparison of acetate- and pyruvate-dependant fatty-acid synthesis by spinach chloroplasts. *Planta* 177: 417–421.

Stitt, M., Lilley, R.McC. and Heldt, H.W. (1982) Adenine nucleotide levels in the cytosol, chloroplasts, and mitochondria of wheat leaf protoplasts. *Plant Physiol.* 70: 971–977.

Stitt, M. and ap Rees, T. (1979) Capacities of pea chloroplasts to catalyze the oxidative pentose phosphate pathway and glycolysis. *Phytochemistry* 18: 1905–1911.

Taylor, A.E., Cogdell, R.J. and Lindsay, J.G. (1992) Immunological comparison of the pyruvate dehydrogenase complexes from pea mitochondria and chloroplasts. *Planta* 188: 225–231.

Thompson, P., Reid, E.E., Lyttle, C.R. and Dennis, D.T. (1977) Pyruvate dehydrogenase complex from higher plant mitochondria and proplastids: Regulation. *Plant Physiol.* 59: 854–858.

Tolbert, N.E. (1983) The oxidative photosynthetic carbon cycle. *Curr. Top. Plant Biochem. Physiol.* 1: 63–77.

Treede, H.-J. and Heise, K.-P. (1986) Purification of the chloroplast pyruvate dehydrogenase from spinach and maize mesophyll. *Z. Naturforsch.* 41C: 149–155.

Turner, S.R., Ireland, S. and Rawsthorne, S. (1992) Purification and primary amino acid sequence of the L subunit of glycine decarboxylase. *J. Biol. Chem.* 267: 7745–7750.

Wexler, I.D., Hemelatha, S.G. and Patel, M.S. (1991) Sequence conservation in the α and β subunits of pyruvate dehydrogenase and its similarity to branched-chain α-keto acid dehydrogenase. *FEBS Lett.* 282: 209–213.

Williams, M. and Randall, D.D. (1979) Pyruvate dehydrogenase complex from chloroplasts of *Pisum sativum*. *Plant Physiol.* 64: 1099–1103.

Wood, C., Masterson, C. and Thomas, D.R. (1992) The role of carnitine in plant cell metabolism. *In*: A.K. Tobin (ed.): *Plant Organelles: Compartmentation of Metabolism in Photosynthetic Tissue*. Cambridge University Press, Great Britain, pp 229–263.

Yeaman, S.J., Hutcheson, E.T., Roche, T.E., Pettit, F.H., Brown, J.R., Reed, L.J., Watson, D.C. and Dixon, G.H. (1978) Sites of phosphorylation on pyruvate dehydrogenase from bovine kidney and heart. *Biochemistry* 17: 2364–2370.

Zeiher, C.A. and Randall, D.D. (1990) Identification and characterization of mitochondrial acetyl-Coenzyme A hydrolase from *Pisum sativum* L. seedlings. *Plant Physiol.* 94: 20–27.

Zeiher, C.A. and Randall, D.D. (1991) Spinach leaf acetyl-Coenzyme A synthetase: purification and characterization. *Plant Physiol.* 96: 382–389.

Alpha-Keto Acid Dehydrogenase Complexes
M.S. Patel, T.E. Roche and R.A. Harris (eds)
© 1996 Birkhäuser Verlag Basel/Switzerland

Regulation of the pyruvate dehydrogenase complex during the aerobic/anaerobic transition in the development of the parasitic nematode, *Ascaris suum*

R. Komuniecki, M. Klingbeil, R. Arnette, D. Walker and F. Diaz

Department of Biology, University of Toledo, Toledo, OH 43606-3390, USA

Introduction

The pyruvate dehydrogenase complex (PDC) occupies a pivotal position in the novel, anaerobic, mitochondrial metabolism of the parasitic nematode, *Ascaris suum* (Kita, 1992; Komuniecki and Komuniecki, 1995). Adult ascarid muscle mitochondria use unsaturated organic acids, instead of oxygen, as terminal electron-acceptors and acetate, propionate, succinate, and the 2-methyl branched-chain fatty acids, 2-methylbutyrate and 2-methylvalerate, accumulate as end products of carbohydrate metabolism. The tricarboxylic acid cycle is not operative and the NADH-dependent reductions of fumarate and 2-methyl branched-chain enoyl CoAs are coupled to site 1, electron-transport associated energy-generation (Kita, 1992; Ma et al., 1993). Most importantly, from the perspective of pyruvate metabolism, intramitochondrial $NADH/NAD^+$ and acyl CoA/CoA ratios appear to be dramatically elevated, when compared to the corresponding aerobic organelles, and serve as the driving force for the reversal of β-oxidation and the synthesis of branched-chain fatty acids (Kita, 1992; Komuniecki and Komuniecki, 1995). Therefore, it was initially surprising to find a functional PDC and PDH_a kinase in these organelles, given the potential for these elevated ratios to reduce PDC activity, either through end product inhibition or stimulation of PDH_a kinase and its subsequent phosphorylation and inactivation of the complex. In fact, the PDC is significantly overexpressed in adult ascarid muscle mitochondria and is present in amounts substantially greater than those reported from other eukaryotic sources (Song and Komuniecki, 1994; Thissen et al., 1986). Not surprisingly, both its subunit composition and regulatory properties differ significantly from complexes isolated from aerobic tissues.

Subunit composition

The subunit composition of the adult *A. suum* muscle PDC differs significantly from both the yeast and mammalian complexes (Fig. 1). Incubation of the ascarid complex with [2-^{14}C]pyruvate results in the acetylation of only a single subunit, dihydrolipoyl transacetylase (E2) and the domain structure of the ascarid E2 appears to be quite similar to that reported for other PDCs (Komuniecki et al., 1992). A second lipoyl domain containing subunit, corresponding to E3BP, is not present in the adult ascarid complex. In contrast, the ascarid complex does contain additional proteins of 43 and 45 kDa (p43 and p45, respectively) which are not visualized on Coomassie blue stained SDS-polyacrylamide gels of the yeast and mammalian complexes (Komuniecki et al., 1992; Komuniecki and Thissen, 1989; Matuda et al., 1987). The function of these additional proteins is not completely clear. However, sequence in the amino terminus of p45 exhibits significant similarity to the putative E3-binding domains of E2 and E3BP (R. Arnette, unpublished). In addition, E2 core substantially depleted of p45 by treatment in 1 M guanidine hydrochloride still retains its ability to fully bind E1 and catalyze the overall PDC reaction, but does not bind additional E3 (M. Klingbeil, unpublished). These results suggest that p45 may play a role in E3-

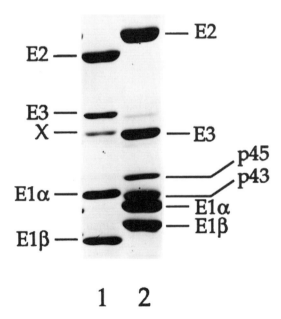

Figure 1. SDS-PAGE of purified PDCs. (Lane 1) bovine kidney PDC; (lane 2) *Ascaris suum* PDC.

binding in the adult ascarid complex. p43 appears to be a mixture of phosphorylated E1α, whose mobility shifts after phosphorylation, and PDH$_a$ kinase (Thissen and Komuniecki, 1988). After separation by two-dimensional gel electrophoresis, phosphorylated E1α focuses at about pH 6.1, while PDH$_a$ kinase activity is associated with one or more of the proteins focusing between pH 6.8 and 7.7 (M. Klingbeil, unpublished).

Regulatory properties

Many of the regulatory properties of the adult ascarid muscle PDC are altered to maintain PDC activity under the reducing conditions present in the vertebrate gut. First, both the intact complex and its associated kinase and phosphatase are much less sensitive to elevated NADH/NAD$^+$ and acetyl CoA/CoA ratios (Thissen et al., 1986). This pattern of reduced sensitivity to NADH coupled to the high expression of the complex is similar to that observed in the anaerobic prokaryote, *Enterococcus faecalis* (Snoep et al., 1992). The mechanism of the reduced sensitivity to NADH is unclear. Whether it resides in an anaerobic-specific E3 or altered E3-binding remains to be determined. Interestingly, the sequence of the conserved redox active thiol domain of E3 is similarly modified in a number of anaerobes, including *Hymenolepis diminuta* and *Parascaris equorum*, members of two diverse helminth phyla, but not in the E3 of the closely related free-living nematode, *Caenorhabditis elegans* (D. Walker, unpublished; Diaz and Komuniecki, 1994). The significance, if any, of this sequence difference remains to be determined. Second, the apparent K_m for pyruvate is higher and the apparent K_m for CoA lower than values reported for the mammalian complexes (Thissen et al., 1986). This allows pyruvate, which is generated intramitochondrially from malate by "malic enzyme", to accumulate *in vivo* to levels which inhibit PDH$_a$ kinase and maintains low free CoA levels, which are essential for the reversal of the thiolase reaction as a prelude to branched-chain fatty acid formation (Komuniecki et al., 1987; Thissen et al., 1986). Third, PDH$_b$ phosphatase activity is stimulated by malate, the major mitochondrial substrate in adult ascarid muscle, but not by calcium or polyamines. PDH$_b$ phosphatase has been purified from adult ascarid muscle mitochondria and is a heterodimer, similar in size to that reported for the mammalian enzyme (Song and Komuniecki, 1994). Malate stimulation requires the presence of the E2 core, in a manner similar to calcium stimulation of the mammalian phosphatase (Song and Komuniecki, 1994). Finally, and perhaps most important, the stoichiometry of phosphorylation and inactivation of the adult ascarid PDC is altered to prevent the complete inactivation of the complex *in vivo* (Thissen and Komuniecki, 1988). Three distinct phosphorylation sites have been identified in the bovine kidney E1α. Inactivation is associated almost exclusively with the phosphorylation of site 1 (Sugden et al., 1979; Yeaman et al., 1978). Additional phosphorylation at sites 2 and 3 appears to lock the complex in an inactive state, since these sites appear to be preferred by the phosphatase (Sale and Randle, 1982a; Sale and Randle, 1982b). In contrast, $E1\alpha$ from the adult ascarid muscle PDC contains only two phosphorylation sites. The ascarid PDH_a kinase exhibits no obvious site preference and both $E1\alpha$ subunits of the $\alpha_2\beta_2$ tetramer are fully phosphorylated at complete inactivation (Thissen and Komuniecki, 1988). To identify the potential role of both the kinase and E1 in this altered pattern of regulation, hybrid complexes have been constructed using an E1-depleted ascarid subcomplex and E1s isolated from either the *A. suum* or bovine kidney PDCs. Both E1s fully restore PDC activity to specific activities observed in the native ascarid complex, but the complex reconstituted with the ascarid E1

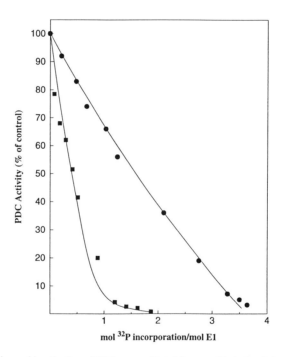

Figure 2. Phosphorylation and inactivation of PDCs reconstituted from an E1-depleted *Ascaris suum* subcomplex and E1s isolated from either *A. suum* or bovine kidney PDC. *A. suum* subcomplex and E1 were prepared as described in Komuniecki et al. (1992). Subcomplex (100 µg) and either *A. suum* or bovine kidney E1 (200 µg) were incubated in 50 mM MOPS (pH 7.4), 1 mM $MgCl_2$, 0.1 mM EDTA, 3 mM dithiothreitol (final volume of 150 µl) at room temperature for 30 min and assayed spectrophotometrically for PDC activity (Thissen et al., 1986). Reconstituted complexes (1.2 mg/ml) were incubated at 30°C in 50 mM MOPS (pH 7.4), 2 mM $MgCl_2$, 3 mM dithiothreitol, 25 mM NaF and 0.5 mM [γ-^{32}P]ATP in a final volume of 250 µl. Aliquots were removed at 0.5, 1, 1.5, 2, 3, 5, 10, 20, 30 and 60 min and assayed for PDC activity and spotted on 2 × 2 cm Whatman P81 paper, washed three times with phosphoric acid and counted for protein-bound radioactivity (Roskoski, 1983). (Circles) *A. suum* E1; (squares) bovine kidney E1.

requires about 3–4 times as much phosphorylation to achieve 90% inactivation (Fig. 2). These data are in strong agreement with results reported for the native complexes and suggest that the altered stoichiometry of phosphorylation/inactivation in the adult *A. suum* PDC is the result of differences in E1. Therefore, even though the ascarid and mammalian E1s are over 50% identical at the amino acid sequence level, they must contain significant differences in structure (Johnson et al., 1992; Wheelock et al., 1991).

Developmental regulation

In contrast to the PDC from the anaerobic adult, much less is known about the PDC from aerobic ascarid larval stages. However, it is clear that both its structure and function differ significantly and it is more similar to PDCs isolated from other eukaryotic sources. During development, *A. suum* undergoes a series of defined larval molts and an aerobic/anaerobic transition in its energy metabolism (Komuniecki and Komuniecki, 1995). Larval development to the second-stage (L2) is complex and accompanied by the conversion of stored triacylglycerides to glycogen catalyzed by a functional glyoxylate cycle (Barrett, 1976). After ingestion by the host, the infective L2 "hatches" in the host gut, migrates to the liver, and molts to the third-stage (L3). In contrast to the anaerobic mitochondrial pathways described for the adult, these larval ascarids are aerobic, cyanide-sensitive and contain a functional tricarboxylic acid cycle (Komuniecki and Vanover, 1987; Vanover-Dettling and Komuniecki, 1989). Under these conditions, the PDC functions, as in its mammalian counterpart, to provide glycolytically generated pyruvate for TCA cycle oxidation. Not surprisingly, the PDC in aerobic larval ascarids appears to differ significantly from that isolated from the adult. For example, the larval PDC does not appear to contain significant amounts of p45, based on the immunoblotting of larval homogenates with affinity purified polyclonal antisera against p45 (Komuniecki, unpublished). In addition, stage-specific, aerobic and anaerobic E1α isozymes have been tentatively identified (Johnson et al., 1992). In contrast to the E1α predominant in anaerobic adult muscle, the larval E1α contains the three phosphorylation sites observed in most mammalian E1αs and recently identified in the closely related aerobic free-living nematode, *C. elegans* (Johnson et al., 1992; Waterson et al., 1992). This suggests that the altered stoichiometry of phosphorylation/inactivation observed in adult muscle may not be operative in these aerobic larvae. Similarly, recent data also suggests that the E1β present in larval homogenates also differs significantly from the E1β found in adult muscle (Fig. 3). Affinity-purified antisera against the adult muscle E1β does not recognize a similar protein on immunoblots of homogenates of unembryonated eggs, L1 and L2, while antisera raised against E1α readily

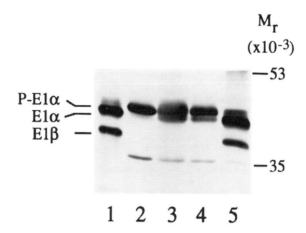

Figure 3. Immunoblot of *A. suum* larval homogenates with antisera against the adult *A. suum* E1. Larval homogenates were prepared in 20 mM MOPS (pH 7.2), 2 mM EDTA, 2 mM EGTA, 1 mM PMSF, 1 mM benzamidine, 2.1 μM leupeptin, 3 μM aprotinin, 1 μM soybean trypsin inhibitor, 66 μM antipain, 33 μM chymostatin and 29 μM pepstatin A by passing unembryonated "eggs", L1, L2 and L3 through a French Press at 20 000 psi (Komuniecki and Vanover, 1987; Vanover-Dettling and Komuniecki, 1989). Triton X-100 then was added to a final concentration of 1% (w/v). After 30 min on ice, samples were centrifuged for 30 min at 10 000 × g and the supernatants frozen in liquid nitrogen. After concentration by ultrafiltration and centrifugation to remove insoluble material, the samples were separated by SDS-PAGE on 10% gels, followed by transfer to nitrocellulose membranes. Affinity purified antisera against adult *A. suum* E1α and E1β were used for immunoblotting as described previously (Johnson et al., 1992; Wheelock et al., 1991). (Lane 1) *A. suum* PDC (0.5 μg); (lane 2) unembryonated "eggs" (60 μg); (lane 3) L1 (60 μg); (lane 4) L2 (60 μg); (lane 5) L3 (30 μg).

crossreacts. Interestingly, the L3, while aerobic, contains an E1β with a mobility identical to that observed in the anaerobic adult (Fig. 3). The presence of anaerobic specific enzymes in this metabolically transitional stage has been noted previously (Duran et al., 1993). Since phosphorylation changes the mobility of E1α during SDS-PAGE, the phosphorylation state of the PDC can be estimated directly from immunoblots (Thissen and Komuniecki, 1988). It appears that the E1α from the unembryonated "egg" is almost completely phosphorylated and inactive, as would be predicted for this quiescent stage (Fig. 3). Clearly, the purification of the PDC from aerobic larvae will be essential to fully understand the relationship between the two complexes. For example, does the aerobic ascarid PDC contain E3BP? In addition, this data suggests that the PDC will be an excellent model to study mitochondrial biogenesis during the aerobic/anaerobic transition which occurs during ascarid development.

Acknowledgements
We would like to thank Emilio Duran for the antisera to E1α and E1β, Karen Hayton for assistance with the immunoblots and Dr. Thomas E. Roche, Kansas State University, for the kind gift of bovine kidney E1.

References

Barrett, J. (1976) Intermediary metabolism in *Ascaris* eggs. *In*: H. Van den Bossche (ed.): *Biochemistry of Parasites and Host-Parasite Relationships*. Elsevier, Amsterdam, pp 117–123.

Diaz, F. and Komuniecki, R. (1994) Pyruvate dehydrogenase complexes from the equine nematode, *Parascaris equorum*, and the canine cestode, *Dipylidium caninum*, helminths exhibiting anaerobic mitochondrial metabolism. *Mol. Biochem. Parasitol.* 67: 289–299.

Duran, E., Komuniecki, R., Komuniecki, P., Wheelock, M.J., Klingbeil, M. and Johnson, K.R. (1993) Isolation and sequence determination of cDNA clones for the 2-methyl branched chain enoyl CoA reductase from *Ascaris suum*. *J. Biol. Chem.* 268: 22391–22396.

Johnson, K.R., Komuniecki, R., Sun, Y. and Wheelock, M.J. (1992) Characterization of cDNA clones for the alpha subunit of pyruvate dehydrogenase from *Ascaris suum*. *Mol. Biochem. Parasitol.* 51: 37–48.

Kita, K. (1992) Electron-transfer complex of mitochondria in *Ascaris suum*. *Parasitol. Today* 8: 155–159.

Komuniecki, P.R. and Vanover, L. (1987) Biochemical changes during the aerobic-anaerobic transition in *Ascaris suum* larvae. *Mol. Biochem. Parasitol.* 22: 241–248.

Komuniecki, R., Campbell, T. and Rubin, N. (1987) Anaerobic metabolism in *Ascaris suum*: acyl CoA intermediates in isolated mitochondria synthesizing 2-methyl branched-chain fatty acids. *Mol. Biochem. Parasitol.* 24: 147–154.

Komuniecki, R. and Thissen, J. (1989) The pyruvate dehydrogenase complex from anaerobic mitochondria of the parasitic nematode *Ascaris suum*: Stoichiometry of phosphorylation and inactivation. *Ann. N.Y. Acad. Sci.* 573: 175–182.

Komuniecki, R., Rhee, R., Bhat, D., Duran, E., Sidawy, E. and Song, H. (1992) The pyruvate dehydrogenase complex from the parasitic nematode *Ascaris suum*: novel subunit composition and domain structure of the dihydrolipoyl transacetylase component. *Arch. Biochem. Biophys.* 296: 115–121.

Komuniecki, R. and Komuniecki, P. (1995) Aerobic-anaerobic transitions in energy metabolism during the development of the parasitic nematode, *Ascaris suum*. *In*: J. Boothroyd and R. Komuniecki (eds): *Molecular Approaches to Parasitology*. Wiley-Liss Inc., New York, pp 109–121.

Liu, S., Baker, J.C. and Roche, T.E. (1995) Binding of the pyruvate dehydrogenase kinase to recombinant constructs containing the inner lipoyl domain of the dihydrolipoyl acetyltransferase component. *J. Biol. Chem.* 270: 793–800.

Ma, Y., Funk, M., Dunham, W.R. and Komuniecki, R. (1993) Purification and characterization of electron-transfer flavoprotein:rhodoquinone oxidoreductase from anaerobic mitochondria of the adult parasitic nematode, *Ascaris suum*. *J. Biol. Chem.* 268: 20360–20365.

Matuda, S., Nakano, K., Uraguchi, Y., Matso, S. and Saheki, T. (1987) Immunogical identification of a new component of *Ascaris* pyruvate dehydrogenase complex. *Biochim. Biophys. Acta* 926: 54–60.

Roskoski, R.J. (1983) Assays of protein kinase. *Meth. Enzymol.* 99: 306–347.

Sale, G.J. and Randle, P.J. (1982a) Occupancy of phosphorylation sites in pyruvate dehydrogenase phosphate complex in rat heart *in vivo*. Relation to proportion of inactive complex and rate of reactivation by phosphatase. *Biochem. J.* 206: 221–229.

Sale, G.J. and Randle, P.J. (1982b) Role of individual phosphorylation sites in inactivation of pyruvate dehydrogenase complex in rat heart mitochondria. *Biochem. J.* 203: 99–108.

Snoep, J.L., Westphal, A.D., Benen, J.A., Texeira de Mattos, M.J., Neijssel, O.M. and de Kok, A. (1992) Isolation and characterization of the pyruvate dehydrogenase complex of the anaerobically grown *Enterococcus faecalis*. *Eur. J. Biochem.* 203: 245–250.

Song, H. and Komuniecki, R. (1994) Novel regulation of pyruvate dehydrogenase phosphatase purified from anaerobic muscle mitochondria of the adult parasitic nematode, *Ascaris suum*. *J. Biol. Chem.* 269: 31573–31578.

Sugden, P.H., Kerbey, A.L., Randle, P.J., Waller, C.A. and Reid, K. (1979) Amino acid sequences around the sites of phosphorylation in the pig heart pyruvate dehydrogenase complex. *Biochem. J.* 181: 419–426.

Thissen, J., Desai, S., McCartney, P. and Komuniecki, R. (1986) Improved purification of the pyruvate dehydrogenase complex from *Ascaris suum* body wall muscle and characterization of PDH$_a$ kinase activity. *Mol. Biochem. Parasitol.* 21: 129–138.

Thissen, J. and Komuniecki, R. (1988) Phosphorylation and inactivation of the pyruvate dehydrogenase from the anaerobic parasitic nematode, *Ascaris suum*. Stoichiometry and amino acid sequence around the phosphorylation sites. *J. Biol. Chem.* 263: 19092–19097.

Vanover-Dettling, L. and Komuniecki, P.R. (1989) Effect of gas phase on carbohydrate metabolism in *Ascaris suum* larvae. *Mol. Biochem. Parasitol.* 36: 29–40.

Waterson, R., Martin, C., Craxton, M., Huynh, C., Coulson, A., Hillier, L., Durbin, R., Green, P., Shownkeen, R., Halloran, N., Metzstein, M., Hawkins, T., Wilson, R., Berks, M., Du, Z., Thomas, K., Thierry-Mieg, J. and Sulston, J. (1992) A survey of expressed genes in *Caenorhabditis elegans*. *Nature Genet.* 1: 114–123.

Wheelock, M.J., Komuniecki, R., Duran, E. and Johnson, K.R. (1991) Characterization of cDNA clones for the beta subunit of pyruvate dehydrogenase from *Ascaris suum*. *Mol. Biochem. Parasitol.* 45: 9–18.

Yeaman, S.J., Hutchenson, E.T., Roche, T.E., Pettit, F.H., Brown, J.R., Reed, L.J., Watson, D.C. and Dixon, G.H. (1978) Sites of phosphorylation on pyruvate dehydrogenase from bovine kidney and heart. *Biochem.* 17: 2364–2370.

Structure, function and assembly of mammalian branched-chain α-ketoacid dehydrogenase complex

R.M. Wynn[1, 2], J.R. Davie[1], M. Meng[1] and D.T. Chuang[1]

Departments of Biochemistry[1] and Internal Medicine[2], University of Texas Southwestern Medical Center, Dallas, TX 75235, USA

Introduction

The mammalian branched-chain α-ketoacid dehydrogenase complex (BCKDC) catalyzes the oxidative decarboxylation of the branched-chain α-ketoacids derived from valine, leucine, and isoleucine. The mammalian BCKDC is a member of the α-ketoacid dehydrogenase complexes comprising pyruvate dehydrogenase (PDC), α-ketoglutarate dehydrogenase (KGDC) and BCKDC with similar structure and function (Reed et al., 1985; Yeaman, 1989). All three of these complexes are located in the mitochondrial matrix compartment and are associated with the inner membrane. BCKDC is a highly assembled macromolecular structure with an estimated molecular mass of 3 to 4 million daltons. BCKDC isolated from bovine kidney and liver has been shown to contain three catalytic components (Pettit et al., 1978; Heffelfinger et al., 1983): a branched-chain α-ketoacid decarboxylase (E1), a dihydrolipoyl transacylase (E2) and a dihydrolipoyl dehydrogenase (E3). The E3 components are shared by PDC, KGDC and BCKDC (Yeaman, 1989). BCKDC also contains two regulatory enzymes, a specific kinase (Shimomura et al., 1990; Lee et al., 1991) and a specific phosphatase (Damuni et al., 1984), which modulate the activity of BCKDC by reversible phosphorylation-dephosphorylation.

Maple syrup urine disease (MSUD) is an inherited metabolic disorder caused by deficient BCKDC activity (for review, see Chuang and Shih, 1995). This metabolic block results in ketoacidosis, neurological derangements and mental retardation. Currently, limited progress has been made in understanding the molecular and biochemical basis of MSUD. The mammalian BCKDC with its intricate subunit structure and architectural design offers a useful genetic model to investigate how disease states perturb structure, function and assembly. This will be a main focus of this review.

Macromolecular organization

The mammalian BCKDC is organized around a cubic core consisting of 24 E2 subunits (Reed et al., 1985; Griffin et al., 1988). The arrangement of the E2 subunits is based on octahedral 432-point group symmetry. The 24 E2 subunits form eight trimers, with each trimer occupying a vertex of a truncated cube. The purified bovine E1-E2 subcomplex is a macromolecule with a sedimentation coefficient ($S_{20, w}$) of 40S. The molecules of E1 appear to be distributed over the surface of the cube, as examined by electron microscopy (Pettit et al., 1978). It has been shown that the E1β subunit, but not the E1α, of the bovine BCKDC binds to the E2 core (Wynn et al., 1992a). The results support the binding order of E2-E1β-E1α, which was suggested by the dark-field electron microscopy of the bovine BCKDC (Hackert et al., 1989). Compared to the PDC, the E3 component apparently has a lower binding affinity for the BCKDC-E2 core (Pettit et al., 1988), and is usually lost during the purification procedure.

Expression and assembly of mammalian E1

The E1 component of BCKDC contains two α and two β subunits with $M_r = 46\,000$ and $35\,000$, respectively, as determined by SDS-PAGE (Pettit et al., 1978). The holo-enzyme of E1 has an apparent $M_r = 170\,000$, as determined by gel filtration (Chuang, 1988). The reconstitution of the heterotetrameric E1 components of the BCKDC and PDC has proven difficult. Dissociation of E1 component of PDC with 8 M urea into its individual E1α and E1β subunits causes irreversible denaturation of the two polypeptide chains (Barrera et al., 1972). Expression of the E1β subunit of the bovine BCKDC in *E. coli* produced largely high molecular weight aggregates (Wynn et al., 1992a). These results have prevented the reconstitution to the E1 component *in vitro*. The initial solution to this problem was to co-express the mature E1α and E1β subunits of the mammalian BCKDC in *E. coli* (Davie et al., 1992). The E1α subunit was fused to the *E. coli* maltose-binding protein (MBP) to facilitate expression and purification. These experiments demonstrated that the E1 component could be stably expressed and assembled into an active E1 heterotetramer. In parallel experiments, separate expressions in respective host cells for the individual MBP-E1α and E1β, followed by mixing experiments, did not result in reconstitution to an active E1 component. The requirement for co-expression of both subunits of the complex in the same *E. coli* host strongly suggest that folding of one subunit into the assembly-competent state is dependent on the presence of the other subunit, and/or on factors in the bacterial cytoplasm. In contrast, Lessard and Perham (1994) have demonstrated that individual E1α and E1β subunits of the *B. stearothermophilus* PDC were stably expressed and capable of yielding an active reconstituted heterotetramer upon mixing the two cell lysates. It is not known if the difference between the bacterial and mammalian BCKDC E1's is due to more rapid folding kinetics of the bacterial PDC E1 than the mammalian BCKDC E1, or if certain cellular components in the bacteria optimize the folding of *B. stearothermophilus* E1. In a separate study, cotransformation of a second plasmid overexpressing bacteria chaperonins GroEL and GroES resulted in an increase of 500 times in the specific activity of E1 (Wynn et al., 1992b). The results strongly suggest bacterial chaperonins promote folding and assembly of heterotetramers ($\alpha_2\beta_2$) of mammalian BCKDC E1 in *E. coli*. Chaperonins GroEL and GroES are homologues of mitochondrial heat-shock proteins Hsp60 and Hsp10 (Hartl et al., 1992). The data establish the role of chaperonin proteins in the biogenesis of mammalian mitochondrial E1 protein. One can envision that chaperonin GroEL may serve as a "scaffold" for the proper folding of assembly-competent E1 subunits, however, the direct involvement of chaperonins in oligomeric assembly is an open question.

Assembly defects in E1 caused by mutations in maple syrup urine disease

The Y393N substitution in the E1α subunit is caused by a T to A transversion in exon 9 of the gene (Zhang et al., 1989; Matsuda et al., 1990; Fisher et al., 1991a). This missense mutation exists both in homozygotes in Mennonites and in compound heterozygotes in other populations. In an earlier study, vectors with E1α cDNAs carrying both normal sequence and the Y393N-α mutation were transfected into E1α-deficient lymphoblasts from another MSUD patient (Fisher et al., 1991b). The transfection with normal E1α cDNA restored decarboxylation activity to intact MSUD lymphoblasts. Transfection with the cDNA carrying the Y393N substitution was unable to correct the deficiency in decarboxylation of branched-chain α-ketoacids. The results offered direct evidence that the Y393N mutation was a cause of MSUD. Western blot analysis of cell lysates showed that transfection with normal E1α cDNA increased the level of both E1α and E1β subunits, the E1β being immunologically absent in the untransfected host cells. The results indicated that the E1β subunit was normally expressed, but was degraded in the E1α deficient host without assembly with the normal E1α subunit. Transfection with the vector carrying the cDNA with Y393N-α mutation failed to restore the E1β subunit to a level detectable by Western blotting. It was suggested that the Y393N-α mutation impeded the assembly of E1α and E1β into stable $\alpha_2\beta_2$ heterotetramers, resulting in the degradation predominantly of the E1β subunit. Constructs containing the Y368C-α (Chuang et al., 1994) and Y364C-α (Chuang et al., 1995)

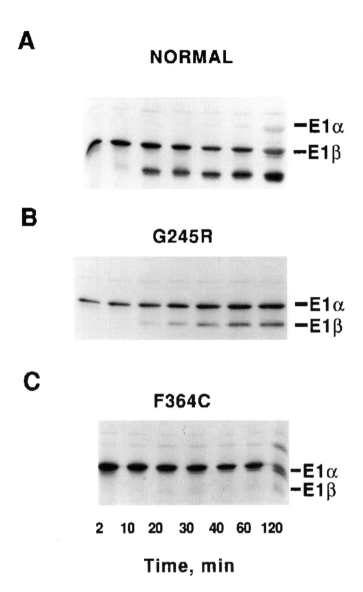

A

NORMAL

−E1α
−E1β

B

G245R

−E1α
−E1β

C

F364C

−E1α
−E1β

2 10 20 30 40 60 120

Time, min

Figure 1. Pulse-chase labeling autoradiogram depicting the kinetics of assembly of normal and mutant His-tagged E1α with untagged E1β subunits. The bacterial expression system overexpressing GroEL and GroES and the pulse-chase labeling of E1 subunits are described in Chuang et al. (1995). Cell lysates containing ^{35}S-labeled His-tagged E1α and associated polypeptides were bound to Ni^{2+}-NTA resin, washed and eluted with an SDS sample buffer. The eluted labeled polypeptides were analyzed by SDS-PAGE followed by fluorography. Assembly at different time points (2–120 min) of the untagged E1β with the His-tagged normal or mutant E1α resulted in the copurification of both subunits as shown in the autoradiograms. (A) Normal His-tagged E1α; (B) G245R His-tagged E1α; (C) F364C His-tagged E1α. (From Chuang et al. (1995) by permission of *J. Clin. Invest.*).

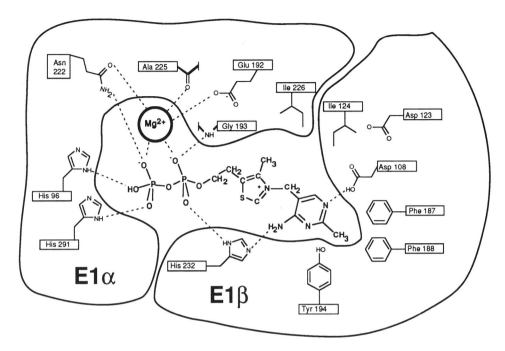

Figure 2. Model of the E1 decarboxylase active site for the human BCKDC with bound thiamine pyrophosphate. Amino acid residues in E1α and E1β subunits of the human BCKDC are assigned based on alignment with the published X-ray structure of the *S. cerevisiae* transketolase (Lindqvist et al., 1992). Computer program BLITZ, Version 1.5 was used in the sequence alignment. The modeling of the TPP-binding pocket refers to that previously described for transketolase and the human PDC-E1 (Robinson and Chun, 1993). A modification is that Mg^{2+} replaces Ca^{2+} in forming the pentameric coordinates involving the pyrophosphate group.

mutations, when transfected into E1α deficient MSUD cells, produced similar results to that obtained with Y393N-α. The results of molecular genetic studies indicate that Tyr-364, Tyr-368 and Tyr-393 in the C-terminal region of the E1α subunit are essential for assembly with the E1β subunit.

The putative E1 assembly defect in MSUD patients was supported by altered kinetics of association between E1β and mutant E1α subunits (Chuang et al., 1995). To investigate the kinetics of E1α and E1β assembly, a method for pulse-chase labeling of E1 subunits in an *E. coli* host was developed. ES^{ts} *E. coli* cells were doubly transformed with a His-tagged E1 expression vector carrying either normal or mutant His-tagged E1α cDNA and normal untagged E1β cDNA, along with a second vector pGroESL that over-expressed bacterial chaperonin proteins GroEL and GroES. The expression of these genes is driven by both an intrinisic heat-shock and a β-glactosidase promoter. The transformed cells were heat-shocked at 42°C for 4 h to induce chaperonins

GroEL and GroES, followed by incubation at 37°C with isopropyl thiogalactoside for 5 min. The cells were pulsed with ^{35}S-cysteine/^{35}S-methionine for 1 min and chased with the corresponding non-radioactive amino acids. Samples were taken at different time intervals from 2–120 min, and lysates purified by Ni^{+2}-NTA affinity chromatography. The eluted radioactive peptides were analyzed by SDS-PAGE and autoradiograms obtained by fluorography. Since E1β is untagged, its copurification with His-tagged E1α is an indication of assembly of the two polypeptides. Figure 1 shows that the assembly of radioactively labeled E1β with labeled normal E1α occurs as early as 10 min after the chase, approaching a plateau in 40 min. The level of the normal labeled His-tagged E1α remains relatively constant during the 2 h chase. The assembly of normal E1β with the labeled G245R E1α appears to be slower than normal E1α, and reached a lower than normal plateau at 60 min after the chase (Fig. 1). In contrast, there was no significant assembly of labeled E1β with His-tagged mutant F364C E1α during the 2 h chase. This was indicated by the absence of the copurified E1β, while the signal of His-tagged E1α remained relatively constant (Fig. 1). Kinetic patterns similar to Y364C-α were obtained with Y393N-α and Y368C-α mutants, confirming impaired assembly caused by these mutations.

Structural relevance of mammalian E1 to other TPP-dependent enzymes

To date, no X-ray structures have been resolved for the E1 components for the enzymes in this multienzyme family of complexes. However, the *Saccharomyces cerevisiae* transketolase (Lindqvist et al., 1992) structure has been determined to 2.5 Å resolution. This structure shows the TPP-binding motif GDGX$_{26/27}$ NN, which was previously assigned by sequence homology (Hawkins et al., 1989). This portion of the TPP-binding regions is only concerned with the pyrophosphate end of the cofactor and its relative position to the metal ion (in the case of transketolase-a calcium ion). On the opposite end of the TPP-binding site, the thiazolium ring structure is held in position by a hydrophobic pocket lined with phenylalanine residues, such that each dimeric transketolase molecule binds two molecules of TPP in a head-to-tail fashion (Lindqvist et al., 1992). It is reasonable to infer that instead of calcium binding to the pyrophosphate structure, magnesium or a similar divalent action is involved in binding TPP to PDC and BCKDC E1 components. The model of human BCKDC E1 shown in Figure 2 proposes that the magnesium is coordinated at the end of the channel to several residues and the pyrophosphate of the TPP. An interesting feature is the coordination of the Asn 222 to the magnesium and the pyrophosphate. This residue is mutated in a patient with MSUD (Chuang et al., 1995) and the catalytic activity of the E1 component is severely decreased compared to wild-type (Griffin, 1989). The alteration of this completely conserved Asn residue could conceivably induce a conformation change in the E1α subunit or decrease the binding efficiency for Mg^{2+} and TPP. In addition, two histidine residues have been identified that coordinate to the pyrophosphate in *S. cerevisiae* transketolase (Lindqvist et al., 1992). Equivalent residues, His-96 and His-291 are present in the E1α subunit of human BCKDC. His-291 is one amino acid away from the site 1 phosphorylation (Ser-292) of the E1α which results in the loss of over 90% of the enzymatic activity when phosphorylated by the kinase. Recent studies from Zhao et al. (1994) suggest that the size of the phosphate bound to site 1 (Ser-292) may prevent binding of TPP, instead of charge repulsion by the negative charged phosphate group interfering with binding. Similar to the transketolase structure, the thiazolium ring in the BCKDC E1 can be enclosed by several hydrophobic groups, Tyr-194, Phe-188, and Phe-187 in the human E1β subunit. The E1β subunit shows significant sequence similarities with other TPP-dependent enzymes, suggesting that E1β may also participate in TPP binding (Zhao et al., 1992). Lindqvist et al. (1992) have suggested that His-481, which is approximately 33 amino acids from the thiazolium binding pocket, serves to abstract a proton from the C-4' amino group of the thiazolium ring (Lindqvist et al., 1992). A possible candidate in the E1β subunit of the BCKDC E1 is His-232 (Fig. 2). This residue is thought to create a reactive thiazole carbanion for receipt of the acyl group in BCKDC E1. This mechanism is in contrast to the one proposed for the yeast pyruvate decarboxylase (Dyda et al., 1993). In the latter enzyme, the active site histidine that creates the reactive thiazole carbanion for transketolase (His-481) is absent.

Domain structure and expression of mammalian E2

The mature E2 subunit consists of three folded domains (N- to C-terminus) i.e., lipoyl-bearing, E1/E3 binding, and inner-core domains (Griffin et al., 1988; Lau et al., 1988; Hummel et al., 1988). These independently folded domains are linked together *via* flexible hinge regions rich in proline or negatively charged amino acid residues. The E2 domain structure is highly conserved in the 16 E2 proteins of the α-ketoacid dehydrogenase complexes from bacteria, yeast, and mammals that were cloned (Russell and Guest, 1991a). The only exception is the number of N-terminal lipoate-bearing domains, which varies from one to three. In the mammalian BCKDC, a single lipoyl-binding domain carries one lipoic acid moiety (Hu et al., 1986; Heffelfinger et al., 1983). The E2 subunit of bovine BCKDC migrates as a 52000-dalton species on SDS polyacrylamide gels (Pettit et al., 1978). The molecular mass based on the deduced amino acid composition is 46158 daltons for the mature bovine E2 subunit (Griffin et al., 1988). The higher apparent molecular mass for the E2 subunit on SDS-PAGE may be attributed to the acidic lipoic acid domain or the hinge regions which contain excess proline and charged amino acid residues (Guest et al., 1985). The C-terminal inner-core domain possesses the active site and the subunit binding regions that confer the 24-mer assembly (Chuang et al., 1984; Hu et al., 1986). The molecular mass of the bovine E2 24-mer is 1.1×10^6 daltons, as measured by sedimentation equilibrium analysis (Griffin et al., 1988; Reed et al., 1985).

Mature bovine E2 subunits of the bovine BCKDC complex were expressed in *E. coli* and shown by electron microscopy to assemble into a cubic 24-mer (Griffin et al., 1990). The recombinant E2 possessed transacylase activity when assayed in the presence of exogenous dihydrolipoamide and branched-chain acyl CoA. However, the recombinant 24-mer is essentially devoid of lipoic acid. It was concluded that the apo-E2 expressed in *E. coli* is capable of assembly into the 24-mer independent of the lipoylation process. Mitochondrial extracts prepared from beef liver catalyze the incorporation of DL-[2-^{3}H] lipoic acid into recombinant apo-E2 (Griffin et al., 1990).

Characterization of the active site of BCKDC-E2

The cloning of full-length mammalian E2 cDNAs and efficient expression of recombinant E2 domains have facilitated probing of the E2 active site. A recombinant protein corresponding to the residues 167 to 421, which contains the entire inner-core domain and a portion of the 5' linker region was stably expressed in *E. coli* (Griffin and Chuang, 1990). The recombinant inner core domain was capable of assembling into a 24-mer structure and exhibited full transacylase activity. Deletion of four residues (Thr-Ile-Pro-Ile) corresponding to residues 175 to 178 drastically reduced the E2 activity. Further deletion up to residue 209 resulted in complete loss of E2 activity without affecting the 24-mer assembly. This result confirms the conclusions derived from limited proteolysis (Chuang et al., 1985) that the segment between residues 175 and 209 of the inner core domain is an integral part of the E2 active site.

The atomic structure at 2.6 Å of the inner core domain of the related PDC from *Azotobacter vinelandii* has been recently solved (Mattevi et al., 1992). This structure shows that trimers represent the basic functional unit of E2 with 29 Å long active site channels formed at the interface between two three-fold related subunits. This topology is similar to that of the bacterial chloramphenicol acetyltransferase (CAT), as seminally suggested by Guest (1987). Based on the *Azotobacter* PDC-E2 structure, the segment of residues 175 to 209 of the BCKDC inner-core domain comprises a β strand and an α-helix that form an extended N-terminal arm connecting the two three-fold related subunits. Removal of this region of BCKDC-E2 by limited proteolysis

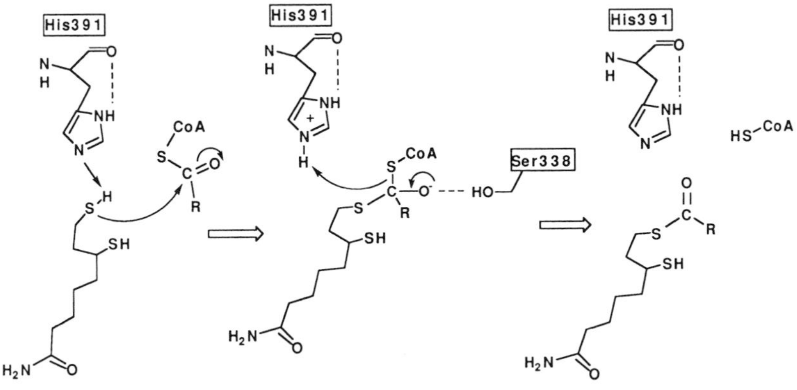

Figure 3. Schematic illustration of the proposed mechanism for the E2 inner-core domain-catalyzed acyltransfer reaction. His391 and Ser338 are active-site residues of the inner-core domain of the bovine BCKDC. Substrate acyl-CoA and dihydrolipoamide are indicated. The dotted line represents the hydrogen bond. (From Meng and Chuang (1994) by permission of *Biochemistry*).

(Chuang et al., 1985) or genetic reconstruction (Griffin and Chuang, 1990) apparently disrupt the integrity of the active site channel, thereby impeding the catalytic activity.

The crystal structure of E2 inner-core domain from *A. vinelandii* shows that the long active-site channel has two entrances; acetyl-CoA approaches from one end, while dihydrolipoamide enters from the other (Mattevi et al., 1993). The active-site histidine and serine, located in the middle of the channel, were proposed to catalyze the acyltransfer reaction by a mechanism analogous to that of CAT (Guest, 1987). Based on the alignment of amino acid sequence of the E2 family, the active site histidine and serine of bovine inner-core domain BCKDC are His-391 and Ser-338. The substitution of either residues with alanine resulted in a drastic decrease in k_{cat} by $10^3 - 10^4$ fold, but a marginal increase in K_M (Meng and Chuang, 1994). The small change in K_M implies that the mutation does not cause significant alterations in the conformation of the active-site. The drastic decreases in activity in H391A and S338A compared to the wild-type inner-core domain agree with the proposed catalytic roles of His-391 and Ser-338 in the actyltransfer reaction. The change in energy of interaction (ΔG_a) between the enzyme and transition state, due to the mutation of S338→Ala, was calculated to be –4.5 kcal/mole based on the equation, $\Delta G_a = RT \ln \{(k_{cat}/K_M)_{mutant}/(k_{cat} K_M)_{wild-type}\}$ (Fersht, 1988). It has been suggested that, in the protein-ligand interaction, 3–6 kcal/mole is the expected range of binding energy contributed by a hydrogen bond involving a charge donor or acceptor (Fersht, 1988). Thus, the active-site serine may act as a hydrogen bond donor to the putative negatively charged transition state. Moreover, in the spontaneous acyltransfer reaction, the rate is higher when the R group is a more electron-withdrawing group (e.g. acetoacetyl-CoA) or lower when the R group is a more electron-donating group (e.g.

isovaleryl-CoA) (Meng and Chuang, 1994). Based on the transition state theory, the reaction rate depends on the energy difference (G_a) between the transition state and the ground state. The reduction of the energy difference (G_a) by the electron-withdrawing ability of the R group indicates the reaction involves an increase of negative charge in the transition state.

A simplified mechanism is shown in Figure 3. The $N^{\epsilon 2}$ of imidazole of the active-site histidine abstracts a proton from the reactive thiol group of dihydrolipoamide. The nucleophilic sulfur, strengthened by the action of proton abstraction, attacks the carbonyl carbon of acyl-CoA and forms the putative tetrahedral intermediate. The energy of transition state is lowered by the active-site serine, which provides a hydrogen bond to the charged transition state. The acyltransfer reaction is completed following the breakdown of the tetrahedral intermediate. This reaction mechanism is analogous to the one described for CAT of *E. coli* (Lewendon et al., 1990). This mechanism may also occur in the E2 transacetylase of the *E. coli* PDC complex, as suggested by previous site-directed mutagenesis studies (Russell and Guest, 1990; Russell and Guest, 1991b). It remains unclear whether the formation or the breakdown of the tetrahedral intermediate is the rate-limiting step.

Steric effects on substrate specificity of E2

Although the amino acid sequences of the various E2 components of the family of α-ketoacid dehydrogenase complexes are highly homologous, each subtype enzyme has its own substrate specificity. For instance, favorable substrates for the E2 of mammalian BCKDC are branched-chain acyl-CoA, whereas E2 of PDC has strong preference for acetyl-CoA. A steric effect may be the key factor responsible for substrate specificity. Recent site-directed mutagenesis studies have identified one such amino acid, which may contact the substrate and partially confer substrate specificity (Meng and Chuang, 1994). When the BCKDC Ala-348 was changed to Val, the catalytic efficiency for isovaleryl-CoA decreased about 10-fold, and that for acetyl-CoA increased about three-fold. Ala-348 presumably contacts the isobutyl group of isovaleryl-CoA in the acyltransfer reaction. Modeling studies based on the structure of the enzyme-substrate complex suggested that Phe-568 of the inner-core domain from *A. vinelandii* (corresponding to Ala-348 in bovine BCKDC) may directly contact the methyl group of acetyl-CoA (Fig. 4). When Ala-348 was changed to Phe, the substrate preference shifted slightly toward acetyl-CoA, and catalytic efficiency (k_{cat}/K_M) remained 4% for isovaleryl-CoA and 6% for acetyl-CoA. The phenyl group in A348F may be too large for maintaining the critical distance between the catalytic residues and the reactive groups of acetyl-CoA and isovaleryl-CoA. Alternatively, the phenyl group in the

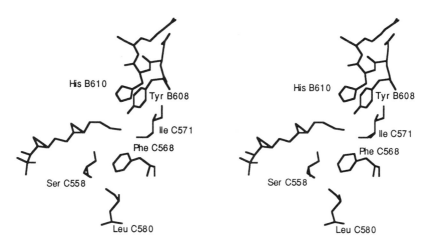

Figure 4. Structure of CoA and the active-site residues of the inner-core domain of PDC from *A. vanelandii* (Mattevi et al., 1992). The active-site channel is formed at the interface of two neighboring subunits (B and C in this figure). HisB610 and TyrB608 are from the B subunit, while SerC558, PheC568, IleC571, and LeuC580 are from the C subunit. Only the pantetheine arm of CoA is shown. The reactive thiol group of the pantetheine arm points toward the active-site amino acids. PheC568 corresponds to Ala 348 in BCKDC-E2 based on sequence alignment (Meng and Chuang, 1994). The coordinates were provided by Dr. Wim G. J. Hol at the University of Washington, Seattle, and visualized on the Silicon Graphics Computer with the aid of the INSIGHT II graphic program. (From Meng and Chuang (1994) by permission of *Biochemistry*).

mutant enzyme may block the approach of acyl-CoA and dihydrolipoamide. The side chain of leucine or valine affects the accommodation of isovaleryl-CoA, but has no adverse impact on acetyl-CoA. These results indicate that substrate preference is sensitive to alterations in the size of the side chain of amino acid 348, and implicates Ala-348 as a major structural determinant for preferential catalysis of isovaleryl-CoA by the bovine E2 of BCKDC. The inability to convert BCKDC-E2 to PDC-E2 with the same catalytic efficiency indicates that certain other structural elements may be important for creating an optimal environment for efficient acyl transfer with acetyl-CoA.

In vitro reconstitution of E2 inner core

In vitro reconstitution of the 24-mer E2 inner-core domain has shown an obligate requirement for chaperonins GroEL and GroES (Wynn et al., 1994). In this study, the recombinant inner-core domain was denatured in 4.5 M guanidine (Gdn).HCl (pH 7.5) at 25°C for 60 min. Under these conditions, the inner-core domain unfolded into a completely "random coil" structure within 10 min as determined by the ultraviolet circular dichroism spectrum. The Gdn.HCl-denatured inner-core domain had no transacylase activity, and migrated as a monomeric species

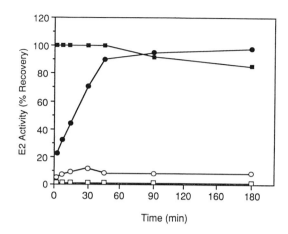

Figure 5. Time course of chaperonin-dependent refolding of denatured bovine inner-core domain (E2c) into active enzyme. Recombinant bovine E2c was denatured in 6 M guanidine.HCl. Refolding was carried out by diluting 100-fold with a buffer in the absence or presence of chaperonins GroEL (600 nM, oligomer) and GroES (600 nM, oligomer) and Mg-ATP (5 mM) at 25°C. The final concentration of monomeric E2c in the refolding reaction mixture was 150 nM in a volume of 0.5 ml. At the indicated times, an aliquot was withdrawn and assayed for E2 activity. E2 activity is expressed as a percentage of the activity of an equal amount of undenatured E2c which was 186 m-units/ml or 100%. (■) Native E2c incubated at 25°C; (●) denatured E2c + GroEL, GroES and Mg-ATP; (○) denatured E2c + GroEL, GroES without Mg-ATP; (□) denatured E2c with no addition. (From Wynn et al. (1994) by permission of *Biochemistry*).

($M_r \approx 27\,000$) in sucrose density gradient centrifugation. The Gdn.HCl-denatured inner-core domain (15 µM, monomer), when diluted 100-fold into a refolding buffer lacking both chaperonins and Mg-ATP, did not result in recovery of detectable enzymatic activity after 3 h at 25°C (Fig. 5). When the denatured inner-core domain (15 µM) was diluted 100-fold into the refolding buffer containing four-fold molar excess of the chaperonins GroEL/GroES multimers minus ATP, there was a slight (10%) recovery of the transacylase activity during the measured period (15 min to 3 h) (Fig. 5). When 5 mM Mg-ATP was included with the chaperonins in the refolding buffer (complete refolding buffer), full (100%) tranacylase activity was recovered in approximately 45 min at 25°C (Fig. 5). Native, non-denatured inner-core domain diluted similarly in the refolding buffer served as a positive control, and showed a slight decline in over 3 h. After 20 min of refolding, a major portion of inner-core domain was dissociated from GroEL and assembled into active 24-mers, as analyzed by sucrose density gradient centrifugation.

The homo 24-mer structure of the reconstituted E2 inner-core domain represents the largest number of subunits that have been refolded and assembled *in vitro* to date with the aid of chaperonin proteins. Other active, partially dissociated E2 species have been previously described. Treatment of the E2 transacetylase of the *E. coli* PDC with N-acetyl-imidazole was previously shown to produce an active smaller species of E2, probably a trimer. The latter was capable of binding E1 and E3 to produce an active PDC with an apparent M_r of about 500 000 (Schwartz and Reed, 1969). The E2 transacetylase of the PDC from *A. vinelandii* was shown to dissociate into a smaller tetrameric species in the presence of Gdn.HCl (Hanemaaijer et al., 1989). It has been speculated that the E2 inner core of α-ketoacid dehydrogenase complexes is assembled from trimers (Mattevi et al., 1992). This was supported by earlier electron microscopy data which showed that the E2 component of *E. coli* PDC (Willms et al., 1967) and bovine BCKDC (Chuang et al., 1985) possess octahedral symmetry with eight trimers occupying the vertices of a truncated cube. This structure is substantiated by the results of the aforementioned crystallographic study of the inner-core domain of the PDC from *A. vinelandii* (Mattevi et al., 1992). Within this structure, the most extensive interactions (involving about 25% of the monomer-accessible surface) occur within individual trimers. Upon oligomerization of the trimers to form a truncated cube, only an additional 8% of the monomer-accessible surface is buried. Active trimers interlock through carboxy-terminal hydrophobic knobs (Mattevi et al., 1992).

During *in vitro* refolding reactions (Wynn et al., 1994) attempts were made to isolate putative trimeric species of inner-core domain of BCKDC by quenching the reaction at different times with trans-1, 2-diaminocyclohexane-N, N, N', N'-tetraacetic acid (CDTA). However, only active inner-core domain 24-mers, but no trimers, were detected by sucrose density ultracentrifugation. An indirect approach was therefore undertaken to partially dissociate the recombinant inner-core domain. This experiment utilized the mild chaotropic conditions of 1.5 M Gdn.HCl, which promoted disruption of the relatively weak interactions between neighboring trimers. The stronger interactions between the two three-fold related subunits required for E2 activity were not perturbed. Results of the radiochemical assay demonstrated unequivocally the existence of an active trimeric species of inner-core domain (apparent $M_r = 84 000$), confirming the basic functional unit deduced from X-ray crystallographic structure (Mattevi et al., 1992). The inclusion of 1.5 M Gdn.HCl in the reaction mixture ensured that reassociation of trimers into active 24-mers did not occur during the enzyme assay. Upon removal of the 1.5 M Gdn.HCl, active inner-core domain 24-mers of BCKDC were reisolated from sucrose density ultracentrifugation (Wynn et al., 1994). The latter results support the model depicted in Figure 6, in which chaperonins GroEL and GroES mediate the folding of assembly-competent monomers. The monomers assemble into an intermediate active trimeric species, where the role of chaperonins is undetermined, as assemble-competent monomers of inner-core domain thus far cannot be isolated for oligomerization studies. The assembly of trimers into 24-mers is apparently spontaneous and independent of chaperones.

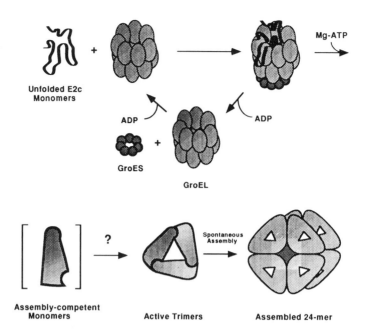

Figure 6. Proposed pathway for *in vitro* reconstitution of the 24-mer structure of the E2 inner-core domain. The refolding of the denatured E2 inner-core domain monomers begins with binding to the open end of the GroEL 14-mer complexed with the GroES 7-mer. Refolding occurs within the central cavity of the GroEL cylinder (Hartl et al., 1992). Multiple rounds of Mg-ATP hydrolysis result in the release of assembly-competent monomers (shown in brackets). Chaperonins GroEL and GroES recycle in the presence of ADP to form a second chaperonin-polypeptide complex. The monomers assemble into an intermediate species of active trimers whereby constitutive polypeptides assume their final conformation. Active trimers interlock through carboxy-terminal hydrophobic knobs (Mattevi et al., 1992) to produce the native 24-mer structure with octahedral symmetry. The mechanism for the assembly of monomers into active trimers is not known, as depicted by a question mark. However, the *in vitro* reassociation experiment establishes that the assembly of the trimeric intermediates into active 24-mers is a spontaneous process independent of chaperonins. (From Wynn et al. (1995) by permission of *Biochemistry*).

Conclusion

Expression of individual E1α and E1β subunits of the mammalian BCKDC in *E. coli* leads to essentially insoluble proteins, and mixing *in vitro* does not reconstitute E1 activity. It is established that coexpression of both subunits with cotransformed GroEL and GroES in the same *E. coli* cells is required for optimal expression and assembly of BCKDC-E1. Using the chaperonin-augmented expression system, impairment in E1 assembly secondary to MSUD mutations in the E1α subunit has been demonstrated by pulse-chase labeling studies. These naturally occurring human mutations may prove a useful genetic model for investigation into steps leading to chaperonin-mediated biogenesis of mitochondrial proteins. Recent structural information of other TPP-dependent enzymes has provided a basis for modeling of mammalian BCKDC-E1. Certain MSUD mutations in the E1α subunit have been localized to this cofactor binding site. Further studies by site-directed mutagenesis of candidate residues in the TPP-binding pocket will undoubtedly shed light on the mechanism of E1 catalysis.

The recent solution of atomic structure of the 24-mer inner core of *Azotobacter* PDC-E2 has greatly facilitated structural and functional studies of other E2 proteins. Site-directed mutagenesis studies have suggested a reaction mechanism, in which His-391 serves as a general base and Ser-338 provide hydrogen bonding to create a tetrahedral transition state intermediate between dihydrolipoamide and branched-chain acetyl-CoA. Moreover, Ala-348 has been proposed to be a key residue contributing to substrate specificity for branched-chain CoAs. This residue presumably contacts the isobutyl group of isovaleryl-CoA in the aceyltransfer reaction, based on modeling of *Azotobacter* PDC-E2. Alteration of Ala-348 to Phe could create a steric hindrance that blocks the approach of isovaleryl-CoA to the active site channel with minimal effects on acetyl CoA.

In vitro reconstitution of the E2 inner core domain has shown a requirement for chaperonins GroEL and GroES. Partial dissociation of the assembled inner core domain in 1.5 M Gdn.HCl resulted in the isolation of active trimers. Removal of the chaotropic reagent spontaneously produced the active 24-mer structure. A model is proposed in which chaperonins mediate folding of assembly-competent monomers. The latter assemble into active trimers, which spontaneously oligomerize into the final 24-mer structure.

Acknowledgements
This work was supported by Grants DK37373 and DK26758 from the National Institutes of Health, and Grant 95G-074 from the American Heart Association, Texas Affiliate, and Grant I-1286 from the Robert A. Welch Foundation.

References

Barrera, C.R., Namihira, G., Hamilton, L., Munk, P., Eley, M.H., Linn, T.C. and Reed, L.J. (1972) α-Keto acid dehydrogenase complexes. XVI. Studies on the subunit structure of the pyruvate dehydrogenase complexes from bovine kidney and heart. *Arch. Biochem. Biophys.* 148: 343–358.

Chuang, D.T., Hu, C.-W.C., Ku, L.S., Niu, W.L., Myers, D.E. and Cox, R.P. (1984) Catalytic and structural properties of the dihydrolipoyl transacylase component of bovine branched-chain α-ketoacid dehydrogenase. *J. Biol. Chem.* 259: 9277–9284.

Chuang, D.T., Hu, C.-W.C., Ku, L.S. and Cox, R.P. (1985) Subunit structure of the dihydrolipoyl transacylase component of branched-chain α-ketoacid dehydrogenase complex from bovine liver. Characterization of the inner transacylase core. *J. Biol. Chem.* 260: 13779–13786.

Chuang, D.T. (1988) Assays for E1 and E2 components of branched-chain ketoacid dehydrogenase complex. *Meth. Enzymol.* 166: 146–154.

Chuang, J.L., Fisher, C.R., Cox, R.P. and Chuang, D.T. (1994) Molecular basis of maple syrup urine disease: Novel mutations at the E1α locus that impair E1 ($\alpha_2\beta_2$) assembly or decrease steady-state E1α mRNA levels of branched-chain α-ketoacid dehydrdogenase complex. *Am. J. Hum. Genet.* 55: 297–304.

Chuang, D.T. and Shih, V.E. (1995) Disorders of branched chain amino acid and ketoacid metabolism. *In*: C.R. Scriver, A.L. Beaudet, W.S. Sly and D. Valle (eds):*The Metabolic and Molecular Bases of Inherited Disease*, 7th Edition. McGraw-Hill, Inc., pp 1239–1277.

Chuang, J.L., Davie, J.R., Chinsky, J.M. Wynn, R.M., Cox, R.P. and Chuang, D.T. (1995) Molecular and biochemical basis of intermediate maple syrup urine disease. Occurrence of homozygous G245R and F364C mutations at the E1α locus of Hispanic-Mexican patients. *J. Clin. Invest.* 95: 954–963.

Damuni, Z., Merryfield, M.L., Humphreys, J.S. and Reed, L.J. (1984) Purification and properties of branched-chain α-ketoacid dehydrogenase phosphatase from bovine kidney. *Proc. Natl. Acad. Sci. USA* 81: 4335–4338.

Davie, J.R., Wynn, M.R., Cox, R.P. and Chuang, D.T. (1992) Expression and assembly of a functional E1 component ($\alpha_2\beta_2$) of mammalian branched-chain α-ketoacid dehydrogenase complex in *Escherichia coli. J. Biol. Chem.* 267: 16601–16606.

Dyda, F., Furey, W., Swaminathan, S., Sax, M., Farrenkopf, B. and Jordan, F. (1993) Catalytic centers in the thiamin diphosphate dependent enzyme Pyruvate Decarboxylase at 2.4-Å resolution. *Biochemistry* 32: 6165–6170.

Fersht, A.R. (1988) Relationships between apparent binding energies measured in site-directed mutagenesis experiments and energetics of binding and catalysis. *Biochemistry* 27: 1577–1580.

Fisher, C.R., Fisher, C.W., Chuang, D.T. and Cox, R.P. (1991a) Occurrence of a Tyr 393→Asn (Y393N) mutation in the E1α gene of the branched-chain α-ketoacid dehydrogenase complex in maple syrup urine disease patients from a Mennonite population. *Am. J. Hum. Genet.* 2: 429–434.

Fisher, C.R., Chuang, J.L., Cox, R.P., Fisher, C.W., Star, R.A. and Chuang, D.T. (1991b) Maple syrup urine disease in Mennonites. Evidence that the Y393N mutation in E1α impedes assembly of the E1 component of branched-chain α-ketoacid dehydrogenase complex. *J. Clin. Invest.* 88: 1034–1037.

Griffin, T.A., Lau, K.S. and Chuang, D.T. (1988) Characterization and conservation of the inner E_2 core domain structure of branched-chain α-ketoacid dehydrogenase complex from bovine liver. Construction of a cDNA encoding the entire transacylase (E_2b) precursor. *J. Biol. Chem.* 263: 14008–14014.

Griffin, T.A. (1989) *Molecular Studies of the Mammalian Branched-chain α-Ketoacid Dehydrogenase Complex*. PhD Thesis, Case Western Reserve University.

Griffin, T.A. and Chuang, D.T. (1990) Genetic reconstruction and characterization of the recombinant transacylase (E_2b) component of bovine branched-chain α-ketoacid dehydrogenase complex. Implication of histidine 391 as an active site residue. *J. Biol. Chem.* 265: 13174–13180.

Griffin, T.A., Wynn, R.M. and Chuang, D.T. (1990) Expression and assembly of mature apotransacylase (E_2b) of bovine branched-chain α-ketoacid dehydrogenase complex in *Escherichia coli*. Demonstration of transacylase activity and modification by lipoylation. *J. Biol. Chem.* 265: 12104–12110.

Guest, J.R., Lewis, H.M., Graham, L.D., Lloyd, L.D., Packman, L.C. and Perham, R.N. (1985) Genetic reconstruction and functional analysis of the repeating lipoyl domains in the pyruvate dehydrogenase multienzyme complex of *Escherichia coli. J. Mol. Biol.* 185: 743–754.

Guest, J.R. (1987) Functional implications of structural homologies between chloramphenicol acetyltransferase and dihydrolipoamide acetyltranferase. *FEMS Microbiol. Lett.* 44: 417–422.

Hackert, M.L., Xu, W.-X., Oliver, R.M., Wall, J.S., Hainfeld, J.F., Mullinax, T.R. and Reed, L.J. (1989) Branched-chain α-ketoacid dehydrogenase complex from bovine kidney: radial distribution of mass determined from dark-field electron micrographs. *Biochemistry* 28: 6816–6821.

Hanemaaijer, R., Wastphal, A.H., van der Heiden, T., de Kok, A. and Veeger, C. (1989) The quaternary structure of the dihydrolipoyl transacetylase component of the pyruvate dehydrogenase complex from *Azotobacter vinelandii*. A reconsideration. *Eur. J. Biochem.* 179: 287–292.

Hartl, F.U., Martin, J. and Neupert, W. (1992) Protein folding in the cell: The role of molecular chaperones Hsp70 and Hsp60. *Ann. Rev. Biophys. Biomol. Struct.* 21: 293–322.

Hawkins, C.F., Borges, A. and Perham, R.N. (1989) A common structural motif in thiamin pyrophosphate-binding enzymes. *FEBS Lett.* 255: 77–82.

Heffelfinger, S.C., Sewell, E.T. and Danner, D.J. (1983) Identification of specific subunits of highly purified bovine liver branched-chain ketoacid dehydrogenase. *Biochemistry* 22: 5519–5522.

Hu, C.-W.C., Griffin, T.A., Lau, K.M., Cox, R.P. and Chuang, D.T. (1986) Subunit structure of dihydrolipoyl transacylase component of branched-chain α-ketoacid dehydrogenase complex from bovine liver. *J. Biol. Chem.* 1: 343–349.

Hummel, K.B., Litwer, S., Bradford, A.P., Aitken, A., Danner, D.J. and Yeaman, S.J. (1988) Nucleotide sequence of a cDNA for branched-chain acyltransferase with analysis of the deduced protein structure. *J. Biol. Chem.* 263: 6165–6168.

Lau, K.S., Griffin, T.A., Hu, C.-W.C. and Chuang, D.T. (1988) Conservation of primary structure in the lipoyl-bearing and dihydrolipoyl dehydrogenase domain of mammalian branched-chain α-ketoacid dehydrogenase complex. Molecular cloning of human and bovine transacylase (E2) cDNAs. *Biochemistry* 27: 1972–1981.

Lee, H.Y., Hall, T.B., Kee, S.M., Tung, H.Y.L. and Reed, L.J. (1991) Purification and properties of branched-chain α-keto acid dehydrogenase kinase from bovine kidney. *Biofactors* 2: 109–112.

Lessard, I.A. and Perham, R.N. (1994) Expression in *Escherichia coli* of genes encoding the E1α and E1β subunits of the pyruvate dehydrogenase complex of *Bacillus stearothermophilus* and assembly of a functional E1 component (α2β2) *in vitro. J. Biol. Chem.* 269: 10378–10383.

Lewendon, A., Murray, I.A. and Shaw, W.V. (1990) Evidence for transition-state stabilization by Ser-148 in the catalytic mechanism of chloramphenicol acetyltransferase. *Biochemistry* 29: 2075–2080.

Lindqvist, Y., Schneider, G., Ermler, U. and Sundstrom, M. (1992) Three-dimensional structure of transketolase, a thiamine diphosphate dependent enzyme, at 2.5 Å resolution. *EMBO J.* 11: 2373–2379.

Matsuda, I., Nobukuni, Y., Mitsubuchi, H., Indo, Y., Endo, F., Asaka, J. and Harada, A. (1990) A T- to -A substitution in the E1α subunit gene of the branched-chain α-ketoacid dehydrogenase complex in two cell lines derived from Mennonite maple syrup urine disease patients. *Biochem. Biophys. Res. Comm.* 172: 646–651.

Mattevi, A., Obmolova, G., Schulze, E., Kalk, K.H., Westphal, A.H., de Kok, A. and Hol W.G.J. (1992) Atomic structure of the cubic core of the pyruvate dehydrogenase multienzyme complex. *Science* 255: 1544–1550.

Mattevi, A., Obmolova, G., Kalk, K.H., Teplyakov, A. and Hol, W.G.J. (1993) Crystallographic analysis of substrate binding and catalysis in dihydrolipoyl transacetylase (E2p). *Biochemistry* 32: 3887–3901.

Meng, M. and Chuang, D.T. (1994) Site-directed mutagenesis and functional analysis of the acitve-site residues of the E2 component of bovine branched-chain α-ketoacid dehydrogenase complex. *Biochemistry* 33: 12879–12885.

Pettit, F.H., Yeaman, S.J. and Reed, L.J. (1978) Purification and characterization of branched chain α-ketoacid dehydrogenase complex of bovine kidney. *Proc. Natl. Acad. Sci. USA* 75: 4881–4885.

Pettit, F.H. and Reed, L.J. (1988) Branched-chain α-ketoacid dehydrogenase complex from bovine kidney. *Meth. Enzymol.* 166: 309–312.

Reed, L.J., Damuni, Z. and Merryfield, M.L. (1985) Regulation of mammalian pyruvate and branched-chain α-ketoacid dehydrogenase complexes by phosphorylation-dephosphorylation. *Curr. Topics Cell Regul.* 27: 41–49.

Robinson, B.H. and Chun, K. (1993) The relationships between transketolase, yeast pyruvate decarboxylase and pyruvate dehydrogenase of the pyruvate dehydrogenase complex. *FEBS Lett.* 328: 99–102.

Russell, G.C. and Guest, J.R. (1990) Overexpression of restructured pyruvate dehydrogenase complexes and site-directed mutagenesis of a potential active-site histidine residue. *Biochem. J.* 269: 443–450.

Russell, G.C. and Guest, J.R. (1991a) Sequence similarities within the family of dihydrolipoamide acyltransferases and discovery of a previously unidentified fungal enzyme. *Biochim. Biophys. Acta* 1076: 225–232.

Russell, G.C. and Guest, J.R. (1991b) Site-directed mutagenesis of the lipoate acetyltransferase of *Escherichia coli. Proc. R. Soc. Lond. B* 243: 155–160.

Schwartz, E.R. and Reed, L.J. (1969) α-Ketoacid dehydrogenase complexes. XII. Effects of acetylation on the activity and structure of the dihydrolipoyl transacetylase of *Escherichia coli. J. Biol. Chem.* 244: 6074–6079.

Shimomura, Y., Nanaumi, N., Suzuki, M., Popov, K.M. and Harris, R.A. (1990) Purification and partial characterization of branched-chain α-ketoacid kinase from rat liver and rat heart. *Arch. Biochem. Biophys.* 2: 293–299.

Willms, C.R., Oliver, R.M., Henney, H.R., Mukherjee, B.B. and Reed, L.J. (1967) α-Ketoacid dehydrogenase complexes VI. Dissociation and reconstitution of the dihydrolipoyl transacetylase of *Escherichia coli. J. Biol. Chem.* 242: 889–897.

Wynn, R.M., Chuang, J.L., Davie, J.R., Fisher, C.W., Hale, M.A., Cox, R.P. and Chuang, D.T. (1992a) Cloning and expression in *Escherichia coli* of mature E1β subunit of bovine mitochondrial branched-chain α-ketoacid dehydrogenase complex. Mapping of the E1β-binding region on E2. *J. Biol. Chem.* 267: 1881–1887.

Wynn, R.M., Davie, J.R., Cox, R.P. and Chuang, D.T. (1992b) Chaperonins GroEL and GroES promote assembly of heterotetramers (α2β2) of mammalian mitochondrial branched-chain α-ketoacid decarboxylase in *Escherichia coli. J. Biol. Chem.* 267: 12400–12403.

Wynn, R.M., Davie, J.R., Zhi, W., Cox, R.P. and Chuang, D.T. (1994) *In vitro* reconstitution of the 24-meric E2 inner core of bovine mitochondrial branched-chain α-ketoacid dehydrogenase complex: Requirement for chaperonins GroEL and GroES. *Biochemistry* 33: 8962–8968.

Yeaman, S.J. (1989) The 2-oxo acid dehydrogenase complexes: recent advances. *Biochem. J.* 257: 625–632.

Zhang, B., Edenberg, H.J., Crabb, D.W. and Harris, R.A. (1989) Evidence for both regulatory mutation and structural mutation in a family with maple syrup urine disease. *J. Clin. Invest.* 83: 1425–1429.

Zhao, Y., Kuntz, M.J., Harris, R.A. and Crabb, D.W. (1992) Molecular cloning of the E1β subunit of the rat branched chain α-ketoacid dehydrogenase. *Biochim. Biophys. Acta* 1132: 207–210.

Zhao, Y., Hawes, J., Popov, K.M., Jaskiewicz, J., Sjimomura, Y., Crabb, D.W. and Harris, R.A. (1994) Site-directed mutagenesis of phosphorylation sites of the branched chain α-ketoacid dehydrogenase complex. *J. Biol. Chem.* 269: 18583–18587.

Alpha-Keto Acid Dehydrogenase Complexes
M.S. Patel, T.E. Roche and R.A. Harris (eds)
© 1996 Birkhäuser Verlag Basel/Switzerland

Lipoylation of E2 component

Y. Motokawa, K. Fujiwara and K. Okamura-Ikeda

The Institute for Enzyme Research, The University of Tokushima, Tokushima 770, Japan

Summary. Lipoic acid is a prosthetic group of the acyltransferase (E2) components of the pyruvate, α-ketoglutarate, and branched chain α-keto acid dehydrogenase complexes, X component (the dihydrolipoamide dehydrogenase-binding protein) of the eucaryotic pyruvate dehydrogenase complex, and H-protein of the glycine cleavage system. The lipoyl moiety is attached in amide linkage to the ε-amino group of specific lysine residues of the proteins. Although the role of lipoic acid in these proteins has been well studied, intensive research on enzyme(s) concerning the lipoylation of proteins have not been undertaken until recently. In this chapter, we describe the recent development about the lipoylation of E2 component and H-protein as well as the historical view of the studies about lipoylation.

Earlier works on lipoylation of the E2 component

Reed et al. (1958) first described the lipoylation of the pyruvate dehydrogenase complex employing lipoic acid-deficient *Streptococcus faecalis* cells. Cell free extracts of *S. faecalis* grown on synthetic medium were incubated with [^{35}S]lipoic acid. The incubation mixtures were dialyzed to remove free lipoic acid. Assay of the dialyzed preparations revealed a parallel increase in the amount of bound lipoic acid and the pyruvate dehydrogenase activity. Mg^{2+} or Mn^{2+} was required for the conversion of the enzyme to the active form. Two protein fractions were obtained from the cell free extracts with protamine sulfate precipitation, and each fraction was partially purified with ammonium sulfate. The fraction precipitated with protamine sulfate and further with ammonium sulfate at 60% saturation contained the apopyruvate dehydrogenase complex, and the other fraction, supernatant of the protamine precipitation and precipitable with ammonium sulfate at 60–100% saturation, contained a system responsible for protein lipoylation. Activation of lipoic acid through its carboxyl group was catalyzed by the latter fraction when incubated with lipoic acid and ATP. The latter fraction was further separated into two fractions, a heat-labile and a heat-stable fraction, both of which were required for the activation of the apopyruvate dehydrogenase complex. The heat-labile fraction catalyzed the activation of lipoic acid, but the activity of the heat-stable fraction has remained unknown. A protein fraction catalyzing the lipoylation of the apopyruvate dehydrogenase complex from *S. faecalis* was also obtained from *Escherichia coli*, but the fraction was not dissolved into two fractions as observed in *S. faecalis*.

Reed et al. (1958) found that lipoyl-AMP is the intermediate for the activation of the apopyruvate dehydrogenase complex. Synthetic lipoyl-AMP could replace the mixture of lipoic acid and ATP. The result suggested that the reaction proceeds as shown,

$$lipoic\ acid + ATP \longrightarrow lipoyl-AMP + PPi \qquad [1]$$
$$lipoyl\text{-}AMP + apoenzyme \longrightarrow holoenzyme + AMP, \qquad [2]$$

but the formation of AMP was not confirmed.

Subsequently, activities of lipoic acid activation and the transfer of lipoic acid to proteins were detected in extracts of *Azotobacter vinelandii* and mung bean seedling (Mitra and Burma, 1965). Radioactive lipoyl-AMP was enzymatically synthesized either with [^{35}S]lipoic acid and ATP or cold lipoic acid and [^{14}C]ATP and separated by paper chromatography. On mild hydrolysis of lipoyl-AMP isolated from the chromatogram, lipoic acid and AMP were obtained, and the radioactivity resided with the component derived from the radioactive precursor used. Transfer of [^{35}S]lipoic acid to proteins was dependent on ATP and Mg^{2+} and proportional to the amount of extracts added. These results supported the mechanism suggested by Reed et al. (1958).

Reversibility of the lipoic acid activation reaction has been demonstrated by the fact that extracts from *A. vinelandii* and mung bean seedling catalyzed the pyrophosphate-dependent disappearance of lipoyl-AMP (Mitra and Burma, 1965). Apparently, ATP was formed in this reaction since [^{32}P]ATP was produced from lipoyl-AMP and [^{32}P]PPi.

The presence of a lipoic acid activating enzyme has been demonstrated in animal liver (Tsunoda and Yasunobu, 1967). The enzyme was partially purified from the soluble fraction of bovine liver by ammonium sulfate fractionation and adsorption to calcium phosphate gel. The activation of lipoic acid was monitored by the formation of lipoyl-hydroxamate. Absolute requirement of lipoic acid and ATP was demonstrated as previously observed in the bacterial enzyme described above. Octanoic acid appeared to be activated by the same enzyme since incubation of lipoic acid and octanoic acid simultaneously resulted in a decrease in product formation.

Lipoate-protein ligase from *E. coli*

About 30 years have elapsed since the first lipoylation experiment made by Reed et al. (1958) without significant progress in the study of protein lipoylation. It is probably because of a lack of suitable protein substrate for lipoylation reaction. Developments in biotechnology now enable us to obtain apoproteins that accept lipoic acid as a prosthetic group and to investigate the mechanism of lipoylation. Lipoyl domains of the pyruvate dehydrogenase complex of *E. coli* were over-expressed, and holo- and apolipoyl domains were purified (Ali and Guest, 1990). Two indepen-dent lipoate-protein ligase activities in *E. coli,* LPL-A and LPL-B, were separated by chromato-graphy on heparin agarose employing the apolipoyl domain as a substrate (Brookfield et al., 1991). Both enzymes have the same molecular weight (about 45 000) and catalyzed lipoylation of the apolipoyl domain in the presence of lipoic acid, ATP, and Mg^{2+}. Mn^{2+}, Co^{2+}, Zn^{2+}, or Ni^{2+} were effective in place of Mg^{2+}. ATP and lipoic acid could be replaced by lipoyl-AMP. In addition to lipoic acid, octanoic acid was used as a substrate by LPL-B, but not LPL-A. The primary difference between LPL-A and LPL-B is the failure of LPL-A to convert octanoic acid to octanoyl-AMP. Both enzyme fractions activated the apopyruvate dehydrogenase complex obtain-ed from anaerobically grown *E. coli*. However, their specificities for lipoyl domains of different origin have not been investigated.

Lipoate-protein ligase of *E. coli* can lipoylate correct lysine residue of the foreign lipoyl domain (Wallis and Perham, 1994). A sub-gene encoding the lipoyl domain of *Bacillus stearothermo-philus* pyruvate dehydrogenase complex was subjected to mutagenesis and expressed in *E. coli*. When additional lysine residue was introduced just before and after the lysine residue (Lys-42) to be lipoylated (mutation of Glu-41 to Lys or Ala-43 to Lys, cf. Fig. 3), the only lysine residue that received lipoic acid was Lys-42. No doubly lipoylated lipoyl domain was produced. In another experiment, the original target lysine residue was shifted to one position towards the N terminus (mutations of Glu-41 to Lys and Lys-42 to Ala). No lipoylated domain was found. Thus the lipoate-protein ligase of *E. coli* is unable to recognize a lysine residue one position removed from that of the original lysine, even when the original lysine residue has been replaced with an unlipoylatable side-chain.

Morris et al. (1994) isolated the *lplA* gene which encodes an *E. coli* lipoate-protein ligase. In order to identify the gene required for protein lipoylation, they employed transposon mutagenesis to isolate mutants which can grow only under conditions that bypassed any requirement for the lipoate-dependent enzymes. Cells that grew only when supplemented with acetate and succinate (to provide metabolic bypasses of the pyruvate and α-ketoglutarate dehydrogenase, respectively) were isolated. Analysis of these mutants indicated that the lipoate utilization defect was conferred by the Tn*10*dTc insertion in the *lplA* gene. The *lplA* null mutants accumulated almost no ^{35}S-labeled lipoyl-proteins when cultured with [^{35}S]lipoic acid. Activity of lipoate-protein ligase was hardly detectable in extracts of *lplA* mutants.

The position of the *lplA* gene in the *E. coli* linkage map (Kohara et al., 1987) was located at 99.6 min by detecting the insertion point of Tn*10*dTc (Morris et al., 1994). The *lplA* gene was cloned and sequenced. The gene encodes a protein of 338 amino acids, and a potential ribosome-binding site (Shine and Dalgarno, 1974) as well as a potential Rho-independent transcriptional terminator (Rosenburg and Court, 1979) were identified just upstream and downstream of the *lplA* coding sequence, respectively.

The product of *lplA* gene was overexpressed in *E. coli* cells bearing plasmid subclones of the *lplA* gene and purified to homogeneity by chromatographies on heparin-agarose and Mono Q resin (Morris et al., 1994). N-terminal sequencing demonstrated that the initiating methionine of the protein had been cleaved. The molecular mass of the protein was measured by mass spectronic technique as 37 795 Da, which agreed with the deduced molecular mass of 37 794.7 Da. The purified enzyme catalyzed the incorporation of [^{35}S]lipoic acid to apoprotein substrate (an overproduced fusion protein composed of the *E. coli* acyl carrier protein and the lipoate acetyltransferase subunit of pyruvate dehydrogenase complex). The reaction required apoprotein substrate, lipoic acid, ATP, and $MgCl_2$. The conclusion of these experiments was that the enzyme encoded by the *lplA* gene is an ATP-dependent lipoate-protein ligase which catalyzes both the ATP-dependent activation of lipoic acid to lipoyl-AMP [Eq. 1] as well as the transfer of the activated lipoyl species to apoprotein with the concomitant release of AMP [Eq. 2]. Morris et al. (1994) insisted that the two lipoylation activities described by Brookfield et al. (1991) are isoforms of the *lplA* gene product, since both of the lipoylation activities were not detectable in extracts of *lplA* null mutants.

A recent indirect experiment employing *E. coli* mutants blocked in serine biosynthesis (Morris et al., 1995) suggested that lipoylation of the glycine cleavage system with exogenous lipoic acid required the *lplA* gene product.

Morris et al. (1995) recently proposed a second protein lipoylation pathway which did not require the *lplA* gene product. Analysis of mutants of *E. coli* indicated that *lipB*, a gene previously considered to be involved in lipoic acid biosynthesis or metabolism, is responsible for the *lplA*-independent lipoylation. *LplA* null mutants displayed no growth defects unless combined with *lipA* (lipoate synthesis) or *lipB* mutations. A *lplA lipB* double mutant accumulated no detectable lipoylated protein. Expression of *lplA* gene from multicopy plasmid in a *lipB* mutant host restored the ability to grow on minimal glucose medium. However, multicopy expression of *lipB* gene in a *lplA lipA* mutant host failed to suppress the growth defect even in the presence of extremely high concentrations of lipoic acid. These results indicated that the *lipB*-dependent ligase could not utilize the exogenous lipoic acid and was dependent on lipoic acid generated *via* endogenous biosynthesis. Morris et al. (1995) proposed that *E. coli* has evolved two different pathways to utilize the two different forms of lipoic acid that are available to the cell. When presented with free lipoic acid in the medium, LplA enzyme catalyzes the formation of lipoyl-AMP. When lipoic acid is not present in the medium, a pathway generates lipoyl groups by de novo synthesis. They suggested that as yet unidentified species of lipoic acid derivatives such as those with acyl carrier protein, CoA, or some other substances interacts specifically with the *lipB*-dependent lipoylation enzyme. Further work will be needed to clarify the mechanism of protein lipoylation in *E. coli*.

Intramitochondrial lipoylation of mammalian E2 and H-protein

Mature form of acyltransferase of the bovine branched-chain α-keto acid dehydrogenase complex (E2b) was overexpressed in *E. coli* (Griffin et al., 1990). Analysis of extracts obtained from cells grown in the presence of [2-^{3}H]lipoic acid indicated that the only proteins into which radiolabeled lipoic acid was incorporated were the endogenous *E. coli* transacetylase (E2p), transsuccinylase (E2k), and H-protein of the glycine cleavage system. No [2-^{3}H]lipoic acid was incorporated into recombinant mature bovine E2b. On the contrary, lipoylated and unlipoylated lipoyl domains were obtained when a subgene encoding the fusion protein of the inner lipoyl domain of human E2p with glutathione S-transferase was overexpressed in *E. coli* (Quinn et al., 1993). Tow forms of the domain have been purified. The unlipoylated lipoyl inner domain was incubated with the partially purified lipoate-protein ligase B from *E. coli* (Brookfield et al., 1991) in the presence of lipoic acid and ATP. The majority of the unlipoylated form was converted into the lipoylated form after 6 h incubation.

Presence of lipoylation activity in mitochondrial matrix was demonstrated utilizing recombinant mature E2b overexpressed in *E. coli* as a protein substrate (Griffin et al., 1990). The E2b was purified from extracts of *E. coli* harboring the expression vector by chromatography on Sepharose 4B. The purified protein was incubated with [2-^{3}H]lipoic acid, ATP, and a soluble extract from the mitochondrial matrix of bovine liver. SDS-polyacrylamide gel electrophoresis of the radioactive protein showed that the ^{3}H-labeled lipoic acid was incorporated into recombinant mature E2b. No radiolabel was incorporated when ATP was omitted from the reaction mixture.

Intramitochondrial lipoylation of H-protein was investigated in our laboratory. H-protein, whose molecular mass is about 14 kDa, is a component of the glycine cleavage system that catalyzes the oxidative degradation of glycine yielding carbon dioxide, ammonia, methylenetetrahydrofolate, and a reduced pyridine nucleotide (Fujiwara and Motokawa, 1983; Fujiwara et al., 1984; Okamura-Ikeda et al., 1987). The glycine cleavage system consists of four protein components termed P-protein (glycine dehydrogenase, EC 1.4.4.2), H-protein, T-protein (aminomethyltransferase, EC 2.1.2.10), and L-protein (dihydrolipoamide dehydrogenase, EC 1.8.1.4). P-protein catalyzes the pyridoxal phosphate-dependent decarboxylation of glycine and transfers the remaining aminomethyl moiety to one of the sulfhydryl groups of the lipoyl prosthetic group of H-

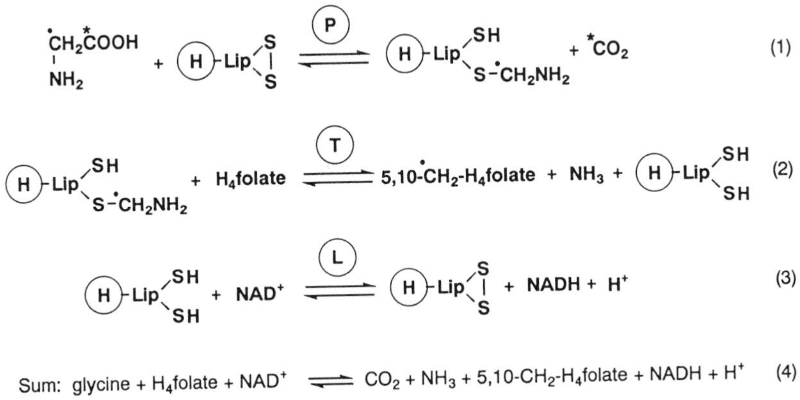

$$\overset{\bullet}{C}H_2\overset{*}{C}OOH \\ | \\ NH_2 \quad + \quad (H)-Lip\overset{S}{\underset{S}{<}} \quad \overset{P}{\rightleftharpoons} \quad (H)-Lip\overset{SH}{\underset{S-CH_2NH_2}{<}} \quad + \quad \overset{*}{C}O_2 \qquad (1)$$

$$(H)-Lip\overset{SH}{\underset{S-\overset{\bullet}{C}H_2NH_2}{<}} \quad + \quad H_4folate \quad \overset{T}{\rightleftharpoons} \quad 5,10\text{-}\overset{\bullet}{C}H_2\text{-}H_4folate \quad + \quad NH_3 \quad + \quad (H)-Lip\overset{SH}{\underset{SH}{<}} \qquad (2)$$

$$(H)-Lip\overset{SH}{\underset{SH}{<}} \quad + \quad NAD^+ \quad \overset{L}{\rightleftharpoons} \quad (H)-Lip\overset{S}{\underset{S}{<}} \quad + \quad NADH \quad + \quad H^+ \qquad (3)$$

Sum: glycine + H_4folate + NAD^+ $\rightleftharpoons$ CO_2 + NH_3 + 5,10-CH_2-H_4folate + NADH + H^+ (4)

Figure 1. Mechanism of the glycine cleavage reaction. P, H, T, and L in the circles represent respective proteins. Lip, H_4folate, 5, 10-CH_2-H_4folate represent lipoyl moiety, tetrahydrofolate, and methylenetetrahydrofolate, respectively.

protein. T-protein catalyzes the release of ammonia and transfer of the 1-carbon unit to tetrahydrofolate. L-protein catalyzes the reoxidation of the dihydrolipoyl residue of H-protein and reduction of NAD^+. The reaction is schematically presented in Figure 1. Thus the lipoic acid prosthetic group of H-protein interacts with the active site of three different enzymes in a manner similar to that found for α-keto acid dehydrogenase complexes.

cDNA encoding the precursor form of bovine H-protein was isolated and sequenced (Fujiwara et al., 1990). *In vitro* transcription and translation of the H-protein cDNA produced a 19-kDa protein recognized by antibody raised to chicken H-protein. When the product was incubated with freshly isolated bovine liver mitochondria, a protein co-migrating with purified H-protein on SDS-polyacrylamide gel electrophoresis was immunoprecipitated. The protein was resistant to trypsin digestion of the mitochondria. The results indicated that the precursor synthesized *in vitro* from cDNA for H-protein could be imported into mitochondria and processed to its mature size. Then we investigated whether the 19-kDa precursor form or the *in vitro* processed mature form of H-protein had received the lipoyl prosthetic group. To distinguish the lipoylated and unlipoylated H-protein, a method was developed to measure the covalent attachment of lipoic acid to the radiolabeled product employing lysyl endopeptidase digestion and resolution of the resultant radioactive peptides by HPLC. Lysyl endopeptidase cleaves the COOH-terminal side of lysyl residues (Masaki et al., 1981), but not if the residue is modified (Fujiwara et al., 1987). As shown in Figure 2 when authentic holoH-protein is digested with lysyl endopeptidase, a peptide containing lipoic acid (K-5 in Fig. 2A) can be separated, whereas two shorter peptides (A and B in Fig. 2B) instead of the lipoyl peptide can be separated from the apoform. Analysis of the radio-

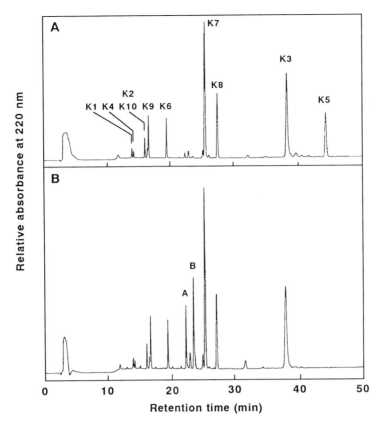

Figure 2. HPLC profile of peptides from bovine H-protein with lysyl endopeptidase digestion. Peptides were numbered according to their location in the sequence of H-protein from NH₂ terminus. (A) Holoform; (B) apoform.

active peptides obtained from the *in vitro* made H-protein revealed that the 19-kDa precursor had no lipoyl prosthetic group, whereas the processed form was modified with lipoic acid, indicating that lipoylation occurred in mitochondria.

The amino acid sequence around the lipoate residue in H-protein shows significant homology with the corresponding region of E2 components (Bradford et al., 1987a, 1987b; Fujiwara et al., 1991a). Highly conserved amino acid residues are indicated in Figure 3. The presence of conserved amino acids in the lipoic acid-attachment site of these proteins suggests that the enzyme responsible for insertion of lipoic acid recognizes some features of the primary structure. In order to determine which residues are essential for lipoylation reaction, codons for Gly-43 and Gly-70 of bovine H-protein were changed to codons for asparagine or serine and codons for Glu-56 and Glu-63 to codons for glutamine, aspartic acid, or alanine by site-directed mutagenesis (Fujiwara et

```
                                                          *
H-protein  Human        41  EVGTKLNKQDEFGALESVKAASPLYSPLSGEV  72
           Bovine       41  EVGTKLNKQEEFGALESVKAASPLYSPLSGEV  72
           Chicken      41  EIGTKLNKDDEFGALESVKAASPLYSPLTGEV  72
           Pea          45  EPGVSVTKGKGFGAVESVKATSDVNSPISGEV  76
           E. coli      46  EVGATVSAGDDCAVAESVKAASDIYAPVSGEI  77

E2p  Human liver        28  KEGDKINEGDLIAEVETDKATVGFESLEECYM  59
     Rat heart          28  KEGEKISEGDLIAEVETDKATVGFESLEECYM  59
     S. cerevisiae      29  KEGDQLSPGEVIAEIETDKAQMDFEFQEDGYL  60
     E. coli            22  KVGDKVEAEQSLITVEGDKASMEVPSPQAGIV  53
     A. vinelandii      21  KTGDLIEVEQGLVVLESAKASMEVPSPKAGVV  52
     B. stearothermophilus 24 KPGDEVNEDDVLCEVQNDKAVVEIPSPVKGKV 55
E2k  Rat heart          25  AVGDAVAEDEVVCEIETDKTSVQVPSPANGII  56
     E. coli            25  KPGDAVVRDEVLVEIETDKVVLEVPASADGIL  56
     A. vinelandii      24  KPGEPVKRDELIVDIETDKVVMEVLAEADGVI  55
E2b  Human and
       bovine liver     26  KEGDTVSQFDSICEVQSDKASVTITSRYDGVI  57
     P. putida          25  KVGDIIAEDQVVADVMTDKATVEIPSPVSGKV  56
                                                          *
```

Figure 3. Comparison of amino acid sequences surrounding the attachment site of lipoic acid. The sequences of human (Fujiwara et al., 1991b; Koyata and Hiraga, 1991), bovine (Fujiwara et al., 1990), chicken (Yamamoto et al., 1991), pea (Kim and Oliver, 1990; Macherel et al., 1990), and *E. coli* (Okamura-Ikeda et al., 1993; Stauffer et al., 1991) H-proteins are compared to the sequences of E2p of human liver (Thekkumkara et al., 1988), rat heart (Matuda et al., 1992), *Saccharomyces cerevisiae* (Niu et al., 1988), *E. coli* (Stephens et al., 1983), *Azotobacter vinelandii* (Hanemaaijer et al., 1988), and *B. stearothermophilus* (Borges et al., 1990), to the sequences of E2k of rat heart (Nakano et al., 1991), *E. coli* (Spencer et al., 1984), and *A. vinelandii* (Westphal and de Kok, 1990), and to the sequences of E2b of human and bovine liver (Lau et al., 1988) and *Pseudomonas putida* (Burns et al., 1988). The lipoyl lysine residues are indicated (*). The conserved amino acid residues are indicated by black boxes. The numbers on the right and left refer to the positions of the amino acids of these proteins.

al., 1991a). In addition, Glu-50, which is less conserved than the above residues, was changed to glutamine for comparison. All the protein products transcribed and translated *in vitro* from the mutated cDNAs co-migrated with the precursor form of H-protein on SDS-polyacrylamide gel electrophoresis. The mutation did not affect the transport into mitochondria and processing of the precursors. The *in vitro* translated products were incubated with mitochondria, and the proteins incorporated and processed were imunoprecipitated and analyzed for their lipoylation. Of three glutamic acid residues examined, only Glu-56 is responsible for the lipoylation. Replacement of Glu-56 by glutamine and alanine, which lack a negative charge, resulted in a great decrease (12% and 17% of that of the wild-type, respectively) in lipoylation. Substitution by the negatively charged aspartic acid residue inhibited the lipoylation by 60%. Gly-43 seems to be not essential for lipoylation, since the substitution of this residue resulted in slight inhibition of lipoylation. On the other hand the lipoylation rate was reduced to 14% of that of the wild-type when Gly-70 was replaced by the asparagine residue. Conversion of the glycine residue to serine, a smaller residue than the asparagine residue, restored the lipoylation partially (54% of the wild-type). Gly-70 is predicted to be situated in a β-turn structure (Fujiwara et al., 1990). The presence of glycine at position 70 may be required to facilitate the lipoylation of the specific lysyl residue by providing a flexible loop.

Lipoyltransferase of bovine liver

As described above, *in vitro* lipoylation studies of H-protein (Fujiwara et al., 1990) and E2b (Griffin et al., 1990) have shown that the mammalian lipoylation enzyme is located within mitochondria. To study the properties of the enzyme, H-protein was chosen for the protein substrate because it is easy to handle, and the assay for the lipoylated H-protein is simple and sensitive. Lipoylated H-protein can be assayed by the P-protein-catalyzed glycine-$^{14}CO_2$ exchange reaction that is a partial reaction of the glycine cleavage reaction (Fujiwara et al., 1992). When glycine and $^{14}CO_2$ are incubated with P-protein and H-protein, the exchange of the carboxyl carbon of glycine with $^{14}CO_2$ is catalyzed. The reaction is proportional to the amount of the lipoylated H-protein if appropriate reaction conditions are provided. The sensitivity of the method is such that we can detect lipoylation of apoH-protein in amounts down to 2 pmol.

A procaryotic expression vector for mature bovine H-protein was constructed (Fujiwara et al., 1992) employing the bacterial T7 system of Studier et al. (1990). *E. coli* BL21(DE3)pLysS cells harboring the expression vector expressed mature bovine H-protein as a soluble form at a level of about 10% of the total cellular protein. About 80% of the recombinant H-protein overexpressed was the unlipoylated apoform, 10% was the lipoylated holoform, and 10% was an inactive octanoylated form even cultured in a medium supplemented with sufficient amounts of lipoic acid. The apoH-protein was separated from the other forms of bovine H-protein and *E. coli* holoH-protein by DEAE-Sepharose column chromatography and purified to homogeneity.

An attempt was made to purify lipoyltransferase from bovine liver mitochondria with apoH-protein as a lipoyl acceptor and chemically synthesized lipoyl-AMP as a donor of lipoyl group (Fujiwara et al., 1994). Lipoyltransferase is localized in mitochondria, and the postmitochondrial supernatant has no lipoyltransferase activity. Extracts of bovine liver mitochondria were applied directly to hydroxylapatite columns. Two peaks of activity could be resolved. Lipoyltransferase in the peak eluting higher phosphate concentration was purified to homogeneity and termed lipoyltransferase II. Lipoyltransferase I eluting at lower phosphate concentration was purified, but the final product contained a minor contaminant. The molecular masses of lipoyltransferase I and II are both 40 kDa either estimated by SDS-polyacrylamide gel electrophoresis or by gel permeation chromatography on Superdex 200-HR.

Lipoyltransferases I and II both catalyzed the lipoylation of apoH-protein. The reaction was absolutely dependent on lipoyl-AMP, apoH-protein, and lipoyltransferase. In contrast with *E. coli* lipoate-protein ligase, the enzymes were unable to use lipoic acid plus MgATP as a lipoyl donor. Lipoyl-AMP inhibited the reaction at concentrations over 50 μM. The inhibition was overcome by the addition of bovine serum albumin. The activity of both enzymes was maximum at pH 7.9 in potassium phosphate buffer. K_m values obtained with lipoyltransferase I and II for lipoyl-AMP were 13 and 16 µM and for apoH-protein were 0.29 and 0.17 µM, respectively. V_{max} values of lipoyltransferase I and II were 135 and 144 nmol/min/mg of protein, respectively.

The lipoylation of H-protein was inhibited by lipoyl-AMP analogues including hexanoyl-, octanoyl-, and decanoyl-AMP to the similar extent. When 40 µM lipoyl-AMP was used as substrate, lipoylation was inhibited to about 30–40% by the addition of 40 µM of acyl-AMPs. Acyl-AMPs described above were utilized as substrates in place of lipoyl-AMP. The lipoyl group and acyl groups were transferred to apoH-protein with similar extent. The acylation was determined by nondenaturing polyacrylamide gel electrophoresis. When specific lysyl residue of H-protein (Lys-59 of bovine H-protein) was modified with an acyl group, H-protein moves faster than apoH-protein because the apoform has an additional positive charge on the unmodified lysyl residue.

Lipoyltransferases I and II were active when the nascent apoH-protein translated *in vitro* in a rabbit reticulocyte lysate was employed as an acceptor protein. The precursor form was modified with lipoyl group similarly as observed with the mature H-protein, indicating that the loosely folded apoH-protein is able to accept the lipoyl moiety.

Lipoyltransferases I and II recognized lipoyl domains of rat E2p and E2k and bovine E2b (K. Fujiwara, K. Okamura-Ikeda and Y. Motokawa, unpublished observation). *In vitro* transcribed and translated apolipoyl domains were incubated with lipoyl-AMP and the enzyme, and the protein products were analyzed by nondenaturing polyacrylamide gel electrophoresis. E2p and E2k were lipoylated with similar extent, but the amount of E2b lipoylated was far less than that of E2p and E2k in the same reaction condition.

It seems likely at present that there are at least two different systems for protein lipoylation utilizing lipoyl-AMP as an intermediate. A system found in *E. coli* requires only one enzyme, lipoate-protein ligase, that catalyzes both activation and ligation of lipoyl groups. The mechanism of lipoylation may be analogous to that of biotin-protein ligase (Eisenberg et al., 1982). The system of bovine liver, on the contrary, consists of two enzymes. Lipoyltransferase purified from mitochondria catalyzes the transfer of lipoyl groups from lipoyl-AMP to apoenzymes. Activation of lipoic acid is catalyzed by another enzyme, lipoic acid activating enzyme. However, properties of the enzyme has not been fully elucidated. Thorough characterization of these two systems requires further study.

References

Ali, S.T. and Guest, J.R. (1990) Isolation and characterization of lipoylated and unlipoylated domains of the E2p subunit of the pyruvate dehydrogenase complex of *Escherichia coli*. *Biochem. J.* 271: 139–145.

Borges, A., Hawkins, C.F., Packman, L.C. and Perham, R.N. (1990) Cloning and sequence analysis of the genes encoding the dihydrolipoamide acetyltransferase and dihydrolipoamide dehydrogenase components of the pyruvate dehydrogenase multienzyme complex of *Bacillus stearothermophilus*. *Eur. J. Biochem.* 194: 95–102.

Bradford, A.P., Howell, S., Aitken, A., James, L.A. and Yeaman, S.J. (1987a) Primary structure around the lipoate-attachment site on the E2 component of bovine heart pyruvate dehydrogenase complex. *Biochem. J.* 245: 919–922.

Bradford, A.P., Aitken, A., Beg, F., Cook, K.G. and Yeaman, S.J. (1987b) Amino acid sequence surrounding the lipoic acid cofactor of bovine kidney 2-oxoglutarate dehydrogenase complex. *FEBS Lett.* 222: 211–214.

Brookfield, D.E., Green, J., Ali, S.T., Machado, R.S. and Guest, J.R. (1991) Evidence for two protein-lipoylation activities in *Escherichia coli*. *FEBS Lett.* 295: 13–16.

Burns, G., Brown, T., Hatter, K. and Sokatch, J.R. (1988) Comparison of the amino acid sequences of the transacylase components of branched chain oxoacid dehydrogenase of *Pseudomonas putida*, and the pyruvate and 2-oxoglutarate dehydrogenases of *Escherichia coli*. *Eur. J. Biochem.* 176: 165–169.

Eisenberg, M.A., Prakash, O. and S.-C. Hsiung (1982) Purification and properties of the biotin repressor. A bifunctional protein. *J. Biol. Chem.* 257: 15167–15173.

Fujiwara, K. and Motokawa, Y. (1983) Mechanism of the glycine cleavage reaction: Steady state kinetic studies of the P-protein-catalyzed reaction. *J. Biol. Chem.* 258: 8156–8162.

Fujiwara, K., Okamura-Ikeda, K. and Motokawa, Y. (1984) Mechanism of the glycine cleavage reaction: Further characterization of the intermediate attached to H-protein and of the reaction catalyzed by T-protein. *J. Biol. Chem.* 259: 10664–10668.

Fujiwara, K., Okamura-Ikeda, K. and Motokawa, Y. (1987) Amino acid sequence of the phosphopyridoxyl peptide from P-protein of the chicken liver glycine cleavage system. *Biochem. Biophys. Res. Comm.* 149: 621–627.

Fujiwara, K., Okamura-Ikeda, K. and Motokawa, Y. (1990) cDNA sequence, *in vitro* synthesis, and intramitochondrial lipoylation of H-protein of the glycine cleavage system. *J. Biol. Chem.* 265: 17463–17467.

Fujiwara, K., Okamura-Ikeda, K. and Motokawa, Y. (1991a) Lipoylation of H-protein of the glycine cleavage system: The effect of site-directed mutagenesis of amino acid residues around the lipoyllysine residue on the lipoate attachment. *FEBS Lett.* 293: 115–118.

Fujiwara, K., Okamura-Ikeda, K., Hayasaka, K. and Motokawa, Y. (1991b) The primary structure of human H-protein of the glycine cleavage system deduced by cDNA cloning. *Biochem. Biophys. Res. Comm.* 176: 711–716.

Fujiwara, K., Okamura-Ikeda, K. and Motokawa, Y. (1992) Expression of mature bovine H-protein of the glycine cleavage system in *Escherichia coli* and *in vitro* lipoylation of the apoform. *J. Biol. Chem.* 267: 20011–20016.

Fujiwara, K., Okamura-Ikeda, K. and Motokawa, Y. (1994) Purification and characterization of lipoyl-AMP:N^ε-lysine lipoyltransferase from bovine liver mitochondria. *J. Biol. Chem.* 269: 16605–16609.

Griffin, T.A., Wynn, R.M. and Chuang, D.T. (1990) Expression and assembly of mature apotransacylase (E_{2b}) of bovine branched-chain α-keto acid dehydrogenase complex in *Escherichia coli*: Demonstration of transacylase activity and modification by lipoylation. *J. Biol. Chem.* 265: 12104–12110.

Hanemaaijer, R., Janssen, A., de Kok, A. and Veeger, C. (1988) The dihydrolipoyltransacetylase component of the pyruvate dehydrogenase complex from *Azotobacter vinelandii*: Molecular cloning and sequence analysis. *Eur. J. Biochem.* 174: 593–599.

Kim, Y. and Oliver, D.J. (1990) Molecular cloning, transcriptional characterization, and sequencing of cDNA encoding the H-protein of the mitochondrial glycine decarboxylase complex in peas. *J. Biol. Chem.* 265: 848–853.

Kohara, Y., Akiyama, K. and Isono, K. (1987) The physical map of the whole *E. coli* chromosome: Application of a new strategy for rapid analysis and sorting of a large genomic library. *Cell* 50: 495–508.

Koyata, H. and Hiraga, K. (1991) The glycine cleavage system: Structure of a cDNA encoding human H-protein, and partial characterization of its gene in patients with hyperglycinemias. *Am. J. Hum. Genet.* 48: 351–361.

Lau, K.S., Griffin, T.A., Hu, C.-W.C. and Chuang, D.T. (1988) Conservation of primary structure in the lipoyl-bearing and dihydrolipoyl dehydrogenase binding domains of mammalian branched-chain α-keto acid dehydrogenase complex: Molecular cloning of human and bovine transacylase (E2) cDNAs. *Biochemistry* 27: 1972–1981.

Macherel, D., Lebrun, M., Gagnon, J., Neuburger, M. and Douce, R. (1990) cDNA cloning, primary structure and gene expression for H-protein, a component of the glycine-cleavage system (glycine decarboxylase) of pea (*Pisum sativum*) leaf mitochondria. *Biochem. J.* 268: 783–789.

Masaki, T., Fujihashi, T., Nakamura, K. and Soejima, M. (1981) Studies on a new proteolytic enzyme from *Achromobacter lyticus* M497–1: II. Specificity and inhibition studies of *Achromobacter* protease I. *Biochim. Biophys. Acta* 660: 51–55.

Matuda, S., Nakano, K., Ohta, S., Shimura, M., Yamanaka, T., Nakagawa, S., Titani, K. and Miyata, T. (1992) Molecular cloning of dihydrolipoamide acetyltransferase of the rat pyruvate dehydrogenase complex: Sequence comparison and evolutionary relationship to other dihydrolipoamide acyltransferases. *Biochim. Biophys. Acta* 1131: 114–118.

Mitra, S.K. and Burma, D.P. (1965) Activation of lipoic acid and its transfer from the free pool to the protein-bound state. *J. Biol. Chem.* 240: 4072–4080.

Morris, T.W., Reed, K.E. and Cronan, J.E. (1994) Identification of the gene encoding lipoate-protein ligase A of *Escherichia coli*: Molecular clonig and characterization of the *lplA* gene and gene product. *J. Biol. Chem.* 269: 16091–16100.

Morris, T.W., Reed, K.E. and Cronan, J.E. (1995) Lipoic acid metabolism in *Escherichia coli*: The *lplA* and *lipB* genes define redundant pathways for ligation of lipoyl groups to apoprotein. *J. Bacteriology* 177: 1–10.

Nakano, K., Matuda, S., Yamanaka, T., Tsubouchi, H., Nakagawa, S., Titani, K., Ohta, S. and Miyata, T. (1991) Purification and molecular cloning of succinyltransferase of the rat α-ketoglutarate dehydrogenase complex: Absence of a sequence motif of the putative E3 and/or E1 binding site. *J. Biol. Chem.* 266: 19013–19017.

Niu, X.-D., Browning, K.S., Behal, R.H. and Reed, L.J. (1988) Cloning and nucleotide sequence of the gene for dihydrolipoamide acetyltransferase from *Saccharomyces cerevisiae*. *Proc. Natl. Acad. Sci. USA* 85: 7546–7550.

Okamura-Ikeda, K., Fujiwara, K. and Motokawa, Y. (1987) Mechanism of the glycine cleavage reaction: Properties of the reverse reaction catalyzed by T-protein. *J. Biol. Chem.* 262: 6746–6749.

Okamura-Ikeda, K., Ohmura, Y., Fujiwara, K. and Motokawa, Y. (1993) Cloning and nucleotide sequence of the *gcv* operon encoding the *Escherichia coli* glycine-cleavage system. *Eur. J. Biochem.* 216: 539–548.

Quinn, J., Diamond, A.G., Masters, A.K., Brookfield, D.E., Wallis, N.G. and Yeaman, S.J. (1993) Expression and lipoylation in *Escherichia coli* of the inner lipoyl domain of the E2 component of the human pyruvate dehydrogenase complex. *Biochem. J.* 289: 81–85.

Reed, L.J., Leach, F.R. and Koike, M. (1958) Studies on a lipoic acid-activating system. *J. Biol. Chem.* 232: 123–142.

Rosenberg, M. and Court, D. (1979) Regulatory sequences involved in the promotion and termination of RNA transcription. *Ann. Rev. Genet.* 13: 319–353.

Shine, J. and Dalgarno, L. (1974) The 3'-terminal sequence of *Escherichia coli* 16S ribosomal RNA: Complementarity to nonsense triplets and ribosome binding sites. *Proc. Natl. Acad. Sci. USA* 71: 1342–1346.

Spencer, M.E., Darlison, M.G., Stephens, P.E., Duckenfield, I.K. and Guest, J.R. (1984) Nucleotide sequence of the *sucB* gene encoding the dihydrolipoamide succinyltransferase of *Escherichia coli* K12 and homology with the corresponding acetyltransferase. *Eur. J. Biochem.* 141: 361–374.

Stauffer, L.T., Steiert, P.S., Steiert, J.G. and Stauffer, G.V. (1991) An *Escherichia coli* protein with homology to the H-protein of the glycine cleavage enzyme complex from pea and chicken liver. *DNA Sequence* 2: 13–17.

Stephens, P.E., Darlison, M.G., Lewis, H.M. and Guest, J.R. (1983) The pyruvate dehydrogenase complex of *Escherichia coli* K12: Nucleotide sequence encoding the dihydrolipoamide acetyltransferase component. *Eur. J. Biochem.* 133: 481–489.

Studier, F.W., Rosenberg, A.H., Dunn, J.J. and Dubendorff, J.W. (1990) Use of T7 RNA polymerase to direct expression of cloned genes. *Meth. Enzymol.* 185: 60–89.

Thekkumkara, T.J., Ho, L., Wexler, I.D., Pons, G., Liu, T.-C. and Patel, M.S. (1988) Nucleotide sequence of a cDNA for the dihydrolipoamide acetyltransferase component of human pyruvate dehydrogenase complex. *FEBS Lett.* 240: 45–48.

Tsunoda, J.N. and Yasunobu, K.T. (1967) Mammalian lipoic acid activating enzyme. *Arch. Biochem. Biophys.* 118: 395–401.

Wallis, N.G. and Perham, R.N. (1994) Structural dependence of post-translational modification and reductive acetylation of the lipoyl domain of the pyruvate dehydrogenase multienzyme complex. *J. Mol. Biol.* 236: 209–216.

Westphal, A.H. and de Kok, A. (1990) The 2-oxoglutarate dehydrogenase complex from *Azotobacter vinelandii*: 2. Molecular cloning and sequence analysis of the gene encoding the succinyltransferase component. *Eur. J. Biochem.* 187: 235–239.

Yamamoto, M., Koyata, H., Matsui, C. and Hiraga, K. (1991) The glycine cleavage system: Occurrence of two types of chicken H-protein mRNAs presumably formed by the alternative use of the polyadenylation consensus sequences in a single exon. *J. Biol. Chem.* 266: 3317–3322.

Alpha-Keto Acid Dehydrogenase Complexes
M.S. Patel, T.E. Roche and R.A. Harris (eds)
© 1996 Birkhäuser Verlag Basel/Switzerland

Pyruvate dehydrogenase phosphatase

L.J. Reed, J.E. Lawson, X.-D. Niu and J. Yan

Biochemical Institute and Department of Chemistry and Biochemistry, The University of Texas at Austin, Austin, TX 78712, USA

Introduction

Four classes of protein serine/threonine phosphatases have been identified in eukaryotic cells on the basis of substrate specificities and sensitivity to activators and inhibitors (Cohen, 1989; Shenolikar and Nairn, 1991). Protein phosphatase 1 is sensitive to the thermostable proteins inhibitor 1 and inhibitor 2, and protein phosphatases 1 and 2A are sensitive to okadaic acid. Protein phosphatase 2B is Ca^{2+}/calmodulin-regulated and protein phosphatase 2C is Mg^{2+}-dependent. The protein serine/threonine phosphatases of known sequence comprise two distinct gene families, a major family that includes protein phosphatases 1, 2A, and 2B and isoforms thereof, and a smaller family that includes protein phosphatase 2C and pyruvate dehydrogenase (PDH) phosphatase.

The physiological substrate for the mammalian PDH phosphatase is the PDH complex, which is localized to mitochondria, within the inner-membrane matrix compartment. The mammalian PDH complex is organized about a 60-mer dihydrolipoamide acetyltransferase (E_2) to which about 30 pyruvate dehydrogenase (E_1) tetramers ($\alpha_2\beta_2$), 12 E_3-binding protein (E_3BP) monomers, and 12 dihydrolipoamide dehydrogenase (E_3) dimers are bound by noncovalent bonds (Reed and Hackert, 1990; Patel and Roche, 1990; Perham, 1991). E_3BP binds inside the central cavity of the pentagonal dodecahedron-like E_2 and anchors the E_3 dimer in a specific manner that is essential for a functional PDH complex (Maeng et al., 1994; J. K. Stoops, T. S. Baker, and L. J. Reed, unpublished data). Activity of the mammalian PDH complex is regulated mainly by a phosphorylation-dephosphorylation cycle (Reed and Yeaman, 1987). Phosphorylation and concomitant inactivation of E_1 (α subunit) is catalyzed by PDH kinase, which is tightly bound to E_2, and dephosphorylation and concomitant reactivation is catalyzed by PDH phosphatase, which associates with E_2 in a Ca^{2+}-dependent manner.

Properties of PDH phosphatase

PDH phosphatase has been purified to apparent homogeneity from bovine heart and kidney mitochondria (Teague et al., 1982; Pratt et al., 1982). The phosphatase has a $s_{20,\ w}$ of about 7.4 S and an M_r of about 150 000 as determined by sedimentation equilibrium. The phosphatase consists of a catalytic subunit (PDPc) with an apparent M_r of about 50 000 and a subunit with an apparent M_r of about 97 000. The latter subunit (PDPr) is a flavoprotein (FAD) of as yet uncertain function. PDH phosphatase is a Mg^{2+} (or Mn^{2+})-dependent and Ca^{2+}-stimulated protein serine/threonine phosphatase. The apparent K_m for Mg^{2+} is ~2 mM, and for Mn^{2+} it is 0.2–0.5 mM. At physiological concentrations of Mg^{2+} (~0.5 mM), PDH phosphatase activity is stimulated markedly by 0.5 mM spermine, apparently by decreasing the apparent K_m for Mg^{2+} (Damuni et al., 1984; Thomas et al., 1986). Spermine can spare but not completely replace Mg^{2+}. Spermine apparently acts directly on the phosphatase (Rahmatullah and Roche, 1988). Protamine (3.6 µg/ml) stimulated the phosphatase activity several-fold at suboptimal levels of ^{32}P-labeled PDH complex, apparently by lowering the apparent K_m of the substrate (Z. Damuni and L. J. Reed, unpublished data). At saturating Mg^{2+} concentration (~10 mM), PDH phosphatase activity toward its physiological substrate, phosphorylated E_1 bound to E_2, is stimulated about 10-fold by micromolar concentrations of Ca^{2+} (Pettit et al., 1972; Denton et al., 1972). However, PDH phosphatase activity toward uncomplexed phosphorylated E_1 and other protein and phosphopeptide substrates is not affected by Ca^{2+} (Davis et al., 1977; C. H. MacGowan, P. Cohen, Z. Damuni, and L. J. Reed, unpublished data). These observations indicate that Ca^{2+} is not directly involved in PDH phosphatase catalysis.

NADH inhibits PDH phosphatase activity toward phosphorylated E_1 bound to E_2, but not toward uncomplexed phosphorylated E_1 (Pettit et al., 1975; Rahmatullah and Roche, 1988). Furthermore, NADH did not reduce the protein-bound FAD in the PDH phosphatase heterodimer (Teague et al., 1982). It has been suggested that NADH does not act directly on PDH phosphatase, but rather reduces E_3 which, in turn reduces the lipoyl moieties on E_2, and that the dihydrolipoyl moieties inhibit the phosphatase (Rahmatullah and Roche, 1988). PDH phosphatase activity is not inhibited by protein phosphatase inhibitor 1 or inhibitor 2, or by BCKDH phosphatase inhibitor protein.

A novel PDH phosphatase has been purified to apparent homogeneity and partially characterized from mitochondria of the adult parasitic nematode, *Ascaris suum* (Song and Komuniecki, 1994). The phosphatase is a heterodimer consisting of subunits with apparent M_r of 50 000 and 89 000. The *A. suum* PDH phosphatase is Mg^{2+}-dependent but, in contrast to the mammalian PDH phosphatase, its activity is not stimulated by spermine or Ca^{2+}. The activity of the *A. suum* PDH phosphatase is stimulated by L-malate, which is the major substrate in the anaerobic muscle mitochondria of this organism. Malate decreased the apparent K_m of the phosphatase for phosphorylated E_1 bound to the E_2 core 4 to 6-fold, but had no effect on uncomplexed phosphorylated E_1.

Molecular basis of Ca^{2+}-stimulation of PDH phosphatase activity

Ca^{2+} stimulates the activity of PDH phosphatase toward complexed phosphorylated E_1 (bound to E_2) but not toward free phosphorylated E_1. In the presence of Ca^{2+}, PDH phosphatase binds to E_2, and its apparent K_m for phosphorylated E_1 is decreased about 20-fold, to 2.9 µM (Pettit et al., 1972). Binding studies with $^{45}Ca^{2+}$ showed that PDH phosphatase possesses an intrinsic Ca^{2+}-binding site with a K_d value of about 8 µM (Teague et al., 1982). A second Ca^{2+}-binding site, with a K_d of about 5 µM, is apparently produced by association of the phosphatase with E_2. The apparent role of Ca^{2+} is to position PDPc on E_2 near its substrate, phosphorylated E_1, and this positioning is apparently an important part of the regulatory mechanism.

Examination of the deduced amino acid sequence of PDPc (Lawson et al., 1993) revealed the presence of a putative helix-loop-helix Ca^{2+}-binding motif (residues 165–192). The sequence in the putative loop is DNDISLEAQVGD (residues 173–184). In a preliminary test of this hypothesis, Asp-173 was replaced by Ala, Asn, or Glu and Asp-175 by Ala by site-directed mutagenesis (J. E. Lawson, J. Yan, and L. J. Reed, unpublished data). The Asp-173→Ala or Asn substitutions eliminated essentially all PDH phosphatase activity, whereas the Asp-173→Glu and Asp-175→Ala substitutions had little effect, if any, on the activity. Ultracentrifugation experiments showed that in the presence, but not in the absence of Ca^{2+}, PDPc and the D173E mutant protein bound to the PDH complex (i.e., the E_2 component), whereas the D173A, N mutant proteins did not bind to the PDH complex, whether or not Ca^{2+} was present. These observations indicate that Asp-173 plays an important role. Whether this role involves the putative Ca^{2+}-binding site, or is catalytic or structural, remains to be determined.

Evolutionary relationship of PDH phosphatase catalytic subunit and protein phosphatase 2C

PDH phosphatase and protein phosphatase 2C exhibit similar substrate specificity (C.H. MacGowan, P. Cohen, Z. Damuni, and L.J. Reed, unpublished data). Protein phosphatase 2C dephosphorylates the PDH complex at about one-third the rate of PDH phosphatase, but dephosphorylation by protein phosphatase 2C is not affected by Ca^{2+} or spermine. Both phosphatases show similar rates of dephosphorylation of phosphorylase kinase and glycogen synthase (sites 2 and 3). Both phosphatases preferentially dephosphorylate the α-subunit of phosphorylase kinase and dephosphorylate phosphorylase a at a very low rate. As observed previously with rabbit protein phosphatase $2C_2$ (Donella-Deana et al., 1990), PDPc shows a striking preference for the phosphopeptide RRATpVA over RRASpVA (J. Yan and L.J. Reed, unpublished data). Other similarities between the PDH phosphatase and protein phosphatase 2C are an absolute requirement for Mg^{2+} and insensitivity toward protein phosphatase inhibitor 1, inhibitor 2 and okadaic acid (Damuni et al., 1985; Haystead et al., 1989).

To gain further insight into the structure, function and regulation of PDPc and its relationship to protein phosphatase 2C, the cloning of cDNA for PDPc was undertaken. All attempts to isolate clones from bovine cDNA libraries were unsuccessful. A PCR-based approach using bovine genomic DNA and cDNA synthesized from bovine total RNA as templates eventually proved to be successful (Lawson et al., 1993). The complete cDNA contains an open reading frame of 1614 bp encoding a putative presequence of 71 amino acid residues and a mature protein of 467 residues with a calculated M_r of 52, 625. Hybridization analysis showed a single mRNA transcript of about 2.0 kilobases in bovine heart, brain and lung total RNA. Alignment of the deduced amino acid sequences of the mitochondrial PDPc and the rat cytosolic protein phosphatase 2C α and β isoforms (Tamura et al., 1989; Wenk et al., 1992) showed 20% and 22% identity between PDPc and protein phosphatase 2C α and β, respectively. The overall pattern indicated that PDPc and protein phosphatase 2C evolved from a common ancestor.

The mature form of PDPc was coexpressed in $E.$ $coli$ with the chaperonin proteins groEL and groES. The recombinant protein was purified to near homogeneity. Its activity toward the bovine ^{32}P-labeled PDH complex was Mg^{2+}-dependent and Ca^{2+}-stimulated and comparable to that of native bovine PDH phosphatase. An active, truncated form of the recombinant PDPc, with M_r ~45 000, was produced in variable amounts during growth of cells and/or during the purification procedure. The truncated PDPc has the same amino-terminal sequence as the unmodified PDPc, indicating that proteolysis occurred in the carboxyl-terminal part of PDPc. A hexahistidine tag has been engineered on the C-terminus of PDPc to facilitate purification of the recombinant protein by Ni^{2+}-nitriloacetic-agarose chromatography (K.M. Popov and R.A. Harris, personal communication).

```
PDPc     astpqkfyltppqvnsilkaneysfkvpefdgknvssvlgfdsnqlpanapiedrrsaatclqtrgm   67
ratPP2c  mgafldkpkmekhnaqgqgnglryglssmqgwrvemedahtaviglpsgletws-------------   54
rabPP2c  mgafldkpkmekhnaqgqgnglryglssmqgwrvemedahtaviglpsgletws-------------   54
humPP2c  mgafldkpkmekhnaqgqgnglryglssmqgwrvemedahtaviglpsglesws-------------   54
yPTC1    msnhseilerpetpydityrvgvaenknskfrrtmedvhtyvknfasrldwg--------------   52
LCPP2c   mgiplpkpvmtqlqerygnaifrcgsncvngyretmedahltyltdswg----------------   49

                  I
PDPc     LLGVFDGHAGCACSQAvserlfyyiavsllphetlleienavesgrallpilqwhkhpndyfskeas   134
ratPP2c  FFAVYDGHAGSQVAKYccehlldhitnnqdfkgsagapsvenvkngirtgf---------------   105
rabPP2c  FFAVYDGHAGSQVAKYccehlldhitnnqdfkgsagapsvenvkngirtgf---------------   105
humPP2c  FFAVYDGHAGSQVAKYccehlldhitnnqdfkgsagapsvenvkngirtgf---------------   105
yPTC1    YFAVFDGHAGIQASKWcgkhlhtiieqniladetrdvrdvlndsf--------------------   97
LCPP2c   FFGVFDGHVNDQCSQYlerawrsaiekesipmtdermkela-----------------------   90

PDPc     klyfnslrtywqelidlntgestdidvkealinafKRLDNDisleaqvgdpnsflnylvlrvafsga   201
ratPP2c  -------------------------------LEIDEHmrvmsekkhgadrsgstavgvlispq   137
rabPP2c  -------------------------------LEIDEHmrvmsekkhgadrsgstavgvlispq   137
humPP2c  -------------------------------LEIDEHmrvmsekkhgadrsgstavgvlispq   137
yPTC1    -------------------------------LAIDEEintklvgnsgctaavcvlrwelpdsv   129
LCPP2c   -------------------------------LRIDQEwmdsgreggstgtffvalkegnkvhl   122

                    II                          III
PDPc     tacvahvdgvdl--HVANTGDSRAMLgvqeedgswSAVTLSNDHNAQNEREVERLklehpkNEAKSV   266
ratPP2c  ht-----------YFINCGDSRGLLcrnr-----KVHFFTQDHKPSNPLEKERI------QNAGGS   181
rabPP2c  ht-----------YFINCGDSRGLLcrnr-----KVHFFTQDHKPSNPLEKERI------QNAGGS   181
humPP2c  ht-----------YFINCGDSRGLLcrnr-----KVHFFTQDHKPSNPLEKERI------QNAGGS   181
yPTC1    sddsmdlaqhqrklYTANVGDSRIVLfrng------NSIRLTYDHKASDTLEMQRV------EQAGGL   185
LCPP2c   -------------QVGNVGDSRVVAcidg-----VCVPLTEDHKPNNEGERQRI------ENCAGR   164

                 IV
PDPc     VKQDRLLGLLMPFRAFGDVkfkwsidlqkrviesgpdqlndneytkfippnyytppyltaepevtyh   333
ratPP2c  VMIQRVNGSLAVSRALGDFdykcvhgkgpteqlvspepevhdiersee------------------   229
rabPP2c  VMIQRVNGSLAVSRALGDFdykcvhgkgpteqlvspepevhdiersee------------------   229
humPP2c  VMIQRVNGSLAVSRALGDFdykcvhgkgpteqlvspepevhdiersee------------------   229
yPTC1    IMKSRVNGMLAVTRSLGDKffdslvvgspfttsveits--------------------------   223
LCPP2c   VENNRVDGSLAVSRAFGDReyklgsgsqleqkvialadvqhkdftfd-------------------   211

                 V
PDPc     rlrpQDKFLVLATDGLWEtmhrqdvvrivgeyltgmhhqqpiavggykvtlgqmhgllterrakmss   400
ratPP2c  ----DDQFIILACDGIWDvmgneelcdfvrsrlevtddlekvcnevvdtclykgsrdnmsvilicfp   292
rabPP2c  ----DDQFIILACDGIWDvmgneelcdfvrsrlevtddlekvcnevvdtclykgsrdnmsvilicfp   292
humPP2c  ----DDQFIILACDGIWDvmgneelcdfvrsrlevtddlekvcnevvdtclykgsrdnmsvilicfp   292
yPTC1    ----EDKFLILACDGLWDviddqdacelikditepneaakvlvryalengttdnvtvmvvfl-----   281
LCPP2c   ----SNDFVLLCCDGVFEgnfpneevvayvkqqletcndlaevagrvceeaiergsrdniscmivqf   274
PDPc     vfedqnaathlirhavgnnefgavdherlskmlslpeelarmyrdditiivvqfnshvvgayqnqeq   467
ratPP2c  napkvsaeavkkeaeldkylenrveeiikkqgegvpdlvhvmrtlasenipslppggelaskrnvie   359
rabPP2c  napkvspeavkkeaeldkylecrveeiikkqgegvpdlvhvmrtlasenipslppggelaskrnvie   359
humPP2c  napkvspeavkkeaeldkylecrveeiikkqgegvpdlvhvmrtlasenipslppggelaskrnvie   359
yPTC1    -------------------------------------------------------------   281
LCPP2c   kdgsdyaaephttvvpgpfsaprnsgfrkayesmadkgnttvgallerrydtlkaaealtpeeteel   341

PDPc     -------------------------------------------------------------   467
ratPP2c  avynrlnpyknddtdsastddmw--------------------------------------   382
rabPP2c  avynrlnpyknddtdststddmw--------------------------------------   382
humPP2c  avynrlnpyknddtdststddmw--------------------------------------   382
yPTC1    -------------------------------------------------------------   281
LCPP2c   sqfengpeakltgaerqkwfsnyfqklceaasngpsdqmerlqslqqqagiplsillslmgeqtq   406
```

Figure 1. Sequence alignment (MACAW) of bovine PDPc and rat, rabbit, human, yeast and *Leishmania* protein phosphatase 2Cs. Capital letters indicate conservative substitutions and boldface letters are invariant residues. Distinct regions of significant sequence identity are denoted by overbars and roman numerals.

Little is known about structure-function relationships in the Mg^{2+}-dependent protein phosphatase 2C family. Only recently have the deduced amino acid sequences of members of this family become available. Sequence alignment (Fig. 1) of bovine PDPcs from rat (Tamura et al., 1989), rabbit and human (Mann et al., 1992), *Saccharomyces cerevisiae* (Maeda et al., 1993) and the protozoan parasite *Leishmania chagasi* (Burns et al., 1983) reveals five distinct regions of significant sequence identity. Site-directed mutagenesis is being carried out with PDPc to ascertain whether the highly conserved residues are involved in Mg^{2+} (or Mn^{2+}) binding, substrate binding, and/or catalysis.

Nature of PDH phosphatase regulatory subunit

The regulatory subunit (PDPr) of bovine PDH phosphatase is a flavoprotein (FAD) with an apparent M_r of about 97 000. To gain further insight into its structure, function, and regulation, the cloning of cDNA encoding PDPr was undertaken. Overlapping cDNA fragments encoding about 70% of PDPr have been isolated from cDNA synthesized from bovine heart total RNA (J.E. Lawson and L.J. Reed, unpublished data). Attempts to isolate a clone from bovine cDNA libraries have been unsuccessful.

A BLAST search of protein data bases revealed about 35% sequence identity between the PDPr fragment and rat liver dimethylglycine dehydrogenase (Lang et al., 1991). The overall pattern indicates that the two proteins evolved from a common ancestor. Dimethylglycine dehydrogenase is also localized to the mitochondrial matrix, has a calculated M_r of 91 600, and contains covalently bound FAD (Wittwer and Wagner, 1980). By contrast, PDPr contains non-covalently bound FAD and does not exhibit dimethylglycine dehydrogenase activity (J. Yan and L. J. Reed, unpublished data). Furthermore, antiserum to dimethylglycine dehydrogenase (kindly furnished by Dr. Conrad Wagner, Vanderbilt University) did not cross-react with PDPr and antiserum to PDPr did not cross-react with dimethylglycine dehydrogenase. Based on the specific activities of dimethylglycine dehydrogenase and PDP in mitochondrial extracts and in the near homogeneous state, there appears to be about 100 times more dimethylglycine dehydrogenase than PDP in mitochondria (rat liver and bovine kidney, respectively).

Because PDPr is a flavoprotein and it is related to dimethylglycine dehydrogenase, it is likely that PDPr is a dehydrogenase and that this activity is involved in its regulatory function. Experiments are in progress to identify a substrate for PDPr, possibly a metabolite or an electron carrier.

Insulin-mediated activation of mammalian PDH phosphatase

PDH complex activity is stimulated by insulin in several tissues, particularly in adipose tissue. A two- to threefold increase in activity occurs within a few minutes of exposure of rat adipose tissue to insulin, is accompanied by a net dephosphorylation of E_1, and apparently results from stimulation of PDH phosphatase activity (Denton et al., 1989). This latter stimulation is characterized by a decrease in the apparent K_m of the phosphatase for Mg^{2+}. A major, and as yet unsolved problem, is how the signal is transferred from the insulin receptor in the plasma membrane to the PDH phosphatase within the mitochondrial inner membrane-matrix compartment. Several investigators have proposed that the insulin-induced activation of the mitochondrial PDH phosphatase is mediated by an inositol phosphate glycan(s) of molecular weight about 1 000, produced from glycolipid precursors in the plasma membrane by a phosphatidylinositol-specific phospholipase C (Larner et al., 1989; Saltiel, 1994). These insulin "mediators" are discussed elsewhere in this volume. Other proposals for the insulin "mediator" include Ca^{2+}, Mg^{2+}, and a postulated large-molecular-weight component in the inner mitochondrial membrane that may be directly involved in the transmission of the insulin signal across this membrane (Reed and Damuni, 1987; Denton et al., 1989).

Evidence is accumulating that other members of the protein phosphatase 2C family play important roles in signaling processes – in the control of plant germination and growth (Meyer et al., 1994), cell growth of *S. cerevisiae* (Maeda et al., 1993), and osmoregulation of fission yeast (Shiozaki and Russell, 1995).

Acknowledgement
We thank Dr. Lawrence Poulsen for assistance with amino acid sequence alignments.

References

Burns, J.M., Jr., Parsons, M., Rosman, D.E. and Reed, S.G. (1993) Molecular cloning and characterization of a 42-kDa protein phosphatase of *Leishmania chagasi. J. Biol. Chem.* 268: 17155–17161.
Cohen, P. (1989) The structure and regulation of protein phosphatases. *Ann. Rev. Biochem.* 58: 453–508.
Damuni, Z., Humphreys, J.S. and Reed, L.J. (1984) Stimulation of pyruvate dehydrogenase phosphatase activity by polyamines. *Biochem. Biophys. Res. Comm.* 124: 95–99.
Damuni, Z., Tung, H.Y. L. and Reed, L.J. (1985) Specificity of the heat-stable protein inhibitor of the branched-chain α-keto acid dehydrogenase phosphatase. *Biochem. Biophys. Res. Comm.* 133: 878–883.
Davis, P.F., Pettit, F.H. and Reed, L.J. (1977) Peptides derived from pyruvate dehydrogenase as substrates for pyruvate dehydrogenase kinase and phosphatase. *Biochem. Biophys. Res. Comm.* 75: 541–549.
Denton, R.M., Randle, P.J. and Martin, B.R. (1972) Stimulation by calcium ions of pyruvate dehydrogenase phosphate phosphatase. *Biochem. J.* 128: 161–163.
Denton, R.M., Midgley, P.J. W., Rutter, G.A., Thomas, A.P. and McCormack, J.G. (1989) Studies into the mechanism whereby insulin activates pyruvate dehydrogenase complex in adipose tissue. *Ann. N.Y. Acad. Sci.* 573: 285–296.

Donella-Deana, A., MacGowan, C.H., Cohen, P., Marchiori, F., Meyer, H.E. and Pinna, L.A. (1990) An investigation of the substrate specificity of protein phosphatase 2C using synthetic peptide substrates; comparison with protein phosphatase 2A. *Biochim. Biophys. Acta* 1051: 199–202.

Haystead, T.A. J., Sim, A.T. R., Carling, D., Honnor, R.C., Tsukitani, Y., Cohen, P. and Hardie, D.G. (1989) Effects of the tumour promoter okadaic acid on intracellular protein phosphorylation and metabolism. *Nature* 337: 78–81.

Lang, H., Polster, M. and Brandsch, R. (1991) Rat liver dimethylglycine dehydrogenase. Flavinylation of the enzyme in hepatocytes in primary culture and characterization of a cDNA clone. *Eur. J. Biochem.* 198: 793–799.

Larner, J., Huang, L.C., Suzuki, S., Tang, G., Zhang, C., Schwartz, C.F. W., Romero, G., Luttrell L. and Kennington, A.S. (1989) Insulin mediators and the control of pyruvate dehydrogenase complex. *Ann. N.Y. Acad. Sci.* 573: 297–305.

Lawson, J.E., Niu, X.-D., Browning, K.S., Le Trong, H., Yan, J. and Reed, L.J. (1993) Molecular cloning and expression of the catalytic subunit of bovine pyruvate dehydrogenase phosphatase and sequence similarity with protein phosphatase 2C. *Biochemistry* 32: 8987–8993.

Maeda, T., Tsai, A.Y. M. and Saito, H. (1993) Mutations in a protein tyrosine phosphatase gene (*PTP2*) and a protein serine/threonine phosphatase gene (*PTC1*) cause a synthetic growth defect in *Saccharomyces cerevisiae*. *Mol. Cell. Biol.* 13: 5408–5417.

Maeng, C.-Y., Yazdi, M.A., Niu, X.-D., Lee, H.Y. and Reed, L.J. (1994) Expression, purification, and characterization of the dihydrolipoamide dehydrogenase-binding protein of the pyruvate dehydrogenase complex from *Saccharomyces cerevisiae*. *Biochemistry* 33: 13801–13807.

Mann, D.J., Campbell, D.G., McGowan, C.H. and Cohen, P.T. W. (1992) Mammalian protein serine/threonine phosphatase 2C: cDNA cloning and comparative analysis of amino acid sequences. *Biochim. Biophys. Acta* 1130: 100–104.

Meyer, K., Leube, M.P. and Grill, E. (1994) A protein phosphatase 2C involved in ABA signal transduction in *Arabidopsis thaliana*. *Science* 264: 1452–1455.

Patel, M.S. and Roche, T.E. (1990) Molecular biology and biochemistry of pyruvate dehydrogenase complexes. *FASEB J.* 4: 3224–3233.

Perham, R.N. (1991) Domains, motifs and linkers in 2-oxo acid dehydrogenase multienzyme complexes: A paradigm in the design of a multifunctional protein. *Biochemistry* 30: 8501–8512.

Pettit, F.H., Roche, T.E. and Reed, L.J. (1972) Function of calcium ions in pyruvate dehydrogenase phosphatase activity. *Biochem. Biophys. Res. Comm.* 49: 563–571.

Pettit, F.H., Pelley, J.W. and Reed, L.J. (1975) Regulation of pyruvate dehydrogenase kinase and phosphatase by acetyl-CoA/CoA and NADH/NAD ratios. *Biochem. Biophys. Res. Comm.* 65: 575–582.

Pratt, M.L., Maher, J.F. and Roche, T.E. (1982) Purification of bovine kidney and heart pyruvate dehydrogenase phosphatase on Sepharose derivatized with the pyruvate dehydrogenase complex. *Eur. J. Biochem.* 125: 349–355.

Rahmatullah, M. and Roche, T.E. (1988) Component requirements for NADH inhibition and spermine stimulation of pyruvate dehydrogenase phosphatase activity. *J. Biol. Chem.* 263: 8106–8110.

Reed, L.J. and Damuni, Z. (1987) Mitochondrial protein phosphatases. *Adv. Prot. Phosphatases* 4: 59–76.

Reed, L.J. and Yeaman, S.J. (1987) Pyruvate dehydrogenase. *In*: P.D. Boyer and E.G. Krebs (eds): *The Enzymes*, Vol. 18. Academic Press, Orlando, FL, pp 77–95.

Reed, L.J. and Hackert, M.L. (1990) Structure-function relationships in dihydrolipoamide acyltransferases. *J. Biol. Chem.* 265: 8971–8974.

Saltiel, A.R. (1994) The paradoxical regulation of protein phosphorylation in insulin action. *FASEB J.* 8: 1034–1040.

Shenolikar, S. and Nairn, A.C. (1991) Protein phosphatases: Recent progress. *Adv. Second Messenger Phosphoprotein Res.* 23: 1–121.

Shiozaki, K. and Russell, P. (1995) Counteractive roles of protein phosphatase 2C (PP2E) and a MAP kinase kinase homolog in the osmoregulation of fission yeast. *EMBO J.* 14: 492–502.

Song, H. and Komuniecki, R. (1994) Novel regulation of pyruvate dehydrogenase phosphatase purified from anaerobic muscle mitochondria of the adult parasitic nematode, *Ascaris suum*. *J. Biol. Chem.* 269: 31573–31578.

Tamura, S., Lynch, K.R., Larner, J., Fox, J., Yasui, A., Kikuchi, K., Suzuki, Y. and Tsuiki, S. (1989) Molecular cloning of rat type 2C (IA) protein phosphatase mRNA. *Proc. Natl. Acad. Sci. USA* 86: 1796–1800.

Teague, W.M., Pettit, F.H., Wu, T.-L., Silberman, S.R. and Reed, L.J. (1982) Purification and properties of pyruvate dehydrogenase phosphatase from bovine heart and kidney. *Biochemistry* 21: 5585–5592.

Thomas, A.P., Diggle, T.A. and Denton, R.M. (1986) Sensitivity of pyruvate dehydrogenase phosphate phosphatase to magnesium ions. *Biochem. J.* 238: 83–91.

Wenk, J., Trompeter, H.-I., Pettrich, K.-G., Cohen, P.T. W., Campbell, D.G. and Mieskes, G. (1992) Molecular cloning and primary structure of a protein phosphatase 2C isoform. *FEBS Lett.* 297: 135–138.

Wittwer, A.J. and Wagner, C. (1980) Identification of folate binding protein of mitochondria as dimethylglycine dehydrogenase. *Proc. Natl. Acad. Sci. USA* 77: 4484–4488.

Alpha-Keto Acid Dehydrogenase Complexes
M.S. Patel, T.E. Roche and R.A. Harris (eds)
© 1996 Birkhäuser Verlag Basel/Switzerland

The mitochondrial α-ketoacid dehydrogenase kinases: Molecular cloning, tissue-specific expression and primary structure analysis

R.A. Harris and K.M. Popov

Department of Biochemistry and Molecular Biology, Indiana University School of Medicine, Indianapolis, IN 46202, USA

Introduction

The protein kinases responsible for phosphorylation of the mitochondrial α-ketoacid dehydrogenase complexes constitute a unique family of eukaryotic protein kinases (Popov et al., 1992; Popov et al., 1993; Popov et al., 1994). The first of these enzymes to be cloned (Popov et al., 1992), the branched chain α-ketoacid dehydrogenase kinase (BDK), phosphorylates and inactivates the branched-chain α-ketoacid dehydrogenase complex (BCKDC), the rate-limiting enzyme in the catabolism of branched chain amino acids. Phosphorylation-mediated inhibition of BCKDC by BDK conserves leucine, isoleucine and valine, the dietary essential branched chain amino acids that must be maintained continuously available for protein synthesis by all tissues. Phosphorylation of Ser^{293} and Ser^{303} of the E1α subunit of BCKDC results in inhibition of the thiamine pyrophosphate-dependent step catalyzed by the E1 component and thereby inhibition of the overall activity. BCKDC phosphatase, an enzyme that has been isolated and partially characterized (Damuni and Reed, 1987) but not cloned or studied in detail with respect to regulation in either the short or the long term, is responsible for dephosphorylation and activation of BCKDC. This phosphatase deserves considerably more study since its activity relative to that of BDK determines the proportion of BCKDC in its active dephosphorylated state. α-Ketoisocaproate, the product of leucine transamination, is a potent inhibitor of BDK (Lau et al., 1982; Paxton and Harris, 1984), an effect believed important in the short term for determining the activity state of BCKDC (for review see: Harris et al., 1986). An excessive amount of blood leucine, originating either from the diet or from proteolysis of tissue protein, results in an increased cellular concentration of α-ketoisocaproate. This inhibits BCK which, because of reduced opposition to the activity of BCKDC phosphatase, results in dephosphorylation of E1α and increased BCKDC activity. Greater BCKDC activity causes irreversible degradation of leucine, along with isoleucine and valine, regardless of whether the latter two amino acids are present in excess. When a decrease in α-ketoisocaproate occurs, as a consequence of dietary leucine insufficiency or unusually rapid rates of protein synthesis, BDK is less inhibited, resulting in a greater degree of phosphorylation and inactivation of BCKDC, thereby preserving leucine, isoleucine and valine for protein synthesis.

Long-term, stable changes in BDK activity have also been described (Espinal et al., 1986; Beggs et al., 1987; Miller et al., 1988; Zhao et al., 1992), most likely caused by changes in the amount of BDK protein expressed and thereby available in the mitochondrial matrix space for association with BCKDC. Alternatively, stable changes could result from a covalent modification of BDK not currently characterized, such as phosphorylation or acylation, which if reversible would be a likely target for regulation. Starvation for protein, for example, increases the amount of hepatic BDK activity (Espinal et al., 1986; Beggs et al., 1987; Miller et al., 1988; Zhao et al., 1992) and contributes, along with less direct inhibition of the kinase by α-ketoisocaproate, to the greater degree of phosphorylation and lower enzyme activity of BCKDC in this nutritional state. As further reviewed below, expression of greater amounts of BDK is certainly part of the mechanism responsible for decreased BCKDC activity during periods of dietary protein insufficiency (Popov et al., 1995).

The second of the mitochondrial protein kinases to be cloned (Popov et al., 1993), the pyruvate dehydrogenase kinase (PDK), phosphorylates and inactivates the pyruvate dehydrogenase complex (PDC), a key regulatory enzyme in the catabolism of glucose. Inhibition of PDC by this kinase diminishes overall glucose utilization and conserves alanine, lactate and pyruvate, metabolic intermediates required for the maintenance of blood glucose levels by hepatic and renal gluconeogenesis. Phosphorylation of Ser^{233}, Ser^{240}, and Ser^{172} of the E1α subunit of PDC results in inhibition of the thiamine pyrophosphate-dependent step catalyzed by the PDC E1 component and, as a consequence, inhibits overall activity of the complex. PDC phosphatase, the catalytic subunit of which was recently cloned (Lawson et al., 1993), catalyzes dephosphorylation and activation of the complex. Activation of PDK by NADH and acetyl-CoA and inhibition by NAD^+, CoA, and pyruvate are important allosteric control mechanisms for the regulation of PDK activity (for reviews see: Roche et al., 1989; Yeaman, 1989; Behal et al., 1993). Stimulation of PDH phosphatase activity by Ca^{2+}, an effect not demonstrable with BCKDH phosphatase, is also of physiological significance for the regulation of the activity state of PDC (Rutter et al., 1989). Physiological conditions, such as starvation, under which acetyl-CoA/CoA and $NADH/NAD^+$ ratios increase because of increased rates of fatty acid oxidation, cause activation of PDK and phosphorylation of PDC. Inactivation of PDC by phosphorylation conserves the three carbon compounds for gluconeogenesis. On the other hand, physiological conditions that promote glycolytic activity, such as over consumption of carbohydrate, tend to increase pyruvate levels and activate PDC, thereby providing more acetyl-CoA for fatty acid synthesis in hepatic tissue and more acetyl-CoA for citric acid cycle activity in peripheral tissues. Long-term, stable changes in PDK activity, most likely due to changes in amount of PDK protein associated with PDC and/or to activation of PDK by an undefined covalent modification, have also been described (Kerbey and Randle, 1982; Kerbey et al., 1984; Denyer et al., 1986; Vary, 1991; Jones et al., 1992). Starvation, diabetes and sepsis have all been demonstrated to increase the amount of PDK activity in one or more tissues, and this likely contributes to the greater degree of phosphorylation and lower amounts of active PDC in specific cell types in these common physiological and pathological states.

The third of the mitochondrial protein kinases to be cloned (Popov et al., 1994) also phosphorylates and inactivates PDC, thereby establishing the existence of PDK isoforms in the tissues of higher eukaryotes. Since changes in PDK activity and altered sensitivity of PDK to allosteric effectors in various physiological and pathological states (Vary, 1991; Feldhoff et al., 1993) may be explained by induction of different PDK isoforms in these states, the discovery that isoenzymes are responsible for the phosphorylation of PDC is of considerable importance to our overall understanding of regulation of the complex.

The purpose of this paper is to review and discuss recent progress that has been made in molecular cloning, primary structure analysis, and characterization of this new family of protein kinases, the mitochondrial protein kinases responsible for the regulation of the mitochondrial α-ketoacid dehydrogenase complexes.

Molecular cloning and sequence analysis of mitochondrial protein kinases

cDNAs encoding three mitochondrial protein kinases have been cloned from rat heart cDNA libraries (Popov et al., 1992; Popov et al., 1993; Popov et al., 1994). All encode mitochondrial protein kinases as evidenced by an exact correspondence between amino acid sequences deduced from the cDNAs and partial sequences of the kinases obtained by direct sequencing of the purified proteins. Furthermore, the protein resulting from expression of the BDK cDNA in *E. coli* specifically phosphorylates the BCKDC E1α component and inhibits BCKDC dehydrogenase activity. Likewise, the proteins resulting from heterologous expression of the two unique PDK cDNAs specifically phosphorylate the E1α component of PDC and inhibit dehydrogenase activity of this complex. The BDK cDNA has an open reading frame encoding a protein of 412 amino acids, including a 30 amino acid mitochondrial leader sequence (Popov et al., 1992). The first PDK cDNA cloned has an open reading frame encoding a protein (PDK1) of 434 amino acids, including a 26 amino acid mitochondrial leader sequence (Popov et al., 1993). The second PDK cDNA cloned has an open reading frame encoding a protein (PDK2) of 407 amino acids, including an unusually short mitochondrial leader sequence of 8 residues (Popov et al., 1994). About 30% amino acid identity exists between BDK and the PDKs; 70% between PDK1 and PDK2 (Fig. 1).

The amino acid sequences deduced from the full-length cDNAs encoding BDK, PDK1 and PDK2 lack motifs usually associated with ser/thr protein kinases in eukaryotes (for review see: Harris et al., 1994). Indeed, detailed sequence analysis indicates that BDK, PDK1 and PDK2 are only distantly related to other ser/thr protein kinases. The catalytic domain of mitochondrial protein kinases consist of five regions of high conservation or subdomains separated by regions of lower conservation (Popov et al., 1992; Fig. 2). These conserved subdomains are likely important for catalytic function, either directly as components of the active site or indirectly by contributing structurally to the formation of the active site. Most consensus sequences lie in the carboxy-terminus, whereas the amino-terminal part is quite variable, except for subdomain II. At the amino-terminus, a putative kinase domain is defined by an invariant histidine residue (His^{155}, numbered according to mature sequence of PDK2). It is separated by a spacer of more than 100 amino acids from subdomain II (defined by invariant Asn^{247} occurring within a highly conserved sequence KNAMRAT). At the carboxyl terminus, the catalytic domain is defined by a glycine-rich loop G^{317}xG^{319}xG^{321} (subdomain V). The central core of the catalytic domain consists of subdomains III (consensus sequence D^{282}rG^{284}gG^{286}) and IV (defined by an aromatic residue, Y^{298} in PDK2).

Considerable sequence identity and similarity are apparent between the mitochondrial protein kinases and members of the prokaryotic histidine protein kinase family, a diverse set of sensing and response systems that are important in regulating numerous processes in bacteria (Stock et al., 1989). The bacterial enzymes are termed histidine protein kinases, not because they phosphorylate target proteins on histidine, but rather because their catalytic mechanism involves autophosphorylation of a histidyl residue. The five subdomains described above for mitochondrial protein kinases are also present in the bacterial histidine protein kinases. With these enzymes, the invariant histidine residue of subdomain I is the site of autophosphorylation, whereas subdomains III, IV and V are likely involved in ATP binding. Subdomain II may serve as a hinge controlling the interaction between the ATP-binding site and the amino-terminal region bearing the catalytic His residue (Parkinson and Kafoid, 1992). In contrast to mitochondrial protein kinases which phosphorylate serine residues, prokaryotic histidine protein kinases phosphorylate aspartate residues of target proteins:

```
PDK1    MRLARLL----RGGTSVRPL--CAVPCASRSLASDSASGSGPASESGVP-      43
PDK2    MRWFRAL----LKNASL-------------------------AGAP-        17
BDK     MILTSVLGSGPRSGSSLWPLLGSSLSLRVRSTSATDTHHVELARERSKTV     50
        *    .  *       ..*.                              . .


PDK1    ------GQVDFYARFSPSPLSMKQFLDFGSVNACEK--TSFMFLRQELPV      85
PDK2    ------KYIEHPSKFSPSPLSMKQFLDFGSSNACEK--TSFTFLRQELPV      59
BDK     TSFYNQSAIDVVAEKPSVRLTPTMMLYSGRSQDGSHLLKSGRYLQQELPV    100
              .. .    .. .*.    .* *.  .. ... .*  .*.*****


PDK1    RLANIMKEISLLPDNLLRTPSVQLVQSWYIQSLQELLDFKDKSAEDAKTI    135
PDK2    RLANIMKEINLLPDRVLSTPSVQLVQSWYVQSLLDIMEFLDKDPEDHRTL    109
BDK     RIAHRIKGFVVFLSSLVATLPYCTVHELYIRAFQKLTDF--PPIKDQADE    148
        *.*.  .*...  .. ..   *  .    *.. *.... ... .*.   .


PDK1    YEFTDTVIRIRNRHNDVIPTMAQGVTEYKESFGVDPVTSQNVQYFLDRFY    185
PDK2    SQFTDALVTIRNRHNDVVPTMAQGVLEYKDTYGDDPVSNQNIQYFLDRFY    159
BDK     AQYCQLVRQLLDDHKDVVVTLLAEGLRESRKHIEDEKL----VRYFLDKTL    194
        . . .. .  *.**.. .*.*. *.   .      .    ..****.


PDK1    MSRISIRMLLNQHSLLFGGKGSPSHRKHIGSINPNCDVVEVIKDGYENAR    235
PDK2    LSRISIRMLINQHTLIFDGSTNPAHPKHIGSIDPNCSVSDVVKDAYDMAK    209
BDK     TSRLGIRMLATHHLALHEDK-----PDFVGIISTRLSPKKIIEKWVDFAR    239
        **..****  ..*    . ...     .* *.... .... .  .  *.


PDK1    RLCDLYYNNSPELELEELNAKSPGQPIQVVYVPSHLYHMVFELFKNAMRA    285
PDK2    LLCDKYYMASPDLEIQEVNATNATQPIHMVYVSPHLYHMLFELFKNAMRA    259
BDK     RLCEHKYGNAPRVRI------NGHVAARFPFIPMPLDYILPELLKNAMRA    283
        **.  *  ..*  ..      .    .   ... .* ...  **.******


PDK1    TMEHH-ADKGVYPPIQVHVTLGEEDLTVKMSDRGGGVPLRKIDRLFNYMY    334
PDK2    TVESH-ESSLTLPPIKIMVALGEEDLSIKMSDRGGGVPLRKIERLFSYMY    308
BDK     TMESHLDTPYNVPDVVITIANNDVDLIIRISDRGGGIAHKDLDRVMDYHF    333
        *.* *  ..      *  . ... .. ** ...******.. ....*...* .


PDK1    STA----PRPRVET----------SRAVPLAGFGYGLPISRLYAQYFQGD    370
PDK2    STA----PTPQPGT----------G-GTPLAGFGYGLPISRLYAKYFQGD    343
BDK     TTAEASTQDPRISPLFGHLDMHSGGQSGPMHGFGFGLPTSRAYAEYLGGS    383
        .**    . *. ..          .  *. ***.***.** **.*. *.


PDK1    LKLYSLEGYGTDAVIYIKALSTESIERLPVYNKAAWKHYRTNHEADDWCV    370
PDK2    LQLFSMEGFGTDAVIYLKALSTDSVERLPVYNKSAWRHYQTIQEAGDWCV    343
BDK     LQLQSLQGIGTD--VYL---------RL--------RHIDGREES-----    383
        *.* *..* *** .*.         **         .*  . .*.


PDK1    PSREPKDMTTFRSS      434
PDK2    PSTEPKNTSTYRVS      407
BDK     ---------FRI-       412
                 .*
```

Figure 1. Alignment of predicted amino acid sequences of PDK1, PDK2 and BDK. Identity is indicated by (*); conservative substitutions by (.). Similarity established by CLUSTAL program of PCGENE.

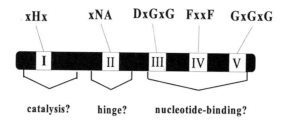

Figure 2. Subdomain structure of the mitochondrial protein kinases. The five sequence motifs characteristic of mitochondrial protein kinases are indicated.

$$PK–His + ATP \longrightarrow PK–His-P + ADP$$
$$PK–His-P + Substrate \longrightarrow PK–His + Substrate-P$$

Whether mitochondrial protein kinases use a similar mechanism of phosphotransfer and substrate recognition is not known.

Additional members of the mitochondrial protein kinase family

Six genes that clearly belong to the mitochondrial protein kinase family have been identified thus far in eukaryotic cells (Fig. 3). As discussed above, three of these genes encode for protein kinases involved in the regulation of BCKDC and PDC activity. Three others, ORF ZK370.5 from *Caenorhabidis elegans* (Sulston et al., 1992), ORF of hypothetical phosphoprotein 3 from *Trypanosoma brucei* (Wirtz et al., 1991), and an ORF from yeast (accession number Z47047, orf YIL042C) have been identified by searching the combined database of GenBank. To date, the protein products of these genes have not been characterized. However, some conclusions may be drawn based on the analysis of the dendrogram (Fig. 3) showing relationships among the members of the family. The topology of the tree indicates that the PDK isoenzymes form one large cluster with BDK and hypothetical protein ZK370.5 from *C. elegans*. It seems reasonable to assume that members of a cluster are more similar to each other than to others in terms of three-dimensional structure, mechanism of substrate recognition and mechanism of phospho-transfer reaction. If this assumption is correct, gene ZK370.5 most likely encodes for a PDK for *C. elegans*. On the other hand, the proteins from *T. brucei* and yeast, even though they have all structural features characteristic of mitochondrial protein kinases, map on the tree very distantly from the main cluster. This suggests that these protein kinases are not closely related to the other

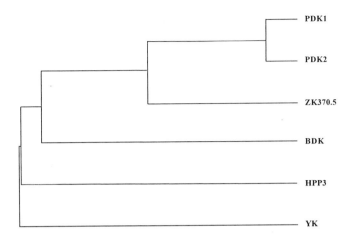

Figure 3. Dendrogram of the alignment of PDK1, PDK2, ZK370.5 hypothetical kinase (ORF ZK370.5 from *C. elegans*), BDK, HPP3 hypothetical kinase (ORF for phosphoprotein 3 from *T. brucei*), and YK hypothetical kinase (ORF from yeast).

members of the family, and may indicate that mitochondria have enzymes other than multienzyme complexes which are controlled by phosphorylation-dephosphorylation cycles.

Phosphorylation sites of BDK

Site-directed mutagenesis has been used to directly assess the roles played by Ser^{293} and Ser^{303}, phosphorylation sites 1 and 2, respectively, of the BCKDC E1α component (Zhao et al., 1994). Mutation of the phosphorylation site 1 serine to glutamate completely inhibits E1 dehydrogenase activity, i.e., produces the same effect as phosphorylation of this residue by BDK. On the other hand, mutation of the phosphorylation site 2 serine to glutamate has no effect on enzyme activity. Replacing phosphorylation site 1 serine with alanine increases the K_m for the α-keto acid substrate of BCKDC but had no effect on V_{max}. Phosphorylation of the open site 2 serine of the latter mutant enzyme does not effect BCKDC activity. Thus, kinase-mediated inactivation of BCKDC E1 is due to the introduction of a negative charge at site 1. Site 2 phosphorylation is silent, in agreement with previous evidence that phosphorylation of site 1 is entirely responsible for kinase-mediated inactivation of BCKDC (Cook et al., 1984; Paxton et al., 1986). The finding that the alanine site 1 mutant enzyme retains enzymatic activity rules out a role for Ser^{293} in the catalytic mechanism of the BCKDC E1 component. The increase in K_m for this mutant suggests that the α-keto acid binding domain may be at or near phosphorylation site 1. Site-directed mutagenesis studies currently in progress (J.W. Hawes, K.M. Popov and R.A. Harris, unpublished studies) indicate that charged residues in the vicinity of phosphorylation site 1 (Arg[288] and Asp[296]) have critical functions in E1 dehydrogenase activity. Arg[288] also appears important for BDK phosphorylation of site 1 on E1α.

Dietary control and tissue specific expression of BDK

The activity state of BCKDC varies widely in different tissues of the rat (for review see: Harris et al., 1986). The activity state is greater in the liver than in any other tissue, and the lowest amount of kinase protein and kinase mRNA are found in this tissue (Popov et al., 1995). In general, the levels of kinase protein and mRNA are highest in tissues with greatest kinase activity and correlate inversely with the activity state of BCKDC in tissues. Starvation for protein, a condition well-established to increase the activity of BDK and decrease the activity state of BCKDC, induces a significant increase in kinase mRNA in liver and the amount of BDK protein associated with the complex (Popov et al., 1995). Thus, the amount of kinase protein expressed in various tissues and tightly associated with the complex under different physiological conditions is an important factor determining the activity state of BCKDC. Whether expression of hepatic BDK is regulated directly by components derived from the diet or indirectly by dietary-induced changes in hormone levels is not known.

Effect of starvation for dietary protein on the amount of BCKDC E1, the substrate for BDK

Starvation of rats for dietary protein decreases the liver content of both BCKDC and mitochondrial protein, resulting in no significant difference in BCKDC activity on a mitochondrial protein basis (Beggs et al., 1987; Miller et al., 1988; Zhao et al., 1994). Nevertheless, by immunoblot analysis the amount of E1 relative to E2 in liver mitochondria is less in protein-starved rats and greater in high-protein-fed rats than that found in normal rats (Zhao et al., 1994). A lesser amount of E1 relative to E2 assures that the dehydrogenase activity of E1 is rate limiting for the complex. This in turn maximizes sensitivity to phosphorylation, since any inhibition of E1 activity by phosphorylation would translate to an identical degree of BCKDC inhibition. On the other hand, the presence of surplus E1 in the liver of high-protein-fed rats may make a different component of the complex rate limiting, and may therefore function to minimize sensitivity to phosphorylation, as would be desirable when excess branched chain amino acids are present.

Tissue-specific expression of PDK1 and PDK2

Northern blot analysis has been used to characterize tissue distribution of PDK1 and PDK2 (Popov et al., 1994). The message for PDK1 is predominantly expressed in heart with only modest expression in other tissues. PDK2 message is best expressed in heart and skeletal muscle, but is also expressed in significant amounts in testes, liver, brain and kidney. Preliminary immunoblot analyses also indicate variations and induced changes in the ratio of PDK1/PDK2 protein expressed in tissues in different physiological and pathological states of rats (J. Sato, K.M. Popov and R.A. Harris, unpublished observations).

Concluding remarks

cDNAs for three mitochondrial protein kinases have been cloned. BDK is encoded by one cDNA, PDK isoforms by the other two. These kinases are related to prokaryotic histidine protein kinases in terms of sequence subdomains but related to eukaryotic ser/thr protein kianses in terms of phosphorylation target sites, making them unique among the protein kinases of both pro- and eukaryotes. That additional protein kinases of this family will be discovered seems likely in view of the finding of *T. brucei* and yeast hypothetical protein kinases that are only distantly related to the mitochondrial α-ketoacid dehydrogenase multienzyme complexes. Inactivation of hepatic BCKDC by phosphorylation conserves branched chain amino acids during dietary protein insufficiency. The message level for BDK and the amount of BDK bound to BCKDC are increased by dietary protein insufficiency, suggesting that increased BDK gene expression is a factor. Changes in gene expression or isozyme shifts may also be responsible for the stable changes in PDK activity induced in various tissues by starvation, diabetes and sepsis.

Acknowledgements
This investigation was supported by United States Public Health Service Grants DK 19259 and DK 47844 (to RAH), a Grant-in-Aid from the National American Heart Association (to KMP), and the Grace M. Showalter Residuary Trust.

References

Beggs, M., Patel, H., Espinal, J. and Randle, P.J. (1987) Temporal relationships in the effects of protein-free diets on the activities of rat liver branched-chain keto acid dehydrogenase complex and kinase. *FEBS Lett.* 215: 13–15.

Behal, R.H., Buxton, D.B., Robertson, J.G. and Olson, M.S. (1993) Regulation of the pyruvate dehydrogenase multienzyme complex. *Ann. Rev. Nutr.* 13: 497–520.

Cook, K.G., Bradford, A.P., Yeaman, S.J., Aitken, A., Fearnley, I.M. and Walker, J.E. (1984) Regulation of bovine kidney branched chain 2-oxoacid dehydrogenase complex by reversible phosphorylation. *Eur. J. Biochem.* 145: 587–591.

Damuni, Z. and Reed, L.J. (1987) Purification and properties of the catalytic subunit of the branched-chain α-keto acid dehydrogenase phosphatase from bovine kidney mitochondria. *J. Biol. Chem.* 262: 5129–5132.

Denyer, G.S., Kerbey, A.L. and Randle, P.J. (1986) Kinase activator protein mediates longer-term effects of starvation on activity of pyruvate dehydrogenase kinase in rat liver mitochondria. *Biochem. J.* 239: 347–354.

Espinal, J., Beggs, M., Patel, H. and Randle, P.J. (1986) Effects of low protein diet and starvation on the activity of branched-chain 2-oxoacid dehydrogenase kinase in rat liver and heart. *Biochem. J.* 237: 285–288.

Feldhoff, P.W., Arnold, J., Oesterling, B. and Vary, T.C. (1993) Insulin-induced activation of pyruvate dehydrogenase complex in skeletal muscle of diabetic rats. *Metabolism* 42: 615–623.

Harris, R.A., Paxton, R., Powell, S.M., Goodwin, G.W., Kuntz, M.J. and Han, A.C. (1986) Regulation of branched-chain α-ketoacid dehydrogenase complex by covalent modification. *Adv. Enz. Regul.* 25: 219–237.

Harris, R.A., Popov, K.M., Zhao, Y., Kedishvili, N.Y., and Crabb, D.W. (1995) A new family of protein kinases – the mitochondrial protein kinases. *Adv. Enz. Regul.* 35: 147–162.

Jones, B.S., Yeaman, S.J., Sugden, M.C. and Holness, M.J. (1992) Hepatic pyruvate dehydrogenase kinase activities during starved-to-fed transition. *Biochim. Biophys. Acta* 1134: 164–168.

Kerbey, A.L. and Randle, P.J. (1982) Pyruvate dehydrogenase kinase/activator in rat heart mitochondria: assay, effect of starvation, and effect of protein-synthesis inhibitors on starvation. *Biochem. J.* 206: 103–111.

Kerbey, A.L., Richardson, L.J. and Randle, P.J. (1984) The roles of intrinsic kinase and of kinase activator protein in the enhanced phosphorylation of pyruvate dehydrogenase complex in starvation. *FEBS Lett.* 176: 115–119.

Lau, K.S., Fatania, H.R. and Randle, P.J. (1982) Regulation of the branched-chain 2-oxoacid dehydrogenase kinase reaction. *FEBS Lett.* 144: 57–62.

Lawson, J.E., Niu, X.-D., Browning, K.S., Trong, H.L., Yan, J. and Reed, L.J. (1993) Molecular cloning and expression of the catalytic subunit of bovine pyruvate dehydrogenase phosphatase and sequence similarity with protein phosphatase 2C. *Biochemistry* 32: 8987–8993.

Miller, R.H., Eisenstein, R.S. and Harper, A.E. (1988) Effects of dietary protein intake on branched-chain keto acid dehydrogenase activity of the rat. Immunochemical analysis of the enzyme complex. *J. Biol. Chem.* 263: 3454–3461.

Parkinson, J.S. and Kafoid, E.C. (1992) Communication modules in bacterial signaling proteins. *Ann. Rev. Genet.* 26: 71–112.

Paxton, R. and Harris, R.A. (1984) Regulation of branched-chain α-ketoacid dehydrogenase kinase. *Arch. Biochem. Biophys.* 231: 48–7.

Paxton, R., Kuntz, M.J. and Harris, R.A. (1986) Phosphorylation sites and inactivation of branched-chain α-ketoacid dehydrogenase isolated from rat heart, bovine kidney, and rabbit liver, kidney, heart, brain, and skeletal muscle. *Arch. Biochem. Biophys.* 244: 187–201.

Popov, K.M., Zhao, Y., Shimomura, Y., Kuntz, M.J. and Harris, R.A. (1992) Branched-chain α-ketoacid dehydrogenase kinase: Molecular cloning, expression, and sequence similarity with histidine protein kinases. *J. Biol. Chem.* 267: 13127–13130.

Popov, K.M., Kedishvili, N.Y., Zhao, Y., Shimomura, Y., Crabb, D.W. and Harris, R.A. (1993) Primary structure of pyruvate dehydrogenase kinase establishes a new family of eukaryotic protein kinases. *J. Biol. Chem.* 268: 26602–26606.

Popov, K.M., Kedishvili, N.Y., Zhao, Y., Gudi, R. and Harris, R.A. (1994) Molecular cloning of the p45 subunit of pyruvate dehydrogenase kinase. *J. Biol. Chem.* 269: 29720–29724.

Popov, K.M., Zhao, Y., Shimomura, Y., Jaskiewicz, J.A., Kedishvili, N.Y., Irwin, J., Goodwin, G.W. and Harris, R.A. (1995) Dietary control and tissue expression of branched-chain α-ketoacid dehydrogenase kinase. *Arch. Biochem. Biophys.* 316: 148–154.

Roche, T.E., Rahmatullah, M., Li, L., Radke, G.A., Chang, C.L. and Powers-Greenwood, S.L. (1989) Lipoyl-containing components of the pyruvate dehydrogenase complex: roles in modulating and anchoring the PDH kinase and the PDH phosphatase. *Ann. N.Y. Acad. Sci.* 573: 168–174.

Rutter, G.A., McCormack, Midgley, P.J.W. and Denton, R.M. (1989) The role of Ca^{2+} in the hormonal regulation of the activities of pyruvate dehydrogenase and oxoglutarate dehydrogenase complexes. *Ann. N.Y. Acad. Sci.* 573: 206–217.

Stock, J.B., Ninfa, A.J. and Stock, A.M. (1989) Protein phosphorylation and regulation of adaptive responses in bacteria. *Microbiol. Rev.* 53: 450–490.

Sulston, J., Du, Z., Thomas, K., Wilson, R., Hillier, L., Staden, R., Halloran, N., Green, P., Thierry-Mieg, J., Qiu, L., Dear, S., Coulson, A., Craxton, M., Durbin, R., Berks, M., Metzstein, M., Hawkins, T., Ainscough, R. and Waterston, R. (1992) The *C. elegans* genome sequencing project: a beginning. *Nature* 356: 37–41.

Vary, T. (1991) Increased pyruvate dehydrogenase kinase activity in response to sepsis. *Am. J. Physiol.* 260: E669-E674.

Wirtz, E., Sylvester, D. and Hill, G.C. (1991) Characterization of a novel developmentally regulated gene from *Trypanosoma brucei* encoding a potential phosphoprotein. *Nucleic Acids Res.* 14: 6745–6763.

Yeaman, S.J. (1989) The 2-oxo acid dehydrogenase complexes: recent advances. *Biochem. J.* 257: 625–632.

Zhao, Y., Jaskiewicz, J. and Harris, R.A. (1992) Effects of clofibric acid on the activity and activity state of the hepatic branched-chain 2-oxo acid dehydrogenase complex. *Biochem. J.* 285: 167–172.

Zhao, Y., Hawes, J.W., Popov, K.M., Jaskiewicz, J.A., Shimomura, Y., Crabb, D.W. and Harris, R.A. (1994) Site-directed mutagenesis of phosphorylation sites of the branched chain α-ketoacid dehydrogenase complex. *J. Biol. Chem.* 269: 18583–18587.

Zhao, Y., Popov, K.M., Shimomura, Y., Kedishvili, N.Y., Jaskiewicz, J., Kuntz, M.J., Kain, J., Zhang, B. and Harris, R.A. (1994) Effect of dietary protein on the liver content and subunit composition of the branched-chain α-ketoacid dehydrogenase complex. *Arch. Biochem. Biophys.* 308: 446–453.

Alpha-Keto Acid Dehydrogenase Complexes
M.S. Patel, T.E. Roche and R.A. Harris (eds)
© 1996 Birkhäuser Verlag Basel/Switzerland

Shorter term and longer term regulation of pyruvate dehydrogenase kinases

P.J. Randle and D.A. Priestman

Nuffield Department of Clinical Biochemistry, University of Oxford, Radcliffe Infirmary, Oxford OX2 6HE, UK

PDH complex and kinases; background material

The mitochondrial PDH complex of animal tissues catalyses the non reversible conversion of pyruvate to acetyl CoA with concomitant reduction of NADH. The reaction connects glycolysis to oxidative metabolism, and to FFA synthesis, resulting in the irretrievable loss of glucose carbon (animals lack the means of converting acetylCoA to glucose). Some cells in the central nervous system are critically dependent upon glucose oxidation for survival and conservation of glucose by substitution of lipid fuels in cells able to oxidise them is of vital importance during carbohydrate deprivation. Regulation of the PDH complex is therefore of crucial importance in glucose homeostasis.

PDH complexes of animal tissues are inactivated by phosphorylation of the E1 (pyruvate dehydrogenase) component by PDH kinase(s). Reactivation requires dephosphorylation catalysed by PDH phosphatase(s); there are no known allosteric activators of the *b* form. In mitochondria and tissues *in vitro* PDH kinases and phosphatases are both active as shown by turnover of phosphate in the complex in the steady state (incorporation of ^{32}P from $^{32}P_i$: Hughes and Denton, 1976; Sale and Randle, 1982a, b). Thus the proportion of PDH complex in the *a* form is determined by the relative activities of kinases and phosphatases.

By purification, N-terminal sequencing, and cDNA cloning and sequencing, rat heart has been shown to contain two PDH kinases which exhibit ~ 70% sequence homology [$M_r = 45$ and 48 kDa by SDS-PAGE (p45 and p48); $M_r = 45\ 031$ and 46 270 from deduced amino acid sequence]. Northern blotting showed that mRNA for p45 is abundant in most rat tissues studied whereas p48 is abundant only in heart (Popov et al., 1993, 1994). PDH kinase purified from bovine kidney by Stepp et al. (1983) yielded two subunits of 45 and 48 kDa; only the 48 kDa subunit exhibited kinase activity and the function of the 45 kDa subunit is not known. Based on limited immunological data bovine kidney kinase may correspond to rat heart p45 (Priestman et al., 1994).

Thus far, work on the regulation of PDH kinases has been largely confined to studies of kinases intrinsic to purified PDH complexes from bovine kidney and heart and porcine heart, and of mitochondria from rat heart, skeletal muscle and liver. Purified PDH kinases phosphorylate three serine residues of E1(α-chain) in the PDH complex, in the purified E1 component, and in synthetic peptides containing phosphorylation sites 1 plus 2 (Mullinex et al., 1985). In bovine PDH complex both E1 and PDH kinase(s) are attached to the E2 component (the kinase more specifically to the inner lipoyl domain of E2). The presence of E2 enhances the rate of phosphorylation of E1 by the kinase (Reed, 1981; Stepp et al., 1983; Liu et al., 1995). The extent to which the two isoforms of rat PDH kinase share identity with respect to regulation and multisite phosphorylations is not clear, and neither is it known whether there is more than one isoform in tissues of species other than the rat.

In bovine kidney and bovine, porcine and rat heart complexes and in mitochondria PDH kinase phosphorylates three seryl residues in E1α. Phosphorylation of sites 1 or 2 (but not of 3) is inactivating and relative rates are sites $1 \gg 2 > 3$. Relative rates of dephosphorylation by PDH phosphatase(s) are sites $2 > 1 >$ or $= 3$ (pig or rat heart) or sites $2 > 3 > 1$ (bovine kidney) (Davis et al., 1977; Sugden and Randle, 1978; Teague et al., 1979; Sugden et al., 1979; Sale and Randle, 1982a, b). In rat heart mitochondria and in heart *in vivo* inactivation is due almost wholly ($> 98.5\%$) to phosphorylation of site 1 (Sale and Randle, 1982a, b). Site 1 dephosphorylation and reactivation are inhibited by phosphorylation (or thiophosphorylation) of sites 2 and 3, i.e PDH kinase(s) regulate reactivation by PDH phosphatase(s) through multisite phosphorylation (Sugden et al., 1978; Kerbey et al., 1981; Tonks et al., 1982; Sale and Randle, 1982a, b).

Shorter term regulation of PDH kinase(s)

PDH kinase(s) have been assayed by the rate of ATP dependent inactivation of PDH complex or of the E1 component and by the rate of [^{32}P] phosphorylation of PDH complex or its E1 component or of synthetic peptides (Hucho et al., 1972; Mullinex et al., 1985). Inactivation, which involves only phosphorylation of a single site, is pseudo first order and hence activity may conveniently be expressed as the apparent first order rate constant (Kerbey and Randle, 1981). [^{32}P] phosphorylation is multisite unless use is made of a peptide with a single site of phosphorylation and hence rate constants are less readily utilised as an index of activity.

The elementary reactions catalysed by the PDH complex are important to understanding regulation and are shown in the scheme below. Reactions 1 to 4 comprise the holocomplex reaction. In the absence of CoA or NAD$^+$ reactions 1 and 1a form acetoin from pyruvate (accelerated by

$$\text{pyruvate} + \text{TPP.E}_1 \longrightarrow \text{hydroxyethylTPP.E}^- + CO_2 + H^+ \quad [1]$$

$$\text{hydroxyethylTPP.E}^- + E_2(\text{lipS.S}) \rightleftharpoons E_2(\text{lipSH.Sacetyl}) + \text{TPP.E}_1 \quad [2]$$

$$E_2(\text{lipSH.Sacetyl}) + \text{CoASH} \rightleftharpoons E_2(\text{lipSH.SH}) + \text{acetyl CoA} \quad [3]$$

$$E_2(\text{lipSH.SH}) + NAD^+ \overset{E3}{\rightleftharpoons} E_2(\text{lipS.S}) + NADH + H^+ \quad [4]$$

$$[2\text{hydroxyethylTPP.E}^- \rightleftharpoons 2\text{TPP.E}_1 + \text{acetoin}] \quad [1a]$$

(TPP, thiamine pyrophosphate; lip, lipoamide)85

the presence of acetaldehyde). Acetoin can also be formed from acetyl CoA plus NADH by reaction 1a plus reverse of reactions 2–4. It is theoretically possible for acetoin to be formed *in vivo* from acetylCoA and NADH generated by oxidation of FFA or ketone bodies. However phosphorylation of the PDH complex by PDH kinase inhibits the elementary reactions involving TPP (i.e., reactions 1 and 2) (Walsh et al., 1976) i.e., it would be expected to inhibit formation of acetoin from pyruvate, and from acetyl CoA + NADH during FFA oxidation.

PDH kinase(s) associated with PDH complexes from bovine, porcine and rat tissues are activated by products of the PDH complex reaction (acetyl CoA and NADH) and inhibited by substrates of the PDH complex reaction (pyruvate, CoA and NAD$^+$); the kinase(s) are also inhibited by ADP (competitive with ATP) (Hucho et al., 1969; Cooper et al., 1974; Pettit et al., 1975; Cooper et al., 1975). Inhibition by pyruvate is enhanced by ADP possibly because pyruvate may bind to the [kinaseADP] intermediate (Pratt and Roche, 1979). The effects are rapid and reversal is rapid following their removal, hence the description "shorter term". These effectors of the kinase influence the rate of phosphorylation of all three sites in E1α (Kerbey et al., 1979). Thus mitochondrial concentration ratios of [ATP]/[ADP], [acetylCoA]/[CoA] and [NADH]/[NAD$^+$] and mitochondrial concentration of pyruvate are major determinants of PDH kinase activity *in vivo*. This has been confirmed rather directly by *in vitro* experiments with rat heart and skeletal muscle mitochondria in which increases in these ratios severally or in combination were associated with decreases in the percentage of complex in the *a* form (Hansford, 1976, 1977; Kerbey et al., 1976, 1977; Ashour and Hansford, 1983; Fuller and Randle, 1984).

PDH complexes are also subject to end product inhibition by increasing concentration ratios of [acetyl CoA]/[CoA], and [NADH]/[NAD$^+$] (Garland and Randle, 1964). This is to be expected from the reversibility of elementary reactions 2–4 (see scheme). As a result, dihydrolipoamide and/or acetylhydrolipoamide and TPPE$_1$ accumulate with consequential inhibition of the irreversible E$_1$ reaction. Studies with rat heart mitochondria have shown that both phosphorylation and end product inhibition mediate inhibitory effects of FFA oxidation on pyruvate oxidation (Hansford, 1977; Hansford and Cohen, 1978). There is now compelling evidence that increasing ratios of [acetylCoA]/[CoA] and of [NADH]/[NAD$^+$] activate PDH kinase by reduction or reductive acetylation of lipoate and not through an allosteric site on the kinase. Thus, PDH kinase is activated as a consequence of slow reduction of lipoamide by thiols such as dithiothreitol or mercaptoethanol (reversed by NAD^+); by pyruvate at low concentration, or by acetoin, in the presence of TPP but not in its absence; by acetylation of lipoate under conditions which result in low residual [acetylCoA]/ [CoA] in the incubate; and following acylation of lipoamide with TPP + α-ketoisovalerate (Cooper et al., 1974, 1975; Walsh et al., 1976; Roche and Cate, 1976; Cate and Roche, 1978, 1979; Rahmatullah and Roche, 1985; Robertson et al., 1990).

It is perhaps important to note the extent to which there is reciprocal regulation of PDH kinase and PDH phosphatase by effectors in common. Ca^{2+} is a major activator of PDH phosphatase at concentrations within the physiological range (10 nM–10 μM; Denton et al., 1972) and it inhibits PDH kinase over the same concentration range (Cooper et al., 1974). PDH phosphatase is inhibited by NADH (reversed by NAD^+) (Reed, 1981) whereas these effectors have the opposite effect on PDH kinase (see above).

Longer term regulation of PDH kinase(s)

The existence of further and longer term mechanism(s) regulating reversible phosphorylation in the PDH complex was first suggested by the observation that the proportion of PDH complex in the *a* form in hearts of alloxan diabetic or starved rats (1–2%) is lower than can be induced *in vitro* by oxidation of FFA or ketone bodies (7%). It was then shown that percent of PDH*a* in rat heart mitochondria oxidising 2-oxoglutarate + L-malate is much lower with heart mitochondria from diabetic or 48 h starved rats, than with those from normal fed controls. This difference persisted in the presence of the PDH kinase inhibitors, pyruvate or dichloracetate, at concentrations sufficient to effect almost complete conversion of PDH*b* to PDH*a* in mitochondria from normal fed rats (Kerbey et al., 1976). Subsequent studies showed that these differences were not explicable in terms of altered mitochondrial concentrations of known effectors of PDH kinase (Kerbey et al., 1977).

It was next shown that the PDH kinase activity in extracts of heart mitochondria from 48 h starved or diabetic rats is increased approximately three fold relative to fed normal controls. In these experiments the mitochondria were incubated under conditions (no substrate, uncoupler of oxidative phosphorylation) which effected complete conversion of PDH*b* to PDH*a* and which removed known activators of PDH kinase (PDH kinase was assayed by rates of ATP dependent inactivation and [^{32}P]-phosphorylation of PDH complex by [γ-^{32}P]ATP; Hutson and Randle, 1978). Similar observations have been made with extracts of mitochondria from rat mammary gland (Baxter and Coore, 1978), rat skeletal muscle (Fuller and Randle, 1984) and rat liver (Denyer et al., 1986). More recently, the effect of starvation to increase PDH kinase activity in extracts of liver mitochondria, has also been demonstrated with an assay based on [^{32}P]-phosphorylation of a synthetic peptide reproducing the amino acid sequence around phosphorylation sites 1 and 2 of PDHE$_1$ (Jones et al., 1992).

It was then shown that a lower M_r protein fraction could be separated from PDH complex in extracts of rat heart and liver mitochondria by high speed centrifugation or by gel filtration on Sephacryl S300 which enhanced the PDH kinase activity of purified pig heart PDH complex. The activity of this protein was enhanced 2–3fold by starvation of the rat (Kerbey and Randle, 1981; Kerbey et al., 1984). This activity was subsequently identified as a free PDH kinase by its ability to phosphorylate and inactivate pig heart PDHE$_1$ devoid of PDH kinase activity; to phosphorylate and inactivate E$_1$ in S. cerevisiae PDH complex (which lacks PDH kinase but is a substrate for it); by its susceptibility to inactivation by thiol reactive reagents, and by fluorosulphonylbenzoyl adenosine which is known to form adducts with the ATP binding site of protein kinases (Mistry et al., 1991); and by its ability to phosphorylate peptide substrates for PDH kinase (Jones and Yeaman, 1991). The starved-fed difference in activity was shown to be maintained with the assays utilising PDHE$_1$ and S. cerevisiae complex (Mistry et al., 1991). A soluble PDH kinase was purified to homogeneity from liver mitochondria of fed and 48 h starved rats. Homogeneity was confirmed by N-terminal amino acid sequencing which showed essentialy only one amino acid residue at each cycle. There were no differences in N-terminal sequences between kinase from fed and starved rats. The specific activity of the kinase purified from starved rats was some 4.5fold greater than that from fed rats (Priestman et al., 1992). Based on N-terminal sequence this kinase corresponds to the p45 PDH kinase (PDH kinase 2) of Popov et al. (1994).

The effect of starvation to increase PDH kinase and decrease percentage of PDHa requires 24–48 h to reach its maximum and complete reversal is effected by a similar period of refeeding (Kerbey and Randle, 1982). More detailed studies have shown that major decreases in percent PDHa occur at 12–18 h in skeletal muscles and at 4–8 h in liver and heart with major increases in percent PDHa on refeeding occurring between 4 and 8 h (Holness and Sugden, 1989; Jones et al., 1992). Evidence concerning factors that may increase PDH kinase activity in starved and diabetic rats has been sought by use of hepatocytes, cardiac myocytes, and soleus muscle strips in tissue culture (Fatania et al., 1986; Marchington et al., 1989, 1990; Stace et al., 1992). These studies showed uniformly that culture of cells from fed rats with agents that increase cAMP or reproduce its effects (glucagon in hepatocytes or dibutyryl cAMP more generally) or with fatty acids (n-octanoate or albumin bound palmitate) or with a combination increases PDH kinase activity 2–3-fold (as measured by the rate of ATP dependent inactivation of PDH complex). In hepatocytes the effect of glucagon was completely reversed by insulin (known to lower cAMP in the presence of glucagon). Insulin had no effect in the presence of dibutyryl cAMP, thus confirming that reversal of the glucagon effect was due to lowering of cAMP. Insulin partially decreased the effect of n-octanoate (by about 20%). The effect of starvation on PDH kinase activity was reversed by about 60% over 24 h of culture; this reversal was blocked by octanoate or dibutyryl cAMP or a combination. In hepatocytes no consistent effect of insulin on reversal of the effect of starvation was seen. It is clear from these studies the action of insulin in longer term regulation of PDH kinase activity *in vivo* is confined to interactive regulation of intracellular cAMP concentration and to regulation of FFA concentration through effects on lipolysis, esterification and possibly carnitine acyl transferase I activity *via* malonyl CoA and acetylCoA carboxylase.

Orfali et al. (1993, 1995) have made the additional observations that high fat diet and also hyperthyroidism increased PDH kinase activity in rat heart, that the effect of the latter was additive with that of starvation, that T3 increased PDH kinase activity in cardiac myocytes in culture, and that the effect of T3 was suppressed by insulin.

The molecular mechanisms involved in longer term regulation of PDH kinase activity are not yet known. Studies in tissue culture showed that the effect of dibutyryl cyclic AMP was blocked by the protein synthesis inhibitor cycloheximide, whereas the effect of FFA (palmitate) was not. The effect of palmitate was blocked by the carnitine acyl transferase I inhibitor Etomoxir whereas that of dibutyryl cAMP was not. By ELISA assays of PDH kinase 2 (p45, the major isoform in liver; Popov et al., 1994) evidence was obtained that the major effects of FFA and dibutyryl cAMP in culture, like those of starvation *in vivo*, were to increase the specific activity of p45 and not its concentration (Priestman et al., 1994). One possibility currently being investigated is that FFA act through acylation of PDH kinase or of some protein associated with its action and that cAMP may increase the concentration of some protein involved in this action.

Physiological and pathological significance of PDH kinase regulation

Glucose oxidation and the percentage of PDH complex in the *a* form in liver, muscles (heart and skeletal) and adipose tissue are decreased by insulin-deficient diabetes (alloxan or streptozotocin), by starvation, by high fat (low carbohydrate) diet, and by obesity in rat and mouse. In man there is clear evidence that glucose oxidation is decreased in uncontrolled Type 1 and Type 2 diabetes, in starvation, in obesity, on feeding high fat (low carbohydrate) diet, and when FFA oxidation is enhanced by triacylglycerol infusion. There is also clear evidence of insulin insensitivity in respect of glucose oxidation in Type 2 diabetes. There is also a small amount of evidence that

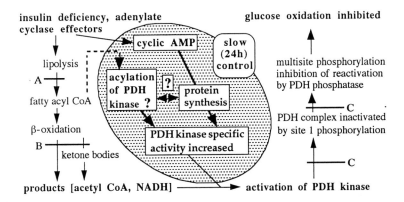

Figure 1. Short term and longer term mechanisms leading to phosphorylation and inactivation of the pyruvate dehydrogenase complex in starvation or insulin-deficient diabetes. (A) Inhibitors of lipolysis, e.g., acipimox, 5-methylpyrazole-3-carboxylic acid: (B) Inhibitors of carnitine acyltransferase, e.g., Etomoxir, 2-tetradecylglycidate: (C) inhibitors of pyruvate dehydrogenase kinase eg dichloroacetate.

percentage of PDHa is decreased in Type 2 diabetes. Use of inhibitors of FFA oxidation (eg tetradecylglycidate) have confirmed the importance of FFA oxidation in the decrease in percentage of PDHa induced by starvation or alloxan diabetes in animals. Use of such inhibitors in man has also shown that inhibition of FFA oxidation in Type 1 and Type 2 diabetes enhances glucose utilization. Use of the PDH kinase inhibitor dichloroacetate (McAllister et al., 1973; Whitehouse and Randle, 1973; Whitehouse et al., 1974) in human diabetics and in experimental animals has confirmed that inhibition of PDH kinase is associated with an increase in glucose disappearance and oxidation. Lack of space precludes detailed bibliography, but this may readily be accessed through recent reviews (Randle et al., 1994a, b) and the chapter in this volume by Sugden and Holness. A summary of our current views on the mechanism of inactivation of PDH complex by phosphorylation in starved or diabetic animals is shown in Figure 1. This incorporates shorter term mechanisms shown in the periphery of the figure and longer term mechanisms regulating PDH kinase in the central shaded area. The inhibition of PDH phosphatase by multisite phosphorylation is envisaged as an important hysteretic mechanism restraining reactivation over short periods of time. It is important to note that the bulk of phosphorylation of sites 2 and 3 in rat heart occurs over the range of 80–100% PDHb, i.e., over the range of the fed to starved or normal to diabetic transitions (Sale and Randle, 1982a, b).

PDH kinase, endocrine functions and glucose sensing

In prolonged carbohydrate deprivation important changes in endocrine function include the development of a decreased insulin secretory response to glucose, and an increase in circulating concentrations of pituitary growth hormone and ACTH. A decreased insulin secretory response to
glucose is also seen in Type 2 diabetes. In the pancreatic β-cell, hyperglycaemia induces insulin
secretion by an enhanced rate of glucose oxidation, the concentration dependence of which is
conferred by the kinetics of glucokinase, and which leads to an increase in the cytosolic
[ATP]/[ADP] ratio. This leads to closure of ATP sensitive K^+ channels with consequential open-
ing of voltage dependent Ca^{2+} channels and a rise in cytosolic and mitochondrial $[Ca^{2+}]$. The
increased mitochondrial Ca^{2+} may further increase the [ATP]/[ADP] ratio (see McCormack et al.,
1991a). The increased cytosolic $[Ca^{2+}]$ induces insulin release (for review see Ashcroft and
Ashcroft, 1992). It is to be expected therefore that PDH complex activity must be of importance
to this overall process. Two factors likely to mediate the increased percentage of PDHa in the islet
β-cell in response to hyperglycaemia (McCormack et al., 1990b) are an increase in pyruvate concentration (inhibitor of PDH kinase) as a result of increased glycolytic flux (the β-cell has low
lactate dehydrogenase activity and a very active glycerophosphate shuttle), and the increase in

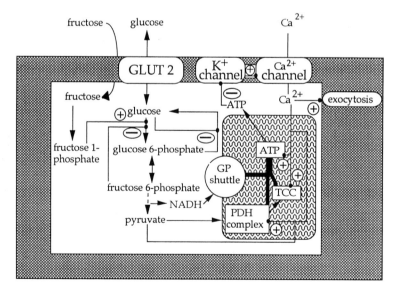

Figure 2. Pathways mediating glucose induced insulin secretion in the pancreatic islet β-cell; (+) Activation, (−)
inhibition, (GP) glycerophosphate and (TCC) tricarboxylate cycle.

[Ca^{2+}] which may activate PDH phosphatase and inhibit PDH kinase. This model for the role of PDH interconversion in glucose-induced insulin secretion is shown in Figure 2. Of particular interest is the recent demonstration, by *in vivo* and *in vitro* studies, that FFA over 24 h can inhibit glucose induced insulin secretion and increase β-cell PDH kinase activity Zhou and Grill, 1994a, b). It seems very likely, therefore, that activation of PDH kinase in the islet β-cell by longer term mechanisms involving FFA (and perhaps cAMP) might mediate inhibitory effects of diabetes and starvation on glucose induced insulin secretion. There is also the beginnings of evidence that glucoregulatory cells in the central nervous system may function through mechanisms similar to those in the islet β-cell. These cells may regulate growth hormone and ACTH secretion through their effects on the secretion of hypothalamic releasing hormones and it thus seems possible that a unifying hypothesis will emerge whereby longer term regulation of PDH kinase might coordinate glucose oxidation in major metabolic tissues, and glucoregulatory effects on the secretion of insulin, growth hormone and ACTH.

References

Ashcroft, F.M. and Ashcroft, S.J.H. (1992) Mechanism of insulin secretion. *In*: F.M. Ashcroft and S.J.H. Ashcroft (eds): *Insulin: Molecular Biology to Pathology*. Oxford University Press, Oxford, pp 97–150.

Ashour, B. and Hansford, R.G. (1983) Effect of fatty acids and ketones on the activity of pyruvate dehydrogenase in skeletal-muscle mitochondria *J. Biol. Chem*.214: 725–736.

Baxter, M.A. and Coore, H.G. (1978) The mode of regulation of pyruvate dehydrogenase of lactating mammary gland. *Biochem. J.* 174: 553–561.

Cate, R.L. and Roche, T.E. (1978) A Unifying mechanism for stimulation of mammalian pyruvate dehydrogenase kinase by reduced nicotinamide adenine dinucleotide, dihydrolipoamide, acetyl coenzyne a, or pyruvate. *J. Biol. Chem*. 253: 496–503.

Cate, R.L. and Roche, T.E. (1979) Function and regulation of mammalian pyruvate dehydrogenase complex. acetylation, interlipoyl acetyl transfer, and migration of the pyruvate dehydrogenase component. *J. Biol. Chem*. 254: 1659–1665.

Cooper, R.H., Randle, P.J. and Denton, R.M. (1974) Regulation of heart muscle pyruvate dehydrogenase kinase. *Biochem. J.* 143: 625–641.

Cooper, R.H., Randle, P.J. and Denton, R.M. (1975) Stimulation of phosphorylation and inactivation of pyruvate dehydrogenase by physiological inhibitors of the pyruvate dehydrogenase reaction. *Nature* 257: 808–809.

Davis, P.F., Pettit, F.H. and Reed, L.J. (1977) Peptides Derived from pyruvate dehydrogenase as substrates for pyruvate dehydrogenase kinase and phosphatase. *Biochem. Biophys. Res. Comm.* 5: 541–549.

Denton, R.M., Randle, P.J. and Martin, B.R. (1972) Stimulation by calcium ions of pyruvate dehydrogenase phosphate phosphatase. *Biochem. J.* 128: 161–163.

Denyer, G.S., Kerbey, A.L. and Randle, P.J. (1986) Kinase activator protein mediates longer term effects of starvation on activity of pyruvate dehydrogenase kinase in rat liver mitochondria. *Biochem. J.* 239: 347–354.

Fatania, H.R., Vary, T.C. and Randle, P.J. (1986) Modulation of pyruvate dehydrogenase kinase activity in cultured hepatocytes by glucagon and *n*-octanoate. *Biochem. J.* 234: 233–236.

Fuller, S.J. and Randle, P.J. (1984) Reversible phosphorylation of pyruvate dehydrogenase in rat skeletal-muscle mitochondria. *Biochem. J.* 219: 635–646.

Garland, P.B. and Randle, P.J. (1964) Control of pyruvate dehydrogenase in the perfused rat heart by the intracellular concentration of acetyl CoA. *Biochem. J.* 91: 6C–7C.

Hansford, R.G. (1976) Studies on the effects of coenzyme a-sh: acetyl coenzyme a, nicotinamide adenine dinucleotide: reduced nicotinamide adenine dinucleotide, and adenosine diphosphate: adenosine triphosphate ratios on the interconversion of active and inactive pyruvate dehydrogenase in isolated rat heart mitochondria. *J. Biol. Chem.* 251: 5483–5489.

Hansford, R.G. (1977) Studies on inactivation of pyruvate dehydrogenase by palmitoylcarnitine oxidation in iso-
    lated rat heart mitochondria. *J. Biol. Chem.* 252: 1552–1560.
Hansford, R.G. and Cohen, L. (1978) Relative importance of pyruvate dehydrogenase interconversion and feed-
    back inhibition in the effect of fatty acids on pyruvate oxidation by rat heart mitochondria. *Arch. Biochem.
    Biophys.* 191: 65–81.
Holness, M.J. and Sugden, M.C. (1989) Pyruvate dehydrogenase activities during the fed to starved transition and
    on refeeding after acute or prolonged starvation. *Biochem. J.* 258: 529–533.
Hucho, F., Randall, D.D., Roche, T.E., Burgett, M.W., Pelley, J.W. and Reed, L.J. (1972) Keto acid dehydrogenase
    complexes. XVII. Kinetic and Regulatory properties of pyruvate dehydrogenase kinase and pyruvate dehydroge-
    nase phosphatase from bovine kidney and heart. *Arch. Biochem. Biophys.* 151: 328–340.
Hughes, W.A. and Denton, R.M. (1976) Incorporation of $^{32}P_i$ into pyruvate dehydrogenase phosphate in mito-
    chondria from control and insulin-treated adipose tissue. *Nature* 264: 471–473.
Hutson, N.J. and Randle, P.J. (1978) Enhanced activity of pyruvate dehydrogenase kinase in rat heart mitochondria
    in alloxan-diabetes or starvation. *FEBS Lett.* 92: 73–76.
Jones, B.S. and Yeaman, S.J. (1991) Longer term regulation of pyruvate dehydrogenase complex. evidence that
    kinase activator protein (KAP) is free pyruvate dehydrogenase kinase. *Biochem. J.* 275: 781–783.
Jones, B.S., Yeaman, S.J., Sugden, M.C. and Holness, M.J. (1992) Hepatic pyruvate dehydrogenase kinase activi-
    ties during the starved to fed transition. *Biochim. Biophys. Acta* 1134: 164–168.
Kerbey, A.L., Randle, P.J., Cooper, R.H., Whitehouse, S., Pask, H.T. and Denton, R.M. (1976) Regulation of pyru-
    vate dehydrogenase in rat heart. mechanism of regulation of proportions of dephosphorylated and phosphorylated
    enzyme by oxidation of fatty acids and ketone bodies and of effects of diabetes; role of coenzyme A, acetyl coen-
    zyme A and reduced and oxidized nicotinamide adenine dinucleotide. *Biochem. J.* 154: 327–348.
Kerbey, A.L., Radcliffe, P.M. and Randle, P.J. (1977) Diabetes and the control of pyruvate dehydrogenase in rat
    heart mitochondria by concentration ratios of ATP/ADP, NADH/NAD+ and acetylCoA/CoA. *Biochem. J.* 164:
    509–519.
Kerbey, A.L., Radcliffe, P.M., Randle, P.J. and Sugden, P.H. (1979) Regulation of kinase reactions in pig heart
    pyruvate dehydrogenase Complex. *Biochem. J.* 181: 427–433.
Kerbey, A.L. and Randle, P.J. (1981) Thermolabile factor accelerates pyruvate dehydrogenase kinase reaction in
    heart mitochondria of starved or alloxan-diabetic rats. *FEBS Lett.* 127: 188–192.
Kerbey, A.L., Randle, P.J. and Kearns, A. (1981) Dephosphorylation of pig heart pyruvate dehydrogenase phos-
    phate complexes by pig heart pyruvate dehydrogenase phosphate phosphatase. *Biochem. J.* 195: 51–59.
Kerbey, A.L. and Randle, P.J. (1982) Pyruvate dehydrogenase kinase/activator in rat heart mitochondria. *Biochem.
    J.* 206: 103–111.
Kerbey, A.L., Richardson, L.J. and Randle, P.J. (1984) The roles of intrinsic kinase and of kinase/activator protein
    in the enhanced phosphorylation of pyruvate dehydrogenase complex in starvation. *FEBS Lett.* 176: 115–119.
Liu, S., Baker, J.C. and Roche, T.E. (1995) Binding of the pyruvate dehydrogenase kinase to recombinant
    constructs containing the inner lipoyl domain of the dihydrolipoyl acetyltransferase component. *J. Biol. Chem.*
    270: 793–800.
McAllister, A. Allison, S.P. and Randle, P.J. (1973) Effects of dichloroacetate on the metabolism of glucose, pyru-
    vate, acetate, 3-hydroxybutyrate and palmitate in rat diaphragm and heart muscle *in vitro* and on extraction of
    glucose, lactate, pyruvate and free fatty acids by dog heart *in vivo*. *Biochem. J.* 134: 1067–1081.
McCormack, J.G., Halestrap, A.P. and Denton, R.M. (1990a) Role of calcium ions in regulation of mammalian
    intramitochondrial metabolism. *Physiol. Rev.* 70, 391–425.
McCormack, J.G., Longo, E.A and Corkey, B. (1990b) Glucose-Induced Activation of Pyruvate Dehydrogenase in
    Islets. *Biochem. J.* 267: 527–530.
Marchington, D.R., Kerbey, A.L., Giardina, M.G., Jones, E.A. and Randle, P.J. (1989) Longer term regulation of
    pyruvate dehydrogenase kinase in cultured rat hepatocytes. *Biochem. J.* 257: 487–491.
Marchington, D.R., Kerbey, A.L. and Randle, P.J. (1990) Longer term regulation of pyruvate dehydrogenase in cul-
    tured rat cardiac myocytes. *Biochem. J.* 267, 245–247.
Mistry, S.C., Priestman, D.A., Kerbey, A.L. and Randle, P.J. (1991) Evidence that rat liver pyruvate dehydrogenase
    kinase activator protein is a pyruvate dehydrogenase kinase. *Biochem. J.* 275: 775–779.
Mullinex, T.R., Stepp, L.R., Brown, J.R. and Reed, L.J. (1985) Synthetic peptide substrate for mammalian pyru-
    vate dehydrogenase kinase and pyruvate dehydrogenase phosphatase *Arch. Biochem. Biophys.* 243: 655–659.
Orfali.K.A., Fryer, L.G.D., Holness, M.J. and Sugden, M.C. (1993) Long-term regulation of pyruvate dehydroge-
    nase kinase by high-fat feeding. *FEBS Lett.* 336: 501–505.
Orfali.K.A., Fryer, L.G.D., Holness, M.J. and Sugden, M.C. (1995) Interactive effects of insulin and triiodothyro-
    nine on pyruvate dehydrogenase kinase activity in cardiac myocytes. *J. Mol. Cardiol.* 27: 901–908.
Pettit, F.H., Humphreys, J. and Reed, L.J. (1982) Regulation of pyruvate dehydrogenase kinase activity by protein
    thiol-disulfide exchange. *Proc. Natl. Acad. Sci. USA* 79: 945–3948.
Popov, K.M., Kedishvili, N.Y., Zhao, Y., Shimomura, Y., Crabb D.W., Harris, R.A. (1993) Primary structure of
    pyruvate dehydrogenase kinase establishes a new family of eukaryotic protein kinases. *J. Biol. Chem.* 268:
    26602–26606.

Popov, K.M., Kedishvili, N.Y., Zhao, Y., Gudi, R. and Harris, R.A. (1994) Molecular cloning of the p45 subunit of pyruvate dehydrogenase kinase. *J. Biol. Chem.* 269: 29720–29724.

Pratt, M.L. and Roche, T.E. (1979) Mechanism of pyruvate inhibition of kidney pyruvate dehydrogenase, kinase and synergistic inhibition by pyruvate and ADP. *J. Biol. Chem.* 254: 7191–7195.

Priestman, D.A., Mistry, S.C., Kerbey, A.L. and Randle, P.J. (1992) Purification and partial characterization of rat liver pyruvate dehydrogenase kinase activator protein (free pyruvate dehydrogenase kinase). *FEBS Lett.* 308: 83–86.

Priestman, D.A., Mistry, S.C., Halsall, A. and Randle, P.J. (1994) Role of protein synthesis and of fatty acid metabolism in the longer-term regulation of pyruvate dehydrogenase kinase. *Biochem. J.* 300: 659–664.

Rahmatullah, M., Roche, T.E., Jilka, J.M. and Kazemi, M. (1985) Mechanism of activation of bovine kidney pyruvate dehydrogenase kinase by malonyl-CoA and enzyme-catalyzed decarboxylation of malonyl-CoA. *Eur. J. Biochem.* 150: 181–187.

Randle, P.J, Priestman, D.A., Mistry, S.C. and Halsall, A. (1994a) Glucose fatty acid interactions and the regulation of glucose disposal *J. Cell Biochem.* 55S: 1–11.

Randle, P.J, Priestman, D.A., Mistry, S.C. and Halsall, A. (1994b) Mechanisms modifying glucose oxidation in diabetes mellitus. *Diabetologia* 37 (Suppl. 2): S155-S161.

Reed, L.J. (1981) Regulation of mammalian pyruvate dehydrogenase complex by a phosphorylation-dephosphorylation cycle. *Curr. Top. Cell. Regul.* 18: 95–106.

Robertson, J.G., Barron, L.L. and Olson, M.S. (1990) Bovine heart pyruvate dehydrogenase kinase stimulation by alpha-ketoisovalerate. *J. Biol. Chem.* 265: 16814–16820.

Roche, T.E. and Cate, R.L. (1976) Evidence for lipoic acid mediated NADH and acetyl-CoA stimulation of liver and kidney pyruvate dehydrogenase kinase. *Biochem. Biophys. Res. Comm.* 72: 1375–1383.

Sale, G.J. and Randle, P.J. (1981) Occupancy of sites of phosphorylation in inactive rat heart pyruvate dehydrogenase phosphate *in vivo. Biochem. J.* 193: 935–946.

Sale, G.J. and Randle, P.J. (1981) Analysis of site occupancies in [^{32}P]phosphorylated pyruvate dehydrogenase complexes by aspartyl-prolyl cleavage of tryptic phosphopeptides. *Eur. J. Biochem.* 120: 535–540.

Sale, G.J. and Randle.P.J. (1982a) Role of individual phosphorylation sites in inactivation of pyruvate dehydrogenase complex in rat heart mitochondria. *Biochem. J.* 203: 99–108.

Sale, G.J. and Randle, P.J. (1982b) Occupancy of phosphorylation sites in pyruvate dehydrogenase phosphate complex in rat heart *in vivo. Biochem. J.* 206: 221–229.

Stace, P.B., Fatania, H.R., Jackson, A., Kerbey, A.L and Randle, P.J. (1992) Cyclic AMP and free fatty acids in the longer-term regulation of pyruvate dehydrogenase kinase in rat soleus muscle. *Biochim. Biophys. Acta* 1135: 201–206.

Stepp, L.R., Pettit, F.H., Yeaman, S.J. and Reed, L.J. (1983) Purification and properties of pyruvate dehydrogenasekinase from bovine kidney. *J. Biol. Chem.* 258: 9454–9458.

Sugden, P.H. and Randle, P.J. (1978) Regulation of pig heart pyruvate dehydrogenase by phosphorylation: studies on the subunit and phosphorylation stoicheiometries. *Biochem. J.* 173: 659–668.

Sugden, P.H., Hutson, N.J., Kerbey, A.L. and Randle, P.J. (1978) Phosphorylation of additional sites on pyruvate dehydrogenase inhibits its reactivation by pyruvate dehydrogenase phosphate phosphatase. *Biochem. J.* 169: 433–435.

Teague, W.M., Pettit, F.H., Yeaman, S.J. and Reed, L.J. (1979) Function of phosphorylation sites on pyruvate dehydrogenase. *Biochem. Biophys. Res. Comm.* 87: 244–252.

Tonks, N.K., Kearns, A. and Randle, P.J. (1982) Pig heart [35S]thiophosphoryl pyruvate dehydrogenase complexes. *Eur. J. Biochem.* 122: 549–551.

Walsh, D.A., Cooper, R.H., Denton, R.M., Bridges, B.J. and Randle, P.J. (1976) The elementary reactions of the pig heart pyruvate dehydrogenase complex; a study of the inhibition by phosphorylation. *Biochem. J.* 137: 41–67.

Whitehouse, S. and Randle, P.J. (1973) Activation of pyruvate dehydrogenase in perfused rat heart by dichloroacetate. *Biochem. J.* 134: 651–653.

Whitehouse, S., Cooper, R.H. and Randle, P.J. (1974) Mechanism of activation of pyruvate dehydrogenase by dichloroacetate and other halogenated carboxylic acids. *Biochem. J.* 141: 761–77.

Zhou, Y.P. and Grill, V. (1994a) Evidence that fatty acids inhibit B-cell function through activation of PDH kinase. *Diabetologia* 37 (Suppl. 1): A11.

Zhou, Y.P. and Grill, V. (1994b) Long term exposure of rat pancreatic islets inhibits glucose-induced insulin secretion and biosynthesis through a glucose fatty acid cycle. *J. Clin. Invest.* 93: 870–876.

Alpha-Keto Acid Dehydrogenase Complexes
M.S. Patel, T.E. Roche and R.A. Harris (eds)
© 1996 Birkhäuser Verlag Basel/Switzerland

Hormonal and nutritional modulation of PDHC activity status

M.C. Sugden and M.J. Holness

Department of Biochemistry (Basic Medical Sciences), Queen Mary and Westfield College, Mile End Road, London E1 4NS, UK

Introduction

In animals, glucose oxidation is decreased by starvation and diabetes (reviewed by Randle, 1986; Patel and Roche, 1990; Sugden and Holness, 1994). As well as short-term effects of starvation and diabetes to suppress glucose oxidation, long-term regulatory mechanisms impose an upper limit on the overall capacity for glucose oxidation and introduce latency of reactivation of glucose oxidation when a high-carbohydrate, low-fat meal is provided, or when insulin is administered after prolonged starvation. These characteristics of whole-body glucose oxidation reflect the activity and regulatory characteristics of the mitochondrial pyruvate dehydrogenase holocomplex (PDHC), which catalyses the oxidative decarboxylation of pyruvate to acetyl CoA. This chapter will review the regulation of PDHC by hormones and nutrients within the context of the competition between fatty acids (FA) and glucose as oxidative substrates.

Regulatory characteristics of PDHC

PDHC (E.C. 1.2.4.1 + E.C. 1.6.4.3 + E.C. 2.3.1.2) contains three component enzymes, namely pyruvate dehydrogenase (E1 or PDH), a heterodimer of subunit composition $\alpha_2\beta_2$, dihydrolipoamide acetyltransferase (E2) and dihydrolipoamide dehydrogenase (E3) (Yeaman, 1989). PDHC activity is regulated by end-product inhibition by NADH and acetyl CoA and by reversible phosphorylation (Randle, 1986; Yeaman, 1989; Patel and Roche, 1990; Sugden and Holness, 1994). End-product inhibition is a short-term regulatory mechanism leading to inhibition of PDH flux; however, the percentage of active PDHC (% PDH_a), which reflects the phosphorylation state of E1, determines the capacity for PDH flux. Changes in % PDH_a are achieved by changes in the relative activities of PDH kinase (phosphorylating, inactivating) (E.C. 2.7.1.99) and PDHP phosphatase (dephosphorylating, activating) (E.C. 3.1.3.43).

PDH kinase is activated by increasing mitochondrial ratios of [ATP]:[ADP], [acetyl CoA]:[CoA] and [NADH]:[NAD$^+$] (Kerbey et al., 1976). The inactivation of PDHC by phosphorylation is important for glucose conservation in states of carbohydrate deprivation, including starvation or the administration of a high-fat/low carbohydrate diet (reviewed Randle et al., 1988; Sugden et al., 1995). Conversely, PDH kinase is inhibited by ADP, pyruvate and TPP (Patel and Roche, 1990). An increased supply of pyruvate secondary to accelerated glycolysis (e.g. in response to insulin) can promote activation of PDHC by opposing its inactivation by PDH kinase (reviewed Sugden and Holness, 1994). PDHP phosphatase is activated by increased mitochondrial Ca^{2+}, a mode of regulation which may be important in exercising muscle and in tissues responding to hormones that can increase mitochondrial Ca^{2+} concentrations (e.g. vasopressin in liver) (Denton and McCormack, 1985; Denton and McCormack, 1990). A further mechanism of activation of PDHP phosphatase appears to operate in lipogenic tissues, adipocytes and possibly liver, namely activation by insulin (reviewed Yeaman, 1989; Sugden and Holness, 1994). The mechanism by which insulin influences PDHP phosphatase activity is thought to be independent of FA oxidation or changes in mitochondrial Ca^{2+}, but is instead achieved *via* a decrease in the apparent Km of PDHP phosphatase for Mg^{2+} (see Thomas et al., 1986).

Pyruvate oxidation in the absorptive state

The conversion of pyruvate to acetyl CoA *via* the PDHC reaction commits pyruvate carbon to oxidation and permits the use of glucose-related fuels as a precursor for lipid synthesis (Sugden and Holness, 1994). In muscle, PDHC has a predominantly bioenergetic function; however, individual muscles differ with respect to their % PDH$_a$ (Caterson et al., 1982; Holness et al., 1989; Stace et al., 1990). These differences reside in differences between individual muscles in terms of their contractile activities and abilities to utilize the alternative lipid-derived fuels as energy substrates. Thus, in the sedentary state, the % PDH$_a$ is greater in heart (whose rhythmic contraction imposes a continual high energy demand) than in slow-twitch oxidative skeletal muscles whereas, despite a lower energy demand, fast-twitch muscles contain relatively high % PDH$_a$, presumably reflecting their reliance on glucose rather than lipid-derived fuels as energy source (Tab. 1). To facilitate high rates of FA synthesis, liver and adipose tissue increase the percentage of PDHC present in the active form from 10–35% PDH$_a$ (Tab. 1) to up to 45–60%. For example, there is a striking correspondence between high PDH$_a$ activities and rapid rates of lipogenesis during in the livers of "meal-fed" rats provided with a single large meal per day (Sugden et al., 1993a). Similarly, brown adipose tissue, which on a fresh weight basis exhibits one of the highest rates of

Table 1. Percentage of active PDHC in tissues of the fed rat

Tissue		PDH$_a$ activity (% of total)	
		Control	Dichloroacetate-treated
Heart		20 – 50	53
Slow – twitch muscles	Soleus	5 – 10	62
	Adductor longus	15 – 25	90
Fast – twitch muscles	Tibialis anterior	20 – 30	77
	Extensor digitorum longus	20 – 30	91
Liver		10 – 20	100
Brown adipose tissue		10 – 35	86

PDH$_a$ activity data are representative for rats with unrestricted access to standard (high carbohydrate/low fat) rodent diet. PDH$_a$ activities *in vivo* after inhibition of PDH kinase with dichloroacetate (Holness et al., 1989; Holness and Sugden, 1990b) are shown for comparison.

FA synthesis de novo of the tissues of the body during insulin stimulation (Holness and Sugden, 1990b), again shows a close correlation between PDH$_a$ activity and lipogenic rate in a variety of situations (Holness and Sugden, 1990b; Sugden and Holness, 1993a). Liver PDHC may have an additional pivotal role in determining the overall direction of hepatic carbon flux (Sugden et al., 1989). For example, the striking temporal relationship between a decline in the rate of hepatic glycogen synthesis and hepatic PDH$_a$ reactivation during refeeding after starvation (Holness et al., 1988) suggests that pyruvate oxidation is of major importance for hepatic carbohydrate disposal when hepatic glycogen concentrations are high. Support for the related concept that suppression of hepatic glucose production (HGP) by insulin in the absence of glycogen storage may be permitted by loss of carbon *via* PDHC is provided from studies with the glycogen-storage-disease (*gsd/gsd*) rat (Holness et al., 1987). In the *gsd/gsd* rat, a constraint on hepatic glycogen synthesis is imposed by the retention of high glycogen concentrations. Starved *gsd/gsd* rats are characterised by rapid reactivation of hepatic PDH in response to glucose loading which is absent in GSD/GSD controls (Holness et al., 1987). Despite a failure to deposit glycogen after a glucose load, there is normal suppression of HGP and the normal positive relationship between net hepatic glucose uptake and portal glycemia (Holness et al., 1987).

Pyruvate oxidation after short-term food deprivation

Although PDHC is traditionally viewed as a conduit for the oxidation of glucose, at least two vital tissues – heart and liver – are avid users of circulating lactate as an oxidative substrate (reviewed Sugden et al., 1993b; Sugden and Holness, 1994)). Both of these tissues respond to food withdrawal with a rapid decline in PDH_a activities (Holness and Sugden, 1989), a response which may be important for sparing C3 derivatives of glucose for gluconeogenesis (and maintenance of glycemia) during the early phase of starvation until adipose-tissue triacylglycerol (TAG) mobilization has been established. The time-courses of PDH_a inactivation in heart and liver thus closely parallel a decline in plasma insulin (Holness and Sugden, 1989; Holness et al., 1991), initiation of hepatic glycogenolysis (Holness and Sugden, 1989; Sugden et al., 1992) and a switch in hepatic carbon flux towards glucose production (Sugden et al., 1992).

In heart, PDH_a inactivation over the first 6 h after food withdrawal can be reversed by inhibition of mitochondrial FA oxidation (Holness and Sugden, 1989; Sugden and Holness, 1993b). The cardiac response to food withdrawal is thus consistent with the idea that cardiac PDHC inactivation is secondary to increased rates of FA oxidation elicited by the fall in insulin. Hepatic PDH_a

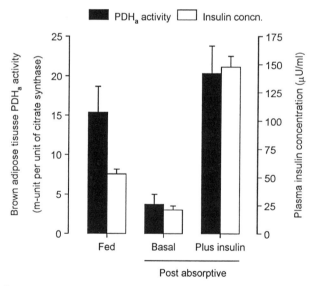

Figure 1. The insulin sensitivity of brown adipose tissue PDH_a activity. PDH_a activity was assayed in homogenates of freeze-clamped tissue obtained from rats sampled in the fed (absorptive) state and in 6 h starved (postabsorptive) rats sampled in the basal state or at the end of a 2 h euglycemic-hyperinsuliemic clamp. PDH_a activities were assayed as described by Caterson et al. (1982). Euglycemic-hyperinsulinemic clamps were performed in unstressed, conscious rats. Human Actrapid insulin (Novo Nordisk, Denmark) was infused at a constant rate of 4.2 $mU \cdot kg^{-1} \cdot min^{-1}$. Blood glucose was clamped at 4 mM by the variable infusion of glucose. Plasma insulin concentrations were determined using a double antibody radioimmunoassay.

inactivation is not reversed by inhibition of mitochondrial FA oxidation (Holness and Sugden, 1989), suggesting that it is not secondary to activation of PDH kinase through increased mitochondrial acetyl CoA/CoA or NADH/NAD$^+$ concentration ratios. This suggests an alternative involvement of a fall in insulin in the response, either direct (through reduced activation of hepatic PDHP phosphatase) or indirect (through diminished rates of pyruvate production and de-suppression of hepatic PDH kinase activity). The early decline in hepatic PDH$_a$ activity is parallelled by suppression of hepatic FA synthesis (Holness and Sugden, 1989; Holness and Sugden, 1990b). A concomitant decline in PDH$_a$ activity and lipogenesis in response to short-term starvation is also observed in brown adipose tissue (Holness and Sugden, 1990b).

We investigated the insulin sensitivity of PDH$_a$ activity in brown fat directly using the euglycemic-hyperinsulinemic clamp technique (Fig. 1). At basal insulin levels (i.e., in the post absorptive state), brown-fat PDH$_a$ activities were significantly lower than in the fed state. Euglycemic hyperinsulinemia (2 h) at an insulin concentration in the physiological range (100 µU/ml) led to a significant increase in PDH$_a$ activities. At the end of the clamp, brown-fat PDH$_a$ activities did not differ significantly from those found in the fed state. The results thus provide strong support for the concept that PDH$_a$ activity in brown fat is directly responsive to the insulin status. Interestingly, euglycemic hyperinsulinemia does not result in brown-fat PDH$_a$ activities that greatly exceed those found in the absorptive state (Fig. 1) c.f. insulin injection without normalization of glycemia (Holness and Sugden, 1990). Furthermore, the response to euglycemic hyperinsulinemia contrasts with that observed *in vivo* in response to the administration of dichloroacetate, an inhibitor of PDH kinase. Exposure to dichloroacetate for 2 h *in vivo* results in almost complete activation of PDHC in brown adipose tissue (Holness and Sugden, 1990b). Overall, the results support the idea that PDHP phosphatase is a target for acute insulin action in this tissue.

The role of PDH inactivation in glucose recycling during more prolonged starvation

During starvation, carbons are transferred from large peripheral storage pools in adipose tissue and skeletal muscle to the liver in order to provide energy and substrates for gluconeogenesis. In man, approximately 60% of the plasma lactate pool is derived from glycolysis of muscle glycogen carbons (Consoli et al., 1990; Virkamaki et al., 1990) and in normal human subjects a decrease in the rate of adipose-tissue lipolysis during fasting regulates HGP by its effect on the availability of glycerol for gluconeogenesis (Jahoor et al., 1992). The mobilization of adipose-tissue TAG and muscle glycogen is inhibited by insulin. The plasma insulin concentrations required for half-maximal suppression of muscle glycogenolysis and adipose-tissue lipolysis are in the low-physiological range (20–50 µU/ml) (Rossetti and Hu, 1993). This is precisely the range over which plasma insulin concentrations vary over the first few hours after food withdrawal: typically, plasma insulin concentrations in the immediately post-absorptive state are around 20–30 µU/ml (Holness et al., 1991).

FA are oxidized in preference to glucose by skeletal muscle *in vivo* (Holness and Sugden, 1989; Holness et al., 1989; Holness and Sugden, 1990a; Sugden and Holness, 1994). The most significant suppression of skeletal-muscle PDH_a activities is observed only after the establishment of high rates of adipose-tissue lipolysis and hepatic ketone body production at 12–15 h after food withdrawal (Holness et al., 1989). Suppression of pyruvate oxidation by skeletal muscle associated with food deprivation is accelerated by an acute elevation in non-esterified FA (NEFA) concentrations (achieved by administration of corn oil + heparin) over the starvation period during which insulin concentrations are declining (Holness et al., 1989; Holness and Sugden, 1990a). This finding suggests that carbohydrate/insulin deprivation (as opposed to calorie deprivation) is important for this response to starvation, and that the mechanism involves increased fat oxidation. In support of the latter suggestion, reactivation of skeletal-muscle pyruvate oxidation after starvation in response to insulin during euglycemic-hyperinsulinemic clamp parallels the suppression of FA supply by insulin (Cooney et al., 1993) and the starvation-induced decline in skeletal-muscle PDH_a activity can be prevented by inhibition of lipolysis (Holness et al., 1989). Taken together, these findings suggest that insulin has an indirect effect on reversible phosphorylation of PDHC in skeletal muscle through lowering plasma FA concentrations and thereby reversing the increased mitochondrial [acetyl CoA]:[CoA] ratio associated with an increased rate of FA β-oxidation.

Inhibition of pyruvate oxidation by skeletal muscle precedes suppression of glucose uptake and phosphorylation by several hours (Holness and Sugden, 1990a). This ensures that pyruvate derived either from muscle glycogen (e.g. in fast-twitch muscles) or from blood glucose and muscle glycogen (slow-twitch muscles and heart) is spared from oxidation, and thereby made available for use as gluconeogenic precursor. Our laboratory has demonstrated that pyruvate sparing resides in greater susceptibility of pyruvate oxidation than glucose uptake/phosphorylation to inhibition by lipid-fuel oxidation (Holness et al., 1989; Holness and Sugden, 1990a). The concept that the loss of normal regulation of adipose tissue plays a central role in both the hyperglycemia and the dyslipidemia that characterizes patients with non-insulin dependent diabetes has recently been reviewed (Reaven, 1995). In relation to the pathogenesis of hyperglycemia in non-insulin dependent diabetes, it is reasonable to suggest that endocrine changes that increase the supply of lipid-derived fuels to muscle (e.g by stimulation of adipose-tissue lipolysis or of hepatic VLDL and ketone body production) would evoke disproportionate suppression of muscle pyruvate oxidation over glucose uptake/phosphorylation, thereby promoting carbon transfer from muscle to liver and supplying additional precursor for HGP.

Regulatory effects of FA oxidation on insulin action

There is convincing evidence that NEFA, acting through short-term mechanisms, inhibit the effects of insulin to suppress HGP and stimulate peripheral glucose disposal (storage and oxidation) in man. Recent work has examined in detail the dose-dependency of the FA effects on HGP in healthy volunteers during euglycemic hyperinsulinemia (plasma insulin of approximately 70 μU/ml) (Boden et al., 1994). Increasing the NEFA concentration from 50 μM to 0.75 mM by infusion of TAG + heparin rapidly opposed suppression of HGP by insulin. The effect of NEFA was maximal at post-absorptive concentrations (approximately 0.5 mM). The effects of elevated FA to inhibit insulin-stimulated carbohydrate oxidation also appeared early and seemed greatest at postabsorptive FA concentrations (Boden et al., 1994). In agreement with our studies demonstrating more rapid suppression of muscle pyruvate oxidation than muscle glucose uptake/phosphorylation by FA during progressive starvation (Holness et al., 1989; Holness and Sugden, 1990a), the FA-mediated inhibition of insulin-stimulated glucose uptake was less acute, developing only after 3–4 h of fat infusion (Boden et al., 1994). Increasing the NEFA concentration from 50 μM to 0.75 mM suppressed stimulation of peripheral glucose uptake by insulin by 55%. Approximately 50% of the decrease in Rd occurred when FA concentrations rose from 50 μM to 0.5 mM, and was accounted for equally by decreases in carbohydrate oxidation and glycogen synthesis (Boden et al., 1994). This suggests a defect in muscle glucose utilization early in the metabolic sequences for disposal, most likely at the level of glucose uptake and/or glucose phosphorylation. The further decline in insulin-stimulated Rd observed when FA concentrations were increased to 0.75 mM was accounted for solely by decreased glycogen synthesis (Boden et al., 1994).

Stable increases in PDH kinase specific activity are observed in response to prolonged (24–48 h) starvation in oxidative tissues, including heart, liver and skeletal-muscle (reviewed by Randle et al., 1994; Sugden et al., 1995). Studies *in vitro* examining the regulation of PDH kinase are consistent with the concept that the response to starvation is secondary to enhanced FA oxidation, possibly in conjunction with a rise in cAMP. There is evidence that insulin can oppose the effects of FA and cAMP to increase PDH kinase in cultured cardiac myocytes prepared from fed rats (Orfali et al., 1995; Sugden et al., 1995). Similarly, the addition of insulin to cultured hepatocytes from starved rats cultured in the absence of FA reversed the stable effects of prolonged starvation to increase hepatocyte PDH kinase activity within 4 h (Marchington et al., 1987). The results suggest that insulin deficiency or resistance to its action may be obligatory for any stable enhancement of PDH kinase activity to be observed in response to a subsequent rise in FA and/or cAMP. However, it has yet to be established whether insulin resistance alone can be a primary event triggering stable increases in tissue PDH kinase activities.

PDH kinase in the pathology of prolonged starvation

Three seryl residues have been identified as phosphorylation sites on E1. Phosphorylation of E1 at sites in addition to the inactivating site (site 1) retards the reactivation of inactive PDHC through the dephosphorylation of site 1 by PDHP phosphatase (Sugden et al., 1978). It is thus possible to modulate the rate of PDHP dephosphorylation by the sustained increase in PDH kinase/PDHP phosphatase activity ratio resulting from starvation-induced increases in PDH kinase specific activity.

The impact of the time-course of reversal of starvation-induced increases in PDH kinase activi-

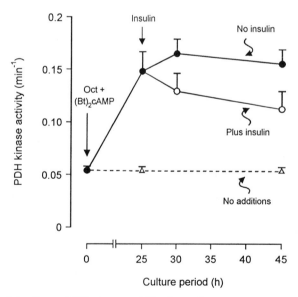

Figure 2. Effects of insulin on PDH kinase activities in cardiac myocytes cultured with n-octanoate and (Bt)$_2$cAMP. Calcium-tolerant ventricular myocytes were isolated by collagenase digestion of hearts from adult female rats as described by Orfali et al. (1993). Cardiac myocytes were cultured at a density of 10^5 cells·ml^{-1} in medium 199 containing 4% (v/v) fetal calf serum, antibiotics and antimycotics at 37°C for up to 45 h. PDH kinase activities were measured in cardiac myocyte extracts as described by Marchington et al. (1989) and computed as the apparent first-order rate constant for ATP-dependent inactivation of active PDH.

ties on the restoration of tissue PDH$_a$ activities during re-feeding after starvation has been examined directly *in vivo* (Kerbey and Randle, 1982; Jones et al., 1992). Suppression of PDH kinase activity to "fed" levels is preceded by a rise in insulin and fall in NEFA concentrations, and occurs before restoration of PDH$_a$ activity (Kerbey and Randle, 1982; Jones et al., 1992). However, although studies with hepatocytes cultured in the absence of (Bt)$_2$cAMP and FA demonstrate reversal of the effects of starvation on PDH kinase by insulin (Marchington et al., 1987), insulin (when present throughout culture) fails to inhibit the effect of (Bt)$_2$cAMP, and only modestly opposes the effect of FA to increase PDH kinase (Marchington et al., 1989). We investigated the possibility that cAMP and/or FA might be dominant effectors using the cultured cardiac myocyte system (Fig. 2). The significant loss of cardiac myocyte PDH kinase activity that occurs when cardiac myocytes from starved rats are cultured in the absence of additions precluded an investigation of whether insulin can reverse the stable effects of prolonged starvation to increase cardiac PDH kinase activity. We therefore examined whether insulin in culture could reverse prior effects of FA and (Bt)$_2$cAMP to increase PDH kinase activity in cultured myocytes from fed rats. The addition of insulin (100 μU/ml) to cardiac myocytes in which PDH kinase activity had already been increased by culture with *n*-octanoate (1 mM) and (Bt)$_2$cAMP (50 μM) for 25 h led to an approximately 18% decline in PDH kinase activity within 5 h; however, the inclusion of insulin for a further 15 h evoked only a modest further decline in PDH kinase activity (Fig. 2). As a consequence, PDH kinase activities in cells pre-cultured with *n*-octanoate and (Bt)$_2$cAMP were only approximately 30% lower than in cells cultured with *n*-octanoate and (Bt)$_2$cAMP alone. Perhaps, more importantly, PDH kinase activities in cell cultured in the presence of insulin, *n*-octanoate and (Bt)$_2$cAMP remained approximately two fold higher than in cells cultured in the absence of additions at the end of the last 20 h of culture (Fig. 2). The clear implication is that any effects of insulin to promote cardiac glucose oxidation will be suppressed if elevated rates of FA oxidation or intracellular cAMP concentrations are sustained. It is thus relevant that *in vivo* in liver, the most dramatic decline in PDH kinase activity (Jones et al., 1992) is coincident with reversal of the effects of starvation to partition FA (which may be derived from plasma NEFA or intracellular TAG) towards oxidation, away from glycerolipid synthesis (see Moir and Zammit, 1992; Moir and Zammit, 1993)). Thus, as suggested from the studies with cardiac myocytes, it appears that a sustained increase in fat oxidation can counteract the effects of a rise in insulin to promote carbohydrate oxidation through its long-term effects to suppress PDH kinase.

PDH kinase in the pathology of insulin resistance

Since glucose modulates the release both of insulin (see Sugden and Holness, 1994) and of lipolytic hormones such as glucagon and the catecholamines (Jensen et al., 1987; Fanelli et al., 1992), mobilization of adipose-tissue TAG is generally geared to the whole-body requirement for glucose conservation. Consequently high NEFA concentrations are not concurrent with high insulin concentrations in healthy individuals eating a balanced diet. In contrast, high insulin concentrations in conjunction with elevated NEFA concentrations are often found in insulin-resistant subjects (reviewed Reaven, 1995).

It has been demonstrated that high-fat feeding leads to impaired whole-body insulin action, assessed as a decreased requirement for glucose infusion to maintain euglycaemia during insulin infusion (euglycemic-hyperinsulinemic clamp) (Storlien et al., 1986; Kraegen et al., 1991). Insulin resistance in terms of glucose metabolism was localised to liver, muscle and brown adipose tissue (Storlien et al., 1986; Kraegen et al., 1991). The long-term administration of a low-carbohydrate diet high in FA also leads to suppression of tissue PDH_a activity (reviewed Randle et al., 1988). Using a high-fat diet containing sufficient carbohydrate (33% of energy intake) to ensure the maintenance of relatively high plasma insulin concentrations, we confirmed that the prolonged (28 day) administration of a fat-enriched diet (47% of energy intake) leads to decreased insulin action and peripheral insulin resistance with respect to glucose uptake and phosphorylation by the muscle mass (unpublished observations). We then examined the association between whole-body insulin resistance and pyruvate oxidation. The administration of the high-fat diet for 28 days elicited significant suppression of PDH_a activities in heart (Orfali et al., 1993; Sugden et al., 1995), liver (Sugden et al., 1995), brown adipose tissue (Holness, 1995) and skeletal muscle (unpublished observations). Plasma insulin concentrations remained relatively high (Orfali et al., 1993). The strong implication is that a modest increase in the provision of dietary fat (at the expense of carbohydrate) leads to insulin resistance with respect to glucose oxidation, as well as with respect to glucose uptake and phosphorylation.

We extended these studies to examine whether high-fat feeding had any long-term stable effects on tissue PDH kinase activities measured in extracts of mitochondria prepared from heart, liver and oxidative skeletal-muscle from rats provided with high-fat diet for 28 days (Sugden and Holness, 1995). In all three tissues, PDH kinase activities in extracts of mitochondria from fat-fed rats were significantly increased relative to controls. The increases were comparable in magnitude (approximately two fold) to those observed after prolonged starvation *in vivo* or in response to FA plus $(Bt)_2cAMP$ *in vitro*. We approached the mechanisms underlying the stable enhancement of tissue PDH kinase activity in two ways. First, based on reports that the replacement of a small proportion of dietary fat with long-chain omega-3 FAs prevents the development of insulin resistance in liver and skeletal muscle (Storlien et al., 1987), we replaced a small percentage of the dietary fat (7%) with long-chain omega-3 FAs from marine oil while holding the total fat content of the diet unchanged. It was found that this dietary manipulation completely prevented the increases in cardiac, hepatic and muscle PDH kinase activities caused by 28 days of high-fat feeding (Fig. 3). These results clearly demonstrate that the precise FA composition of the diet, rather than the absolute fat content of the diet, is a vital aspect of the response of tissue PDH kinase activity to high-fat feeding. In addition, since the high-fat diets contained identical amounts of carbohydrate, the results demonstrate that the response of PDH kinase to high fat feeding is not due to carbohydrate deprivation. Secondly, we investigated the molecular signals which might underlie the long-term changes in PDH kinase activity invoked by high-fat feeding using the cultured cardiac myocyte system. The pattern of response to culture alone, and to $(Bt)_2cAMP$ and FA in culture, of PDH kinase activities in cardiomyocytes prepared from high-fat fed rats was reminiscent of that seen with cardiomyocytes prepared from starved rats (Orfali et al., 1993). In particular, the inclusion of FA alone in culture was unable to elicit any stable increase in PDH kinase (c.f. cardiac myocytes from carbohydrate-fed rats). Taken together, these results support

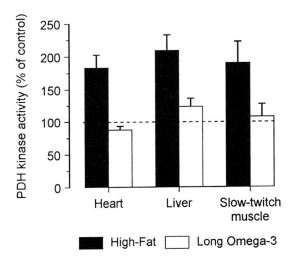

Figure 3. Effects of replacement of 7% of the fat component of high-fat diets with long omega-3 fatty acids on tissue PDH kinase activities. Experimental rats were maintained for 28 days on either high-fat diet or high-fat diet supplemented with long omega-3 fatty acids. PDH kinase activities were measured as described by Kerbey and Randle (1982) in mitochondria prepared from heart, liver or slow-twitch muscles as indicated. PDH kinase activities are expressed as a percentage of corresponding values obtained in rats fed standard (low-fat/high-carbohydrate) diet. PDH kinase activities were computed as the apparent first-order rate constant for ATP-dependent inactivation of active PDH.

the concept that insulin deficiency (starvation) and insulin resistance (high-fat feeding) share common signalling mechanisms, one of which is related to an increased supply of FA. Since PDH kinase activities are increased in response to high-fat feeding *in vivo* despite high insulin concentrations, it is implied that *in vivo* a diet high in fat opposes the action of insulin to suppress PDH kinase. Dominant factors in addition to FA supply may include an altered hormone profile facilitating the partitioning of intracellular fatty acyl CoA towards oxidation.

Concluding remarks

This chapter has reviewed the regulation of PDHC by hormones and nutrients within the context of the competition between FA and glucose as oxidative substrates. The tissue specificity of the mechanisms that underly the inactivation of PDHC by phosphorylation in states of carbohydrate deprivation or lipid excess have been identified. These include a key role of suppression of hepatic PDH_a activity in directing hepatic pyruvate from oxidation/lipogenesis towards glucose production. In addition, we draw attention to a key property of glucose degradation by muscle, namely more rapid suppression of pyruvate oxidation than of glucose uptake and phosphorylation, that facilitates the transfer of carbon from the periphery to the liver to provide energy and substrates for gluconeogenesis. Most evidence points towards a direct role of insulin in directing hepatic carbon flux during short-term starvation; in contrast in muscle, insulin appears to act indirectly, *via* changes in NEFA supply which lead to differential suppression of pyruvate oxidation and glucose uptake, in facilitating carbon transfer from the liver. The direct action of insulin on the liver may thus be viewed as a mechanism to "set" the liver in the gluconeogenic mode before carbon is presented from the periphery as a consequence of indirect (NEFA-mediated) actions of insulin to modify peripheral pyruvate disposition between oxidation and lactate formation. A focus on the short-and long-term regulation of PDH kinase by insulin and FA oxidation may, in addition, be central to an understanding of the molecular mechanisms underlying substrate competition between glucose and lipid in the pathogenesis of insulin resistance developing in response to endocrine malfunction such as type 2 diabetes, or in response to inappropriate diet.

Acknowledgements
We thank L.G. Fryer and K.A. Orfali for their contributions to this work and The British Diabetic Association, The British Heart Foundation and the U.K. Medical Research Council for financial support.

References

Boden, G., Chen, X., Ruiz, J., White, J.V. and Rossetti, L. (1994) Mechanisms of fatty acid-induced inhibition of glucose uptake. *J. Clin. Invest.* 93: 2438–2446.

Caterson, I.D., Fuller, S.J. and Randle, P.J. (1982) Effects of the fatty acid oxidation inhibitor 2-tetradecylglycidic acid on pyruvate dehydrogenase complex activity in starved and alloxan-diabetic rats. *Biochem. J.* 28: 53–60.

Consoli, A., Nurjhan, N., Reilly, J.J., Jr., Bier, D.M. and Gerich, J.E. (1990) Mechanisms of increased gluconeo-genesis in non-insulin dependent diabetes mellitus. *J. Clin. Invest.* 87: 2083–2045.

Cooney, G.J., Denyer, G.S., Jenkins, A.B., Storlien, L.H., Kraegen, E.W. and Caterson, I.D. (1993) *In vivo* insulin sensitivity of the pyruvate dehydrogenase complex in tissues of the rat. *Am. J. Physiol.* 265: E102-E107.

Denton, R.M. and McCormack, J.G. (1985) Ca^{2+} transport by mammalian mitochondria and its role in hormone action. *Am. J. Physiol.* 249: E543-E554.

Denton, R.M. and McCormack, J.G. (1990) Ca^{2+} as a second messenger within mitochondria of the heart and other tissues. *Ann. Rev. Physiol.* 52: 451–466.

Fanelli, C.G., DeFeo, P., Porcellati, F., Perriello, G., Torlone, E., Santeusanio, F., Brunetti, P. and Bolli, G.B. (1992) Adrenergic mechanisms contribute to the late phase of hypoglycemic glucose counterregulation in humans by stimulating lipolysis. *J. Clin. Invest.* 89: 2005–5013.

Fatania, H.R., Vary, T.C. and Randle, P.J. (1986) Modulation of pyruvate dehydrogenase kinase in cultured hepa-tocytes by glucagon and *n*-octanoate. *Biochem. J.* 234: 233–236.

Holness, M.J., Palmer, T.N., Worrall, E.B. and Sugden, M.C. (1987) Hepatic carbon flux after re-feeding in the glycogen-storage-disease (*gsd/gsd*) rat. *Biochem. J.* 248: 969–972.

Holness, M.J., MacLennan, P.A., Palmer, T.N. and Sugden, M.C. (1988) The disposition of carbohydrate between glycogenesis, lipogenesis and oxidation in liver during the starved-to-fed transition. *Biochem. J.* 252: 325–330.

Holness, M.J. and Sugden, M.C. (1989) Pyruvate dehydrogenase activities during the fed-to-starved transition and on re-feeding after acute or prolonged starvation. *Biochem. J.* 258: 529–533.

Holness, M.J., Liu, Y.-L. and Sugden, M.C. (1989) Time-courses of the responses of pyruvate dehydrogenase activities to short-term starvation in diaphragm and selected skeletal muscles of the rat. *Biochem. J.* 264: 771–776.

Holness, M.J. and Sugden, M.C. (1990a) Glucose utilization in heart, diaphragm and skeletal muscle during the fed-to-starved transition. *Biochem. J.* 270: 245–249.

Holness, M.J. and Sugden, M.C. (1990b) Pyruvate dehydrogenase activities and rates of lipogenesis during the fed-to-starved transition in liver and brown adipose tissue of the rat. *Biochem. J.* 268: 77–81.

Holness, M.J., Liu, Y.-L., Beech, J.S. and Sugden, M.C. (1991) Glucose utilization by interscapular brown adipose tissue *in vivo* during nutritional transitions in the rat. *Biochem. J.* 273: 233–235.

Holness, M.J. (1995) Dietary-induced insulin resistance in brown adipose tissue at the level of glucose oxidation. *Diabetic Medicine*; *in press.*

Jahoor, F., Klein, S. and Wolfe, R. (1992) Mechanisms of regulation of glucose production by lipolysis in humans. *Am. J. Physiol.* 262: E353-E358.

Jensen, M.D., Haymond, M.W., Gerich, J.E., Cryer, P.E. and Miles, J.M. (1987) Lipolysis during fasting: decreased suppression by insulin and increased stimulation by epinephrine. *J. Clin. Invest.* 87: 207–213.

Jones, B.S., Yeaman, S.J., Sugden, M.C. and Holness, M.J. (1992) Hepatic pyruvate dehydrogenase kinase activi-ties during the starved-to-fed transition. *Biochim. Biophys. Acta* 1134: 164–168.

Kerbey, A.L., Randle, P.J., Cooper, R.H., Whitehouse, S., Pask, H.T. and Denton, R.M. (1976) Regulation of pyru-vate dehydrogenase in rat heart. Mechanism of regulation of proportions of dephosphorylated and phosphorylated enzyme by oxidation of fatty acids and ketone bodies and of effects of diabetes: role of coenzyme A, acetyl-coenzyme A and reduced and oxidized and oxidized nicotinamide-adenine dinucleotide. *Biochem. J.* 154: 327–348.

Kerbey, A.L. and Randle, P.J. (1982) Pyruvate dehydrogenase kinase/activator in rat heart mitochondria. Assay, effect of starvation, and effect of protein-synthesis inhibitors in starvation. *Biochem. J.* 206: 103–111.

Kraegen, E.W., Clark, P.W., Jenkins, A.B., Daley, E.A., Chisholm, D.J. and Storlien, L.H. (1991) Development of muscle insulin resistance after liver insulin resistance in high-fat-fed rats. *Diabetes* 40: 1397–1403.

Marchington, D.R., Kerbey, A.L., Jones, A.E. and Randle, P.J. (1987) Insulin reverses effects of starvation on the activity of pyruvate dehydrogenase kinase in cultured hepatocytes. *Biochem. J.* 246: 233–236.

Marchington, D.R., Kerbey, A.L., Giardina, M.G., Jones, A.E. and Randle, P.J. (1989) Longer-term regulation of pyruvate dehydrogenase kinase in cultured rat hepatocytes. *Biochem. J.* 257: 487–491.

Moir, A.M. B. and Zammit, V.A. (1992) Selective labelling of hepatic fatty acids *in vivo*. Studies on the synthesis and secretion of glycerolipids in the rat. *Biochem. J.* 283: 145–149.

Moir, A.M. B. and Zammit, V.A. (1993) Monitoring of changes in hepatic fatty acid and glycerolipid metabolism during the starved-to-fed transition *in vivo*. *Biochem. J.* 289: 49–55.

Orfali, K.A., Fryer, L.G. D., Holness, M.J. and Sugden, M.C. (1993) Long-term regulation of pyruvate dehydro-genase kinase by high-fat feeding. *FEBS Lett.* 336: 501–505.

Orfali, K.A., Fryer, L.G. D., Holness, M.J. and Sugden, M.C. (1995) Interactive effects of insulin and triiodothyronine on pyruvate dehydrogenase kinase activity in cardiac myocytes. *J. Mol. Cell. Cardiol.* 27: 901–908.

Patel, M.S. and Roche, T.E. (1990) Molecular biology and biochemistry of pyruvate dehydrogenase complexes. *FASEB J.* 4: 3224–3233.

Randle, P.J. (1986) Fuel selection in animals. *Biochem. Soc. Trans.* 14: 799–806.

Randle, P.J., Kerbey, A.L. and Espinal, J. (1988) Mechanisms decreasing glucose oxidation in diabetes and starvation: role of lipid fuels and hormones. *Diabetes Metab. Rev.* 4: 623–638.

Randle, P.J., Priestman, D.A., Mistry, S. and Halsall, A. (1994) Mechanisms modifying glucose oxidation in diabetes mellitus. *Diabetologia* 37: S155-S161.

Reaven, G.M. (1995) The fourth Musketeer – from Alexandre Dumas to Claude Bernard. *Diabetologia* 38: 3–13.

Rossetti, L. and Hu, M. (1993) Skeletal muscle glycogenolysis is more sensitive to insulin than is glucose transport/phosphorylation. Relation to the insulin-mediated inhibition of hepatic glucose production. *J. Clin. Invest.* 92: 2963–2974.

Stace, P.B., Marchington, D.R., Kerbey, A.L. and Randle, P.J. (1990) Long-term culture of rat soleus muscle *in vitro*. Its effects on glucose utilization and insulin sensitivity. *FEBS Lett.* 273: 91–94.

Stace, P.B., Fatania, H.R., Jackson, A., Kerbey, A.L. and Randle, P.J. (1992) Cyclic AMP and free fatty acids in the longer-term regulation of pyruvate dehydrogenase kinase in rat soleus muscle. *Biochim. Biophys. Acta* 1135: 201–206.

Storlien, L.H., James, D.E., Burleigh, K.M., Chisholm, D.J. and Kraegen, E.W. (1986) Fat feeding causes widespread *in vivo* insulin resistance, decreased energy expenditure, and obesity in rats. *Am. J. Physiol.* 251: E576-E583.

Storlien, L.H., Kraegen, E.W., Chisholm, D.J., Ford, G.L., Bruce, D.G. and Pascoe, W.S. (1987) Fish oil prevents insulin resistance induced by high-fat feeding in rats. *Science* 237: 885–888.

Sugden, P.H., Hutson, N.J., Kerbey, A.L. and Randle, P.J. (1978) Phosphorylation of additional sites on pyruvate dehydrogenase inhibits its reactivation by pyruvate dehydrogenase phosphate phosphatase. *Biochem. J.* 169: 433–435.

Sugden, M.C., Holness, M.J. and Palmer, T.N. (1989) Fuel selection and carbon fluxes during the starved-to-fed transition. *Biochem. J.* 263: 313–323.

Sugden, M.C., Howard, R.M. and Holness, M.J. (1992) Variations in hepatic carbon flux during unrestricted feeding. *Biochem. J.* 284: 721–724.

Sugden, M.C. and Holness, M.J. (1993a) Physiological modulation of the uptake and fate of glucose in brown adipose tissue. *Biochem. J.* 295: 171–176.

Sugden, M.C. and Holness, M.J. (1993b) Control of muscle pyruvate oxidation during late pregnancy. *FEBS Lett.* 321: 121–126.

Sugden, M.C., Grimshaw, R.M. and Holness, M.J. (1993a) The regulation of hepatic carbon flux by pyruvate dehydrogenase and pyruvate dehydrogenase kinase during long-term food restriction. *Biochem. J.* 296: 217–223.

Sugden, M.C., Howard, R.M., Munday, M.R. and Holness, M.J. (1993b) Mechanisms involved in the co-ordinate regulation of strategic enzymes of glucose metabolism. *Adv. Enz. Regul.* 33: 71–95.

Sugden, M.C. and Holness, M.J. (1994) Interactive regulation of the pyruvate dehydrogenase complex and the carnitine palmitoyltransferase system. *FASEB J.* 8: 54–61.

Sugden, M.C. and Holness, M.J. (1995) Mechanisms underlying suppression of glucose oxidation in insulin-resistant states. *Biochem. Soc. Trans.* 23: 314–320.

Sugden, M.C., Orfali, K.A. and Holness, M.J. (1995) The pyruvate dehydrogenase complex: nutrient control and the pathogenesis of insulin resistance. *J. Nutrition* 125: 1746S–1752S.

Thomas, A.P., Diggle, T.A. and Denton, R.M. (1986) Sensitivity of pyruvate dehydrogenase phosphate phosphatase to magnesium ions. Similar effects of spermine and insulin. *Biochem. J.* 238: 83–91.

Virkamaki, A., Puhaikainen, I., Nurjhan, N., Gerich, J.E. and Yki-Jarvinen, H. (1990) Measurement of lactate formation from glucose using [6-^{3}H] and [6-^{14}C]glucose in humans. *Am. J. Physiol.* 259: E397-E404.

Yeaman, S.J. (1989) The 2-oxo acid dehydrogenases: recent advances. *Biochem. J.* 257: 625–632.

Regulation of branched-chain α-keto acid dehydrogenase complex in rat liver and skeletal muscle by exercise and nutrition

Y. Shimomura

Department of Bioscience, Nagoya Institute of Technology, Gokiso-cho, Showa-ku, Nagoya 466, Japan

Summary. Branched-chain α-keto acid dehydrogenase complex catalyzes oxidative decarboxylation of branched-chain α-keto acids. This reaction is the rate-limiting step in the catabolism of branched-chain amino acids. The enzyme complex is subject to covalent modification; a phosphorylation / dephosphorylation cycle regulates the enzyme activity. It has been reported that dietary protein has a great effect on the activity state of the enzyme complex in rat liver. We describe here effects of exercise and high-fat diet intake on the activity state of the hepatic enzyme complex. The activity state of the enzyme complex in rat skeletal muscle is very low under rested conditions and was significantly elevated by a bout of exercise or muscle contractions elicited by electrical stimulation. Inhibition of branched-chain α-keto acid dehydrogenase kinase by branched-chain α-keto acids, which are increased in the muscle cell by exercise, is suggested to be the mechanism responsible for the enzyme activation induced by exercise. Physical training enhanced activation of the enzyme complex in skeletal muscle by starvation and leucine administration, suggesting that the kinase is sensitized by training. Only long-term training increased the amount of the enzyme complex in the muscle, which was associated with mitochondrial biogenesis.

Introduction

Branched-chain α-keto acid dehydrogenase complex (BCKADC) is a rate-limiting enzyme in the catabolism of branched-chain amino acids (Harris et al., 1986). BCKADC is a mitochondrial multi-enzyme complex consisting of 2-oxoisovalerate dehydrogenase (E1; EC 1.2.4.4), dihydrolipoamide acyltransferase (E2; no EC number), and dihydrolipoamide dehydrogenase (E3; EC 1.8.1.4). The activity of the complex is regulated by covalent modification; branched-chain α-keto acid dehydrogenase kinase (no EC number) is responsible for phosphorylation and inactivation of the BCKADC (Popov et al., 1992; Shimomura et al., 1990a, 1991) and branched-chain α-keto acid dehydrogenase phosphatase (no EC number) is responsible for dephosphorylation and activation of the complex (Damuni et al., 1984; Damuni and Reed, 1987). Phosphorylation of the enzyme is an important mechanism for regulating the enzyme activity in response to alterations in physiological conditions. The enzyme activity is also regulated by genetic enzyme expression, but this mechanism appears to take longer for regulation than phosphorylation.

The catabolism of branched-chain amino acids has been characterized using rats (Harper et al., 1984). Due to lack of significant branched-chain aminotransferase (EC 2.6.1.42) activity in the liver (Hutson, 1988; Hutson et al., 1992), this tissue has practically no capacity to directly catabolize the branched-chain amino acids, but has markedly high activity of BCKADC (Harris et al., 1990), resulting in a significant contribution toward disposal of branched-chain α-keto acids produced by other organs. On the other hand, skeletal muscle has a very high activity of branched-chain aminotransferase (Hutson, 1988; Hutson et al., 1992), but contains only a minor activity of BCKADC (Fujii et al., 1994; Kasperek et al., 1985; Wagenmakers et al., 1984). Since skeletal muscle makes 35–40% of total body weight, the muscle contributes to the production of branched-chain α-keto acids from excess quantities of branched-chain amino acids. From these findings, interorgan cooperation for the catabolism of branched-chain amino acids is proposed (Harper et al., 1984).

It has been demonstrated that the BCKADC activity is affected by nutrition and exercise. The effect of protein starvation (e.g., feeding a low protein diet) on liver BCKADC activity has been most intensively studied in rat, and the results showed that feeding low protein diets decreases the activity state as well as the amount of BCKADC in rat liver (Harris et al., 1990; Zhao et al., 1994). Since regulation of liver BCKADC by dietary protein has been described elsewhere (Harris et al., 1990, 1991; Zhao et al., 1994), effects of exercise and a combination of exercise and nutrition on BCKADC in liver and skeletal muscle are the main topics of this review.

Effects of exercise and nutrition on liver BCKADC

It has been demonstrated that endurance training affects fat, carbohydrate and protein metabolism (Holloszy, 1988; Poortmans, 1988). Animals adapted to regular physical training have a higher capacity for gluconeogenesis in liver (Huston et al., 1975), and amino acids and their metabolites are used as substrates for gluconeogenesis. Rat liver contains a very high activity of BCKADC (more than 1000 munits/g wet wt. and about 100% of the enzyme in the dephosphorylated/active form) in normal conditions (Harris et al., 1990) and is believed to actively catabolize branched-chain α-keto acids obtained from the circulation in rats.

It has also been reported that the gluconeogenic capacity in rat liver is enhanced by feeding a high-fat diet (Huston et al., 1975), suggesting that the high-fat diet may affect liver BCKADC activity. We compared the activity states of liver BCKADC in trained rats fed a high-fat diet or a low-fat diet (Shimomura et al., 1990b). The high-fat diet provided 40, 35 and 25% of calories as fat, carbohydrate and protein, respectively; the low fat diet 5, 70 and 25%. Food consumption of the rats was isoenergetically adjusted between the two diet groups. The activity state of liver BCKADC in rested rats tended to be higher in rats fed the high-fat diet than in those fed the low-fat diet (Fig. 1). The activity state only in the high-fat diet group was significantly increased by a

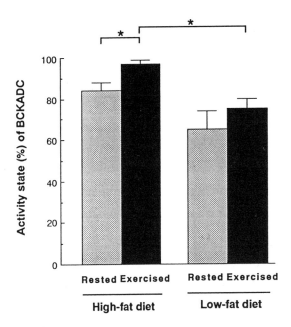

Figure 1. The activity state of liver BCKADC in rats fed a high-fat diet or a low-fat diet. Values are means ± SE for 7–9 rats. (*) Significantly different between indicated two groups (P < 0.05). Data from Shimomura et al. (1990b). All rats were run on a motor-driven treadmill for 5 weeks. The intensity of exercise was gradually increased during 5 weeks, with the final speed and duration of 35 m/min (up an 8° incline) and 30 min/day. The rats were exercised every other day (i.e., 3 to 4 times per week). On the final day of the experiment, approximately half of rats in each diet group was run on the treadmill at 30 m/min (up an 8° incline) for 2 h. Immediately after running, the rats were rapidly killed by cervical dislocation. The rested rats in each diet group were killed before the exercised rats. Liver was rapidly removed and freeze-clamped at liquid nitrogen temperature.

bout of 2 h-exercise and the activity state after exercise was significantly higher in the high-fat diet group than in the low-fat diet group (Fig. 1), suggesting that the high-fat diet promotes branched-chain α-keto acid catabolism in rat liver.

It has been reported that the activity state of liver BCKADC in rats fed a low protein diet (8% protein) is very low (~10%), presumably, for saving dietary branched-chain amino acids for protein synthesis (Harris et al., 1990). We recently observed that the activity state of liver BCKADC was significantly elevated by a bout of exercise (running speed 30 m/min for 85 min), even in rats fed the low protein diet.

The increase in the activity of liver BCKADC should contribute to enhanced energy expenditure during exercise. Since valine and isoleucine are glycogenic amino acids, these branched-chain amino acids may also be used as substrates for gluconeogenesis in liver during exercise. Branched-chain amino acid concentrations in rat liver are increased by exercise (Shimomura et al., 1990b), but it is not known whether branched-chain α-keto acid concentrations are affected.

However, since it has been reported that the α-keto acid concentrations in rat serum are increased by exercise (Kasperek, 1989), the α-keto acids are suggested to be involved in the mechanism responsible for activation of liver BCKADC during exercise. This mechanism is similar to that for activation of muscle BCKADC during exercise, which is described below.

Activation of skeletal muscle BCKADC by exercise

Small amounts of branched-chain amino acids are oxidized in rested skeletal muscle and this is consistent with the results of *in vivo* assays that have determined only a small percentage (4–6%) of muscle BCKADC is in the active form in rested skeletal muscle of normal rats (Fujii et al., 1994; Kasperek et al., 1985; Wagenmakers et al., 1984). It has been demonstrated that contractions of skeletal muscle promote leucine oxidation in the muscle. BCKADC in the muscle plays an important role in enhancement of branched-chain amino acid catabolism caused by exercise (Hood and Terjung, 1991; Shimomura et al., 1993). Kasperek et al. (1985) have clearly demonstrated that rat muscle BCKADC is greatly activated by exercise and that the muscle BCKADC activity elevated by exercise is rapidly decreased after stopping exercise and reaches the level of rested rats 10 min after exercise. Kasperek and Snider (1987) pointed out that rapid removal of skeletal muscle after death (within ~20 s) is very important to arrest *in vivo* BCKADC activity elevated by exercise.

We have examined activation of skeletal muscle BCKADC by a bout of running exercise in rats fed a high-fat diet or a low-fat diet (Shimomura et al., 1990b). In this experiment, rats were trained for 5 weeks to keep running for 2 h on the final day of the experiment. The muscle BCKADC was activated by the 2 h exercise and the activity state was elevated to 70–80%. However, no significant difference was observed in the activation of muscle BCKADC between the high-fat diet and the low-fat diet groups.

In order to examine the mechanism responsible for activation of muscle BCKADC by exercise, we used a model of tetanic contractions elicited by electrical stimulation (Shimomura et al., 1993), which was developed by Hood and Terjung (1987) for the study on leucine metabolism in perfused rat skeletal muscle. It has been demonstrated that muscle contractions in this model increase leucine oxidation in the muscle (Hood and Terjung, 1987). According to the findings of Kasperek's group, we took care to remove and then to freeze-clamp muscle samples within 20 s for analyzing the activity state of BCKADC in skeletal muscle. The muscle contractions caused by electrical stimulation for 5–60 min significantly increased the activity state of the BCKADC at all the time points examined (Fig. 2). Since, in this model, contractions of hindlimb muscles were

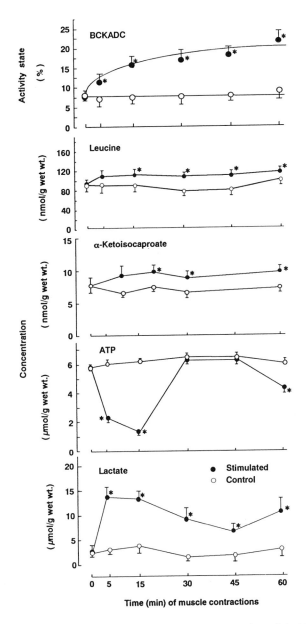

Figure 2. Activation of BCKADC in rat skeletal muscle by tetanic contractions elicited by electrical stimulation and alterations of leucine, α-ketoisocaproate, ATP and lactate concentrations in the muscle during contractions. Values are means ± SE; n = 5 or 6 per time point. (*) Significantly different from control muscle (p < 0.05). Data from Shimomura et al. (1993). Untrained rats were used in this experiment. The animals starved for 24 h before the experiment were anesthetized with sodium pentobarbital and tetanic contractions of the left hindlimb muscles were elicited *via* supramaximal square-wave pulses (6 V, 0.05 ms) at tetanic frequency of 60 tetani/min (100 ms train, 100 Hz). The right hindlimb muscles were not stimulated and served as a control. At the indicated time points, gastrocnemius muscles were removed and immediately freeze-clamped at liquid nitrogen temperature.

made in only the left leg and the enzyme activity in the contralateral muscle was not elevated during contractions, we concluded that blood components are not involved in the mechanism for activation of muscle BCKADC. The total activity of BCKADC in the stimulated muscle was not altered during contractions, indicating that dephosphorylation of phosphorylated BCKADC in the muscle is responsible for the activation.

Since branched-chain α-keto acid dehydrogenase kinase exists in a form tightly bound to BCKADC (Shimomura et al., 1990a), regulators of the kinase activity are possible factors

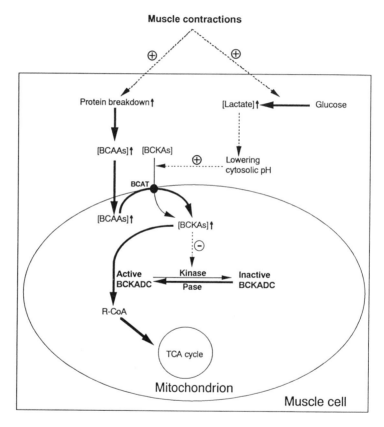

Figure 3. Schematic model of the mechanism for activation of branched-chain α-keto acid dehydrogenase complex (BCKADC) in skeletal muscle by exercise. This is a modified version of the model reported previously (Shimomura et al., 1995). Muscle contractions (exercise) stimulate protein breakdown, thereby increasing the concentrations of branched-chain amino acids (BCAAs) in the muscle cell. The BCAAs are transaminated in muscle mitochondria by branched-chain aminotransferase (BCAT) and branched-chain α-keto acids (BCKAs; especially α-keto acid from leucine) inhibit branched-chain α-keto acid dehydrogenase kinase, resulting in elevation of the activity state of BCKADC. Muscle contractions also increase the lactate concentration in the muscle cell by stimulation of glycolysis, reflecting a lower cytosolic pH. The acidification of the cytosol in the cell may promote accumulation of BCKAs in mitochondria by inhibition of an efflux of BCKAs from mitochondria and/or by stimulation of mitochondrial uptake of BCKAs from the cytosol. (Pase = phosphatase).

involved in the mechanism for activation of BCKADC. α-Ketoisocaproate derived from leucine and ATP are considered as candidates for the regulators, because the former is a potent inhibitor of the kinase activity (Lau et al., 1981; Paxton and Harris, 1984) and the latter is used in the kinase-mediated inactivation of BCKADC (Lau et al., 1983; Paxton and Harris, 1982). The concentrations of leucine and α-ketoisocaproate in the muscle were increased by muscle contractions and the profile of the increases were correlated with the activity state of BCKADC (Fig. 2). On the other hand, the altered pattern of ATP concentration in the muscle was obviously different from the pattern of BCKADC activation (Fig. 2). The lactate concentration in the stimulated muscle was increased three- to five-fold during contractions compared to that in the contralateral muscle (Fig. 2), resulting in acidification of muscle cells which promotes accumulation of intramitochondrial branched-chain α-keto acids (Hutson, 1987). These findings suggest that α-ketoisocaproate is the most likely regulator of BCKADC activity in skeletal muscle. A schematic model for the activation of BCKADC by muscle contractions is shown in Figure 3. In this model, branched-chain α-keto acids act as feedforward activators of BCKADC by inhibiting kinase activity.

Table 1. Effects of starvation and leucine administration on the activity state of muscle BCKADC in untrained and trained rats

Condition	Rat	Muscle BCKADC in active state (%)
Fed	Untrained	3.8 ± 1.2
	Trained	3.4 ± 0.5
Starved	Untrained	8.6 ± 1.6
	Trained	14.8 ± 1.4*
Leucine administration		
Control (saline)	Untrained	4.7 ± 0.5
	Trained	3.8 ± 0.7
Leucine	Untrained	14.7 ± 0.9
	Trained	19.7 ± 1.3*

Values are means ± SE for 4 – 5 rats. (*) Significantly different from untrained rats (p < 0.05). Data from Fujii et al. (1994). Rats in the trained group were run for 5 weeks, 5 days/week, 30 min/day. The intensity of exercise was gradually increased during the 5-week exercise program, with a final speed of 35 m/min up an 8° incline. (1) Experiment for starvation (upper panel). On the final day of the experiment, rats under fed or 24 h-starved conditions were anesthetized with sodium pentobarbital, and gastrocnemius muscles were removed and immediately freeze-clamped at liquid nitrogen temperature. (2) Experiment for leucine administration (lower panel). Untrained and trained rats under fed conditions were used in this experiment. The animals were anesthetized with sodium pentobarbital and were injected into the tail vein with a 125 mM leucine solution (0.25 mmol/kg body wt.) prepared in saline or with saline as control. The time of the injection was ~24 h after final bout of exercise. Muscle samples were prepared 10 min after injection as described above.

Effects of training on skeletal muscle BCKADC

Effects of training on skeletal muscle BCKADC were examined using rats trained for 5 weeks by running exercise (Fujii et al., 1994). The training significantly enhanced the activation of BCKADC by starvation; starvation for 24 h elevated the activity state to ~9% in the untrained rats but to ~15% in the trained rats (Tab. 1). This training effect was not observed in rats under fed conditions. These results suggest that endurance training does not activate muscle BCKADC in rested rats, but predisposes the BCKADC to activation by starvation. This hypothesis was supported by the leucine administration experiment, in which leucine administration caused greater activation of muscle BCKADC in trained rats compared to untrained rats (Tab. 1). These findings further suggest that the branched-chain α-keto acid dehydrogenase kinase is sensitized to inhibition by the α-keto acid derived from leucine by exercise training. A mechanism to sensitize the kinase to the inhibitor by training is not known, but the hypothesis is supported by the findings of Hood and Terjung (1991) that activation of BCKADC in skeletal muscle by perfusion of α-chloroisocaproate, a potent inhibitor of the kinase (Harris et al., 1982), was greater in trained rats compared to untrained rats.

It has been demonstrated that long-term (12 weeks) exercise training stimulates mitochondrial biogenesis in rat skeletal muscle (Murakami et al., 1994). We determined actual and total activities of BCKADC and mitochondrial DNA in skeletal muscle of rats trained for relatively short (5 weeks) and long (12 weeks) periods (Fujii et al., 1995). The parameters examined were not

Table 2. BCKADC activities and mitochondrial DNA content in skeletal muscle of untrained and trained rats

Rat	BCKADC (nmol/min/g wet wt.)		Mitochondrial DNA
	Actual	Total	(ng/μg total DNA)
5 weeks-			
Untrained	1.2 ± 0.3	30.3 ± 2.5	10.1 ± 0.9
Trained	1.2 ± 0.2	34.4 ± 2.3	11.7 ± 0.6
12 weeks-			
Untrained	1.0 ± 0.2	32.8 ± 2.9	10.0 ± 0.7
Trained	$1.8 \pm 0.2*$	$45.3 \pm 3.5*$	$13.3 \pm 1.2*$

Values are means $\pm$ SE for 8 rats. (*) Significantly different from untrained rats ($p < 0.05$). Data from Fujii et al. (1995). Rats in the trained group were run for either 5 or 12 weeks. The training program was as described in Table 1, except that the running period was 30 min/day in the initial 5 weeks and 60 min/day in the following 7 weeks. On the final day of the experiment (24 h after the final bout of exercise), all rats under fed conditions were anesthetized with sodium pentobarbital and gastrocnemius muscles were removed and immediately freeze-clamped at liquid nitrogen temperature.

affected by the short-term training (Tab. 2). The actual activity of BCKADC in the muscle of rested rats was increased to 180% by the long-term training, suggesting that branched-chain amino acid catabolism in the muscle under rested conditions is enhanced by long-term training (Tab. 2). The total activity of BCKADC and mitochondrial DNA in the muscle were increased to a similar level (130–140%) by long-term training (Tab. 2). The total activity of BCKADC corresponded to the enzyme amount in the muscle, which was confirmed by Western blotting analysis. The mitochondrial DNA content corresponded to the mitochondrial number in the muscle (Murakami et al., 1994). These findings suggest that biosynthesis of BCKADC in skeletal muscle is stimulated by mitochondrial biogenesis induced by long-term training.

mRNA levels for E1α, E1β and E2 subunits of BCKADC were determined in the muscle. E1α and E2 mRNA levels were not affected by either short- or long-term training, but E1β mRNA level was slightly increased by both durations of training. In contrast to the training effects, mRNA levels for E1α and E2, but not E1β, were significantly increased by starvation for 24 h, although the amount of BCKADC in skeletal muscle was not affected by starvation (Shimomura et al., 1995). These divergent alterations of the message levels for the subunits of BCKADC in skeletal muscle suggest that coordinate regulation of gene expression of the subunits in skeletal muscle does not occur in response to exercise training and starvation and also that posttranslational regulatory mechanisms determine the amount of BCKADC in skeletal muscle. Such complex and divergent changes in mRNA levels for the subunits was also observed in rat liver BCKADC in response to dietary protein content (Zhao et al., 1994).

References

Damuni, Z., Merryfield, M.L., Humphreys, J.S. and Reed, L.J. (1984) Purification and properties of branched-chain α-keto acid dehydrogenase phosphatase from bovine kidney. *Proc. Natl. Acad. Sci. USA* 81: 43356–4338.

Damuni, Z. and Reed, L.J. (1987) Purification and properties of the catalytic subunit of the branched-chain α-keto acid dehydrogenase phosphatase from bovine kidney mitochondria. *J. Biol. Chem.* 262: 5129–5132.

Fujii, H., Shimomura, Y., Tokuyama, K. and Suzuki, M. (1994) Modulation of branched-chain 2-oxo acid dehydrogenase complex activity in rat skeletal muscle by endurance training. *Biochim. Biophys. Acta* 1199: 130–136.

Fujii, H., Tokuyama, K., Suzuki, M., Popov, K.M., Zhao, Y., Harris, R.A., Nakai, N., Murakami, T. and Shimomura, Y. (1995) Regulation by physical training of enzyme activity and gene expression of branched-chain 2-oxo acid dehydrogenase complex in rat skeletal muscle. *Biochim. Biophys. Acta* 1243: 277–281.

Harper, A.E., Miller, R.H. and Block, K.P. (1984) Branched-chain amino acid metabolism. *Ann. Rev. Nutr.* 4: 409–454.

Harris, R.A., Paxton, R. and DePaoli-Roach, A.A. (1982) Inhibition of branched-chain α-ketoacid dehydrogenase kinase activity by α-chloroisocaproate. *J. Biol. Chem.* 257: 13915–13918.

Harris, R.A., Paxton, R., Powell, S.M., Goodwin, G.W., Kuntz, M.J. and Han, A.C. (1986) Regulation of branched-chain α-ketoacid dehydrogenase complex by covalent modification. *In*: G. Weber (ed.): *Advance in Enzyme Regulation*, Vol. 25. Pergamon, New York, pp 219–237.

Harris, R.A., Zhang, B., Goodwin, G.W., Kuntz, M.J., Shimomura, Y., Rougraff, P., Dexter, P., Zhao, Y., Gibson, R. and Crabb, D.W. (1990) Regulation of the branched-chain α-ketoacid dehydrogenase and elucidation of a molecular basis for maple syrup urine disease. *In*: G. Weber (ed.): *Advance in Enzyme Regulation*, Vol. 30. Pergamon, New York, pp 245–263.

Harris, R.A., Shimomura, Y., Popov, K., Zhao, Y., Hu, H. and Crabb, D.W. (1991) Regulation of hepatic branched-chain amino acid catabolism. *In*: N. Grunnet and B. Quirstorff (eds): *Regulation of Hepatic Function*. Munksgaard, Copenhagen, pp 374–385.

Holloszy, J.O. (1988) Metabolic consequences of endurance exercise training. *In*: E.S. Horton and R.L. Terjung (eds): *Exercise, Nutrition, and Energy Metabolism*. Macmillan, New York, pp 116–131.

Hood, D.A. and Terjung, R.L. (1987) Leucine metabolism in perfused rat skeletal muscle during contractions. *Am. J. Physiol.* 253: E636-E647.

Hood, D.A. and Terjung, R.L. (1991) Effect of α-ketoacid dehydrogenase phosphorylation on branched-chain amino acid metabolism in muscle. *Am. J. Physiol.* 261: E628-E634.

Huston, R.L., Weiser, P.C., Dohm, G.L., Askew, E.W. and Boyd, J.B. (1975) Effects of training, exercise and diet on muscle glycolysis and liver gluconeogenesis. *Life Sci.* 17: 369–376.

Hutson, S.M. (1987) pH Regulation of mitochondrial branched chain α-keto acid transport and oxidation in rat heart mitochondria. *J. Biol. Chem.* 262: 9629–9635.

Hutson, S.M. (1988) Subcellular distribution of branched-chain aminotransferase activity in rat tissues. *J. Nutrition* 118: 1475–1481.

Hutson, S.M., Wallin, R. and Hall, T.R. (1992) Identification of mitochondrial branched chain aminotransferase and its isoforms in rat tissues. *J. Biol. Chem.* 267: 15681–15686.

Kasperek, G.J., Dohm, G.L. and Snider, R.D. (1985) Activation of branched-chain keto acid dehydrogenase by exercise. *Am. J. Physiol.* 248: R166-R171.

Kasperek, G.J. and Snider, R.D. (1987) Effect of exercise intensity and starvation on activation of branched-chain keto acid dehydrogenase by exercise. *Am. J. Physiol.* 252: E33-E37.

Kasperek, G.J. (1989) Regulation of branched-chain 2-oxo acid dehydrogenase activity during exercise. *Am. J. Physiol.* 256: E186-E190.

Lau, K.S., Fatania, H.R. and Randle, P.J. (1981) Inactivation of rat liver and kidney branched-chain 2-oxoacid dehydrogenase complex by adenosine triphosphate. *FEBS Lett.* 126: 66–70.

Lau, K.S., Phillips, C.E. and Randle, P.J. (1983) Multi-site phosphorylation in ox-kidney branched-chain 2-oxoacid dehydrogenase complex. *FEBS Lett.* 160: 149–152.

Murakami, T., Shimomura, Y., Fujitsuka, N., Nakai, N., Sugiyama, S., Ozawa, T., Sokabe, M., Horai, S., Tokuyama, K. and Suzuki, M. (1994) Enzymatic and genetic adaptation of soleus muscle mitochondria to physical training in rats. *Am. J. Physiol.* 267: E388-E395.

Paxton, R. and Harris, R.A. (1982) Isolation of rabbit liver branched chain α-ketoacid dehydrogenase and regulation by phosphorylation. *J. Biol. Chem.* 257: 14433–14439.

Paxton, R. and Harris, R.A. (1984) Regulation of branched-chain α-ketoacid dehydrogenase kinase. *Arch. Biochem. Biophys.* 231: 48–57.

Poortmans, J.R. (1988) Protein metabolism. *In*: J.R. Poortmans (ed.): *Principles of Exercise Biochemistry, Medicine and Sport Sciences*, Vol. 27. Karger, Basel, pp 164–193.

Popov, K.M., Zhao, Y., Shimomura, Y., Kuntz, M.J. and Harris, R.A. (1992) Branched-chain α-ketoacid dehydrogenase kinase; molecular cloning, expression, and sequence similarity with histidine protein kinases. *J. Biol. Chem.* 267: 13127–13130.

Shimomura, Y., Nanaumi, N., Suzuki, M., Popov, K.M. and Harris, R.A. (1990a) Purification and partial characterization of branched-chain α-ketoacid dehydrogenase kinase from rat liver and rat heart. *Arch. Biochem. Biophys.* 283: 293–299.

Shimomura, Y., Suzuki, T., Saitoh, S., Tasaki, Y., Harris, R.A. and Suzuki, M. (1990b) Activation of branched-chain α-keto acid dehydrogenase complex by exercise: effect of high-fat diet intake. *J. Appl. Physiol.* 68: 161–165.

Shimomura, Y., Nanaumi, N., Suzuki, M. and Harris, R.A. (1991) Immunochemical identification of branched-chain 2-oxo acid dehydrogenase kinase. *FEBS Lett.* 288: 95–97.

Shimomura, Y., Fujii, H., Suzuki, M., Fujitsuka, N., Naoi, M., Sugiyama, S. and Harris, R.A. (1993) Branched-chain 2-oxo acid dehydrogenase complex activation by tetanic contractions in rat skeletal muscle. *Biochim. Biophys. Acta* 1157: 290–296.

Shimomura, Y., Fujii, H., Suzuki, M., Murakami, T., Fujitsuka, N. and Nakai, N. (1995) Branched-chain α-keto acid dehydrogenase complex in rat skeletal muscle: regulation of the activity and gene expression by nutrition and physical exercise. *J. Nutrition* 125: 1762S–1765S.

Wagenmakers, A.J.M., Schepens, J.T.G. and Veerkamp, J.H. (1984) Effect of starvation and exercise on actual and total activity of the branched-chain 2-oxo acid dehydrogenase complex in rat tissues. *Biochem. J.* 223: 815–821.

Zhao, Y., Popov, K.M., Shimomura, Y., Kedishvili, N.Y., Jaskiewicz, J., Kuntz, M.J., Kain, J., Zhang, B. and Harris, R.A. (1994) Effect of dietary protein on the liver content and subunit composition of the branched-chain α-ketoacid dehydrogenase complex. *Arch. Biochem. Biophys.* 308: 446–453.

Dephosphorylation of PDH by phosphoprotein phosphatases and its allosteric regulation by inositol glycans

S. Abe[1], L.C. Huang and J. Larner

Department of Pharmacology, University of Virginia School of Medicine, Charlottesville, Virginia 22908, USA
[1]*Present address: Department of Legal Medicine, Fukushima Medical College, Fukushima 960-12, Japan*

Pyruvate dehydrogenase (PDH) is a key rate-limiting enzyme responsible for insulin-stimulated oxidative glucose metabolism. Insulin action leads to an increase in the dephosphorylated (active) form of the enzyme (Jungas, 1971; Linn, Pettit and Reed, 1969). MaCauley and Jarett (1985) demonstrated that the specific interconverting enzyme under insulin control was its phosphatase. They clearly demonstrated that a substance was produced when insulin interacted with plasma membranes which stimulated PDH phosphatase in mitochondria. Independently, Larner et al. (1979) presented evidence of an insulin-generated heat-stable factor(s) which inhibited cAMP-dependent protein kinase and which also stimulated crude glycogen synthase phosphatase. In 1979; Jarett and Seals (1979) reported that the factor(s) isolated in Larner's laboratory could also stimulate PDH phosphatase. We then further separated from the original fraction by anion exchange, two separate fractions; one contained the cAMP-dependent protein kinase inhibitor, while the other contained the PDH phosphatase stimulator (Larner et al., 1988a, b). The former contained myo-inositol and glucosamine after acid hydrolysis; the latter contained D-chiro-inositol and galactosamine after acid hydrolysis.

Using BC_3H_1 myocytes, Romero et al. (1988) studied the formation of the PDH phosphatase stimulator. When insulin was added to the cell medium, an increase in activity of the PDH stimulator was observed in the medium which coincided with the time course of an increase of alkaline phosphatase released into the medium. Both of these increases were blocked by the addition of a protease inhibitor, p-aminobenzamidine indicating the involvement of a protease in regulating the release of alkaline phosphatase and the PDH phosphatase stimulator (Fig. 1). On the other hand, insulin action also led to an increase of [^{14}C] myristate-labeled diacylglycerol (DAG) in the cell membranes coinciding with the increase in PDH stimulator activity in the medium. Pretreatment of cells with pertussis toxin abolished the insulin-dependent labeled DAG accumulation and the release of the stimulator as well. Thus, the formation of the PDH stimulator required the action of a protease as well as a pertussis toxin sensitive enzyme, presumably a phos-

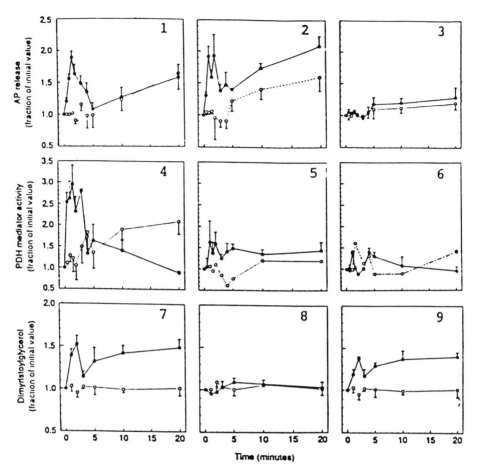

Figure 1. The time course of the effects of insulin on the release of (AP) (1 to 3), release of insulin mediator to the incubation medium (4 to 6), and generation of dimyristoylglycerol (7 to 9). BC$_3$H1 myocytes were cultured, serum starved for 20 h, and stimulated with 100 nM insulin in serum-free medium containing 0.1% bovine serum albumin. All experiments were carried out at 37°C at least in triplicate. Results are expressed as fraction of the initial value of the parameter measured in order to facilitate comparisons among the various experiments. The inhibitors used (PAB and PT) had no significant effect on basal activities. In all the graphs (○) indicates the presence of insulin, and (●) the absence of insulin. (1) The release of AP to the incubation medium. At the times indicated, portions of the culture medium were withdrawn and frozen in liquid nitrogen. The activity of AP in the medium was estimated from the rate of hydrolysis of p-nitrophenylphosphate (2 mg/ml) at 37°C in 0.1 M tris (pH 9.2) containing 2 mM MgCl$_2$. The reaction was monitored spectrophotometrically at 410 nm, and the rate of hydrolysis of the substrate was determined from the slope of the time-course of the reaction. (2) The release of AP from PT-treated BC$_3$H1 cells. Cells were exposed to PT (100 ng/ml) for 20 h before insulin stimulation. (3) The effect of PAB on the release of AP. The inhibitor (0.8 mM) was added to the culture medium 2 to 4 min before exposure to insulin. (4) The activity of insulin mediator released to the extracellular medium. Mediator activity was determined in 0.2 ml portions of the culture medium collected at the indicated times with a PDH stimulation assay as described elsewhere. The samples were chromatographed in 2-ml AG-1 columns eluted successively with 10 ml of distilled water and three solutions of aqueous HCl adjusted at pH 3, 2, and 1.3, respectively. The eluates were concentrated by lyophilization, suspended in water, and lyophilized. Mediator acitvity was found in the pH 2 and pH 3 fractions. (5) The effect of PT treatment on the release of mediator to the culture medium. (6) The effect of PAB (0.8 mM) treatment (2 to 4 min) on the release of insulin mediator. (7) The production of dimyristolglycerol by myocytes. Dimyristoylglycerol production was determined. (8) The effect of PT on the production of dimyristoylglycerol. (9) The effect of PAB in the generation of dimyristoylglycerol.

pholipase activated *via* a G protein (Fig. 1). These results strongly indicate that the PDH stimulator was derived from a glycosyl phosphatidylinositol precursor.

Using an antibody raised against the phosphoinositol glycan from *Trypanosoma brucei brucei*, Romero et al. (1990) demonstrated that this PDH phosphatase stimulator formed in the presence of added insulin was released into the medium and was bound by the antibody which was also added to the medium. The antibody completely blocked insulin action to activate PDH in the cells. The PDH stimulator could then be released from the immunoprecipitate by mild acid treatment (Fig. 2). These data further demonstrated that the stimulator belonged to the inositol glycan family and was released by insulin from the outside of the cells. Furthermore, they demonstrated

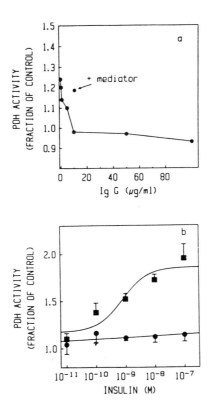

Figure 2. Inhibition of the insulin-induced stimulation of PDH in intact myocytes by anti-inositolglycan glycan antibodies. (a) Cells were stimulated with 10 nM insulin. Protein A-Sepharose-purified IgG was used at the concentrations indicated. Where indicated, a purified mediator fraction was added to the culture dishes to a final concentration of 0.1 nmol of organic phosphate per ml of incubation medium. These cells were also incubated with insulin in the presence of antibodies. (b) Effect of anti-inositolglycan antibodies on the PDH stimulation response to insulin. Protein A-purified antibodies were used at a concentration of 20 µg/ml (●). The insulin dose-response curve in the absence of antibodies is included for comparative purposes (■). The data represent the average of five separate experiments carried out in duplicate.

the importance of intracellular entry of the stimulator to activate PDH phosphatase. The antibody, thus, did not inhibit stimulator generation by insulin but prevented its access to its intracellular site of action.

Saltiel and Cuatrecassas (1986) and Mato et al. (1987) demonstrated that phospholipase C treatment of purified glycophospholipids generated water soluble substances termed modulators, with similar properties to this stimulator. Thus, this or similar stimulators may originate from membrane glycophospholipids, or alternatively from glycophospholipid proteinated species.

By monitoring its PDH phosphatase stimulatory action, the stimulator was purified from rat liver from animals treated *in vivo* with and without insulin. Rat liver was extracted under acidic conditions, boiled and charcoal-treated to remove nucleotides. The extract was then absorbed onto AG1x8 anionic exchanger. The PDH phosphatase stimulator was eluted with 0.015 M HCl and was found to be present with higher activity in deproteinized insulin-treated liver extracts compared to controls (Larner et al., 1988a). The rat liver (and beef liver) stimulators were further purified to homogeneity by Chelex 100, thin layer chromatography, Bio-gel P4 gel filtration and finally HPLC with protein pak-60 (Nestler et al., 1991; Farese et al., 1994). The purified sample gave rise to a single ninhydrin positive spot on TLC chromatography. Only one stimulatory activity was found during the purification scheme. This purification procedure was scaled up to prepare this stimulator from bovine liver (Nestler et al., 1991; Farese et al., 1994). After complete

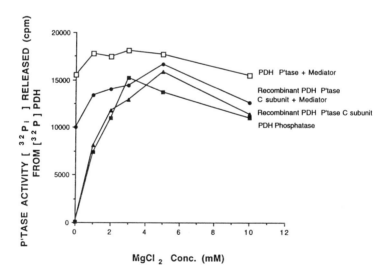

Figure 3. Mg^{2+} sensitivity of pyruvate dehydrogenase (PDH) phosphatase holoenzyme, recombinant catalytic subunit in the absence and presence of D-chiro-inositol containing mediator (stimulator) from rat skeletal muscle. Note the increased sensitivity to Mg^{2+} of both the holoenzyme and recombinant catalytic subunit in the presence of the mediator.

hydrolysis with 6N HCl, D-chiro-inositol was the only inositol present. With mild acid hydrolysis (4 M trifluoroacetic acid, 100°, 4 h), galactosamine and pinitol were established by Dionex HPLC at a ratio of 1.0 to 1.2. Thus, it appears that the beef liver PDH phosphatase stimulator is a pseudo-disaccharide composed of galactosamine and pinitol (Larner et al., 1994). The detailed structure of this pseudo-disaccharide has been established and will be presented in a subsequent publication.

This chiro-inositol containing PDH phosphatase stimulator also has been shown to mimic insulin action. One-half to two nanomoles of this stimulator injected *via* the tail vein to streptozotocin-diabetic rats decreased plasma glucose 30 to 40% (Huang et al., 1993). The effect was seen as early as 15 min after injection and lasted up to 2–3 h. Similar nanomole amounts increased [^{14}C] glucose incorporation into glycogen when injected intraperitoneally to normal rats (Huang et al., 1993). Mechanistically of interest is that, *in vitro* this stimulator decreased the Mg^{++} requirement of purified Mg^{++} dependent PDH phosphatase in the same manner as insulin-treated adipose tissue (Larner et al., 1994; Larner et al., l989, and Denton et al., 1989). Thus, this chiro-inositol containing PDH stimulator may be a second messenger produced following insulin binding to its receptor which regulates PDH phosphatase *in vivo*.

A similar PDH phosphatase stimulator was also purified from rat muscle. Rat muscle was extracted, boiled and charcoal treated the same way as rat liver (S. Abe, J. Larner and L.C. Huang, unpublished data). The lyophilized extract was subjected to Sephadex G-25 gel filtration followed by Sephadex G-15 chromatography. The active fraction was applied to a AG1x8 column eluted, and purified by thin layer chromatography as above. The stimulator which had the same mobility as the stimulator from rat or bovine liver was eluted from thin layer plates and applied onto a Sephadex G-15 column to remove silica. The active fraction yielded a single ninhydrin positive spot on TLC chromatography.

Upon strong acid hydrolysis it also contained chiro-inositol. Myo-inositol, hexosamine amino acids and ethanolamine were also present. The significance of their presence is unclear at present, but indicate heterogeneity. The muscle stimulator is probably larger and more active than the liver stimulator based on its chiro-inositol content. The structure of the muscle PDH phosphatase stimulator is under investigation.

Like the PDH phosphatase stimulator from liver, the muscle stimulator also sensitizes the PDH phosphatase to Mg^{++}. More importantly the stimulator can increase the phosphatase activity in the presence of maximal Mg^{++} concentration (10 mM Mg^{++}). Thus the stimulatory effect seen with the stimulator cannot be explained by comtamination of free Mg^{++} ion. In fact, the Mg^{++} concentration in the stimulator fraction has been determined to be 20 μM. The same amount of Mg^{++} added to the phosphatase assay mixture exerted no significant effect on its activity.

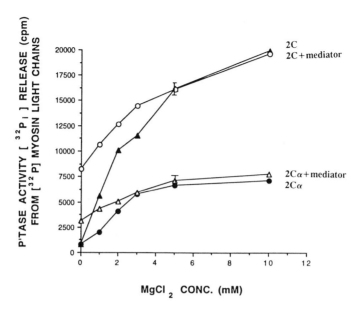

Figure 4. Mg^{2+} sensitivity of phosphoprotein phosphatase 2C and recombinant 2C-α dephosphorylating [^{32}P] myosin light chains in the presence and absence of D-chiroinositol IPG containing mediator (stimulator) from rat skeletal muscle.

The catalytic subunit of the PDH phosphatase has been cloned and shown to have 20% homology in amino acid sequence to phosphatase 2C, another Mg^{++} requiring enzyme (Lawson et al., 1993; Tamura et al., 1989). The muscle stimulator was tested for its effect on the recombinant PDH phosphatase C subunit (a kind gift from Dr. L. Reed), PDH phosphatase holoenzyme from beef heart and phosphatase 2C from smooth muscle (a kind gift from Dr. M. Pato) as well as on cloned phosphatase 2Cα (a kind gift from Dr. Tamura). As shown in Figure 3, the stimulator activates the PDH phosphatase and its C subunit as well as phosphatase 2Cα using [^{32}P]PDH as substrate. The muscle stimulator also activates phosphatase 2C isolated from smooth muscle acting on myosin light chains (Fig. 4), although the smooth muscle phosphatase 2C did not dephosphorylate PDH as did the cloned phosphatase 2Cα. In all these cases, the muscle stimulator decreases the requirement of Mg^{++} for the four species of enzyme (Figs 3, 4). These data suggest that the stimulator interacts with the phosphatases at or near the Mg^{++} binding sites. Greater activity with the stimulator is seen in both cases (PDH phosphatase and phosphatase 2C) with the holoenzymes isolated from tissue as compared to the cloned catalytic subunits. With the holoenzymes, significant activity is now observed in the absence of Mg^{++} and Ca^{++}. This indicates that the accompanying "regulatory" subunits sensitize the catalytic subunits to the action of the muscle stimulator. This proposed mechanism of action was supported by the fact that Mg^{++}

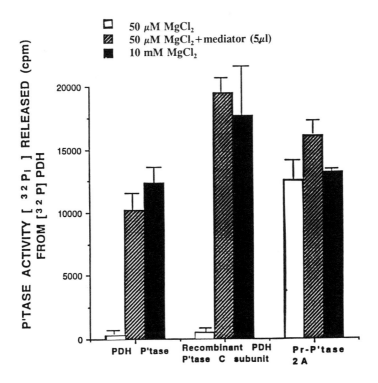

Figure 5. Pyruvate dehydrogenase (PDH) phosphatase activity assayed by [^{32}P] release from [^{32}P] PDH. PDH phosphatase (holoenzyme), PDH phosphatase recombinant catalytic subunit and cytosolic phosphatase 2A are assayed with low Mg^{2+} (left-hand column; with low Mg^{2+} + 5λIPG (middle column) and with high Mg^{2+} (right-hand column). The D-chiro-inositol mediator (stimulator) was prepared from rat skeletal muscle.

independent phosphatase 2A (a kind gift from Dr. Villar-Palasi) was not affected by the stimulator.

It was interesting to note that when the phosphatase activity was determined by the increase of PDH activity, the cloned phosphatase 2Cα failed to activate the inactive PDH. On the other hand, phosphatase 2A both activated and dephosphorylated the PDH complex (Figs 5 and 6). Since the inactive PDH has been shown to be phosphorylated at three serine sites, it became important to determine which site(s) was dephosphorylated by each phosphatase. The product of each phosphatase treatment was subjected to sequence determination by LC/MS. (Collaboration with Drs. D. Hunt and R. Salchetto). It was noted that PDH phosphatase, its cloned C subunit and phosphatase 2A dephosphorylated all three sites, whereas phosphatase 2Cα only dephosphorylated sites 2 and 3. Thus, these data strongly support the previous results that site 1 phosphorylation is responsible for inactivation of the enzyme.

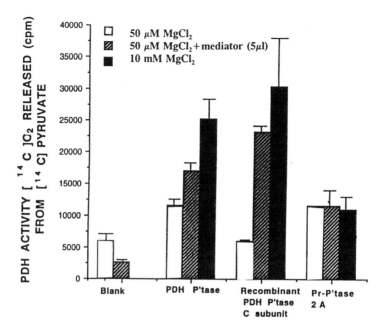

Figure 6. Pyruvate dehydrogenase (PDH) phosphatase activity assayed by [^{14}C]0$_2$ release from [^{14}C] pyruvate. PDH phosphatase (holoenzyme), PDH phosphatase recombinant catalytic subunit and cytosolic phosphatase 2A are assayed. Columns as in Figure 5. The blank without added enzyme is also shown. The D-chiro-inositol mediator (stimulator) was prepared from rat skeletal muscle.

In summary, PDH phosphatase is stimulated by a pseudo-disaccharide containing D-chiro-inositol and galactosamine. Isolated from liver, the pseudo-disaccharide and a similar inositol containing compound from rat muscle act by sensitizing the enzyme to its required metal Mg^{++}. This pseudo-disaccharide and similar members of the class of inositol glycans may be responsible for mediating insulin's action in sensitive cells to activate PDH phosphatase and phosphatase 2C both Mg^{++} requiring enzymes, thus activating rate-limiting enzymes of oxidative and non-oxidative glucose metabolism.

Acknowledgements
We are grateful to Drs. M. Pato, L. Reed, S. Tamura and C. Villar-Palasi for their gifts of phosphatase and to Dr. T. Haystead for supplying myosin light chain and its kinase. We would also like to thank Drs. D. Hunt and R. Salchetto for their excellent analysis of the phosphorylation sites. This work is supported by grants from NIH-USPHS-AM14334, Thomas F. and Kate Miller Jeffress Memorial Trust-J-255, The Center of Innovative Technology, and Insmed Pharmaceuticals, Inc.

References

Denton, R.M., Midgley, P.J.W., Rutter, G.A., Thomas, A.P. and Mc Cormack, J.G. (1989) Studies into the mechanism whereby insulin activates pyruvate dehydrogenase complex in adipose tissue. *Ann. N.Y. Acad. Sci.* 573: 285–296.

Farese, R.V., Standaert, M.L., Yamada, K., Huang, L.C., Zhang, C., Cooper, D.R., Wang, Z., Yang, Y., Suzuki, S., Toyota, T. and Larner, J. (1994) *Proc. Natl. Acad. Sci. USA* 91: 11040–11044.

Huang, L.C., Fontelles, M.C., Houston, D.B., Zhang, C. and Larner, J. (1993) Acute glycogenic and hypoglycemic effects of two inositol-phosphoglycan insulin mediators in normal and streptozotocin-diabetic rats *in vivo. Endocrinol.* 132: 652–657.

Jarett, L. and Seals, J.R. (1979) Pyruvate dehydrogenase activation in adipocyte mitochondria by an insulin-generated mediator from muscle. *Science* 206: 1406–1408.

Jungas, R.L. (1971) Hormonal regulation of pyruvate dehydrogenase. *Metabolism* 20: 43–53.

Larner, J., Galasko, G., Cheng, K., DePaoli, A.A., Huang, L., Daggy, P. and Kellogg, J. (1979) Generation by insulin of a chemical mediator that controls protein phosphorylation and dephosphorylation. *Science* 206: 1408–1410.

Larner, J., Huang, L.C., Schwartz, C.F.W., Oswald, A.S., Shen, T.-Y., Kinter, M., Tang, G. and Zeller, K. (1988a) Rat liver mediator which stimulates pyruvate dehydrogenase phosphatase contains galactosamine and D-chiroinositol. *Biochem. Biophys. Res. Comm.* 151: 1416–1426.

Larner, J., Huang, L.C., Tang, G., Suzuki, S., Schwartz, C.F.W., Romero, G., Roulidis, Z., Zeller, K., Shen, T.-Y., Oswald, A.S. and Luttrell, L. (1988b) Insulin mediators: Structure and function. *Cold Spring Harb. Symp. LIII* 965–971.

Larner, J., Huang, L.C., Suzuki, S., Tang, G., Zhang, C., Schwartz, C.F.W., Romero, G., Luttrell, L. and Kennington, A.S. (1989) Insulin mediators and the control of pyruvate dehydrogenase complex. *Ann. of N.Y. Acad. Sci.* 573: 297–305.

Larner, J., Abe, S., Zhang, C., Rule, G., Price, J. and Huang, L.C. (1994) Action mechanisms and partial structures of insulin mediators-special poster session. Late-breaking topics. *Am. Soc. Biochem. Molec. Biol.*, Washington, D.C.

Lawson, J.E., Niu, X.-D., Browning, K.S., Trong, H.L., Yan, J. and Reed, L.J. (1993) Molecular cloning and expression of the catalytic subunit of bovine pyruvate dehydrogenase phosphatase and sequence similarity with protein phosphatase 2C. *Biochem.* 32: 8987–8993.

Linn, T.C., Pettit, F.H., and Reed, L.J. (1969) Regulation of the activity of the pyruvate dehydrogenase complex from beef kidney mitochondria by phosphorylation and dephosphorylation. *Proc. Natl. Acad. Sci. USA* 62: 234–241.

MaCauley, S.L. and Jarett. L. (1985) Insulin mediator causes dephosphorylation of the α subunit of pyruvate dehydrogenase by stimulating phosphatase activity. *Arch. Biochem. Biophys.* 237: 142–150.

Mato, J.M., Kelly, K.L., Abler, A., Jarett, L., Corkey, B.E., Cashel, J.A. and Zopf, D. (1987) Partial structure of an insulin-sensitive phospholipid. *Biochem. Biophys. Res. Comm.* 146: 764–770.

Nestler, J.E., Romero, G., Huang, L.C., Zhang, C. and Larner, J. (1991) Insulin mediators are the signal transduction system responsible for insulin's actions on human placental steroidogenesis. *Endocrinol.* 129: 2951–2956.

Romero, G., Luttrell, L., Rogol, A., Zeller, K., Hewlett, E. and Larner, J. (1988) Phosphatidylinositol-glycan anchors of membrane proteins: Potential precursors for insulin mediators. *Science* 240: 509–511.

Romero, G., Games, G., Huang, L.C., Lilley, K. and Luttrell, L. (1990) Anti-inositol glycan antibodies selectively block some of the action of insulin in intact BC_3H_1 cells. *Proc. Natl. Acad. Sci.* 88: 1476–1480.

Saltiel, A.R. and Cuatrecassas, P. (1986) Insulin stimulates the generation from hepatic plasma membranes of modulators derived from an inositol glycolipid. *Proc. Natl. Acad. Sci. USA* 83: 5793–5797.

Tamura, S., Lynch, K.R., Fox, J., Yasui, A., Kikuchi, K., Suzuki, Y. and Tsuiki, S. (1989) Molecular cloning of rat type 2C (1A) protein phosphatase mRNA. *Proc. Natl. Acad. Sci. USA* 86: 1796–1800.

Long-term regulation and promoter analysis of mammalian pyruvate dehydrogenase complex

M.S. Patel[1], S. Naik[2], M. Johnson[2] and R. Dey[1]

[1]*Department of Biochemistry, School of Medicine and Biomedical Sciences, State University of New York at Buffalo, Buffalo, NY 14214, USA*
[2]*Departments of Biochemistry and Genetics, Case Western Reserve University, School of Medicine, Cleveland, OH 44106, USA*

Introduction

The mammalian pyruvate dehydrogenase complex (PDC) plays a key role in the irreversible decarboxylation of pyruvate derived from glucose and amino acids to form acetyl-CoA in the mitochondria. Acetyl-CoA is then utilized for either energy production by the tricarboxylic acid cycle or energy storage by the lipogenic pathway. This enzyme complex contains multiple copies of three catalytic components: pyruvate dehydrogenase (E1), dihydrolipoamide acetyltransferase (E2) and dihydrolipoamide dehydrogenase (E3), two regulatory components (E1-kinase, phospho-E1 phosphatase) and one non-catalytic protein X (also referred to as E3-binding protein) (for review see: Reed, 1974; Patel and Roche, 1990; Behal et al., 1993).

Although the short-term regulation of the complex by covalent modification has been investigated extensively (for review see: Patel and Roche, 1990; Behal et al., 1993; Sugden and Holness, 1994), the long-term regulation of PDC by either nutritional or hormonal changes has not been subjected to vigorous investigation. In recent studies, transcriptional and/or pretranslational regulation induced by diet has been implicated in the coordinated modulation of all the component proteins of the complex (MacDonald et al., 1991; Da Silva et al., 1993). Changes in total PDC activity during differentiation and development also suggest coordinate regulation at the pretranslational and/or translational level(s) (Hu et al., 1983; Malloch et al., 1986).

c-DNA clones for the mammalian PDC components have been characterized (for review see Roche and Patel, 1989; Patel and Harris, 1995). Genomic organization and the promoter regulatory regions for the human E1α, E1β and E3 genes have recently been reported (Johanning et al., 1992; Chang et al., 1993; Madhusudan et al., 1995). Availability of cDNAs and genomic clones have facilitated the analyses of changes in mRNA levels as well as promoter regulatory regions of these genes. The aim of this chapter is two fold: (i) to discuss the recent findings on hormonal and dietary regulation of PDC and (ii) to present the emerging information on the analysis of promoter regions of the PDC genes.

Long-term regulation of PDC

Due to its importance in carbohydrate metabolism, PDC is subjected to both short-and long-term regulation. Short-term regulation of PDC is accompanied by a combination of end-product inhibition and reversible covalent phosphorylation (Patel and Roche, 1990; Behal et al., 1993). Both intrinsic (metabolites) and extrinsic (hormones) effectors regulate the active state of PDC. Intrinsic factors, like increase in the intramitochondrial ratios of NADH/NAD$^+$, acetyl-CoA/CoA and ATP/ADP ratios, stimulate kinase activity (for reviews see Patel and Roche, 1990; Behal et al., 1993). The activity of PDC in most tissues is maintained at constant levels; however the levels of active PDC (dephosphorylated) in most tissues are influenced by nutritional and hormonal modifications.

In progressive starvation, the short-term effects of oxidation of fatty acids and ketone bodies to decrease the percent active PDC in heart is enhanced by a more long-term mechanism involving a stable increase in the specific activity of E1-kinase appearing after 24 to 48 h (Kerbey and Randle, 1982). The increase in the E1-kinase specific activity may be mediated by fatty acids and cAMP. In both liver and heart, the inactivation of PDC after prolonged starvation has been correlated with an induction in E1-kinase activity. Similarly, the provision of a high fat diet (47% of calories as fat) for 28 days evoked a decline in "active" PDC activity together with an increase in the activity of E1 kinase in adult rats (Orfali et al., 1993). In contrast, in fed condition increased PDC activity is an important regulatory step leading to enhancement of *de novo* hepatic lipogenesis associated with high-fructose feeding. This activation is achieved *via* stable and long-term downregulation of E1-kinase as well as by acute activation of the phospho-E1 phosphatase (Sugden and Holness, 1994).

Insulin activates PDC *via* phospho-E1 phosphatase activation (Wieland et al., 1989), although it is not known how insulin communicates with the phosphatase. One hypothesis to explain the mechanism of insulin action proposes that the generation of small molecules referred to as "insulin mediator" initially and more recently "inositolglycan/inositol phosphoglycan mediator" in response to insulin, which in turn, modulate the activity of cellular protein kinases and phosphatases (for review see Romero and Larner, 1993). Such an insulin mediator isolated from rat liver membranes stimulated the most abundant phospho-E1 phosphatase, the divalent cation dependent phospho-E1 phosphatase, by decreasing the phosphatase's metal requirement. A metal independent phospho-E1 phosphatase was isolated from bovine heart mitochondria which is also stimulated by the insulin mediator but is insensitive to the effects of okadaic acid and spermine (Lilley et al., 1992). Recent observations indicate that the signal transduction pathway whereby insulin activates PDC in several cell lines involves protein kinase C (Gottschalk, 1991; Farese et al., 1985; Benelli et al., 1994).

Modulation by insulin of expression of several genes involved in glucose metabolism appears to be dependent on the presence of glucose. Recent evidence indicates that aerobic glycolysis is required for glucose-induced insulin secretion (Gottschalk, 1991). Pyruvate carboxylase and E1 were significantly altered by glucose in primary pancreatic islets indicating the important role of these enzymes in the signal transduction for insulin release in the β-cells of the pancreas (MacDonald et al., 1991). The exact biochemical nature of the signals that link glucose metabolism and other nutrient stimuli to insulin secretion is not well defined. Leucine, a potent physiological insulin secretagogue stimulates insulin release in the absence of extracellular glucose. The steady-state mRNA levels of the E1α subunit of branched-chain keto acid dehydrogenase complex and its enzymatic activity were decreased by 90% in islets cultured in 20 mM glucose (MacDonald et al., 1991). Conversely, at 20 mM glucose, the glucose-induced insulin release was preserved with increase in the mRNA of PDC E1α and pyruvate carboxylase. Changes in the level of these mRNAs in response to glucose could involve regulation occurring at the gene expression level *via* "glucose response elements". Such a putative carbohydrate response element has been identified in the L-type pyruvate kinase gene whose expression is induced after feeding a diet rich in carbohydrate (Thompson and Towle, 1991). Similar glucose response elements have been identified in the insulin gene, whose expression increased in primary cultures of rat islet cells with increasing concentrations of glucose (Goodison et al., 1992).

A significant increase in total PDC activity in liver was reported in rats fed a high-sucrose diet but little or no changes were observed in the adipose tissue (Chicco et al., 1986). Basal PDC activity (both active and total enzyme activity) was decreased in liver homogenates from rats fed high-fat *versus* high-glucose diets (Begum et al., 1983). Hepatic total and active PDC was 50–80% higher in rats fed a high-sucrose diet for 7 or 15 days (DaSilva et al., 1992). Dietary polyunsaturated fat caused a marked inhibition on total and active PDC in the liver (DaSilva et al., 1993). Total activity in adipose tissue increased three- to four-fold in animals fed a high-sucrose diet as compared with chow-fed animals (Fig. 1). In contrast, there was a significant decrease in total activity (60%) after 2 weeks of feeding a high-fat diet (22% corn oil) (DaSilva et al., 1992). These changes in total activity in the liver were closely correlated with changes in the enzyme content measured by immunological analysis (Fig. 1). Changes in total PDC activity result from alterations in the amount of enzyme present per mitochondrion. Long-term dietary regulation of

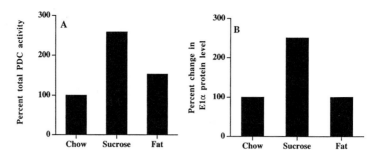

Figure 1. Effects of high-sucrose and high-fat diets on "total" PDC activity and on the level of E1α protein determined by Western analysis in rats fed for 2 weeks. The results are adapted from Da Silva et al. (1993).

PDC may involve transcriptional and/or posttranscriptional regulation. No effects on total and active PDC were observed in heart and skeletal muscle of animals fed the high-sucrose or high-fat diets (DaSilva et al., 1992).

The activity of PDC is not affected by hyperthyroidism. However, total PDC activity in rat liver was decreased by 33% in the hypothyroid state (Weinberg and Utter, 1979). Immunological experiments suggested that the decreased activity of PDC in the hypothyroid state was accompanied by decreased amounts of E1, the first component of the complex. These studies provide no evidence concerning the effects of thyroid status on the rates of synthesis or breakdown of other enzymes associated with the complex, although it is likely that they are coordinately regulated.

Development and differentiation

During normal mammalian development, each differentiating cell type must contend with changing energetic and synthetic needs while concurrently adapting to changing supplies of nutrients. Questions concerning the role of PDC in the development of the mammalian central nervous system have recently resurfaced in the light of clinical studies that have reported a broad range of neuropathologic findings associated with PDC deficiency (Cross et al., 1994; Robinson et al.,1987). The activity of active and total PDC during the postnatal development of the rat brain increases approximately three- to four-fold to attain adulthood levels by weaning (Cremer and Teal, 1974; Wilbur and Patel, 1974). The high levels of PDC at weaning are viewed as being critical for brain metabolism at this stage when this tissue becomes almost exclusively dependent on glucose as an energy source. By Western blot and ELISA analyses, Malloch et al. (1986) have shown that the changes seen in enzymatic activity during rat brain development were paralleled by

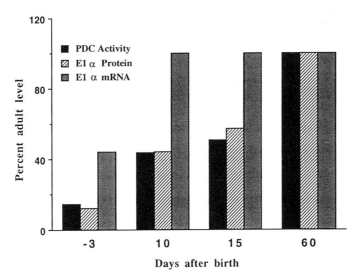

Figure 2. Regulation of PDC expression during pre- and postnatal rat brain development. The activity of total PDC and the E1α protein abundance were determined in rat brain homogenates and were adapted from Malloch et al. (1986). The steady-state levels of E1α mRNA determined by RNAse protection assays using rat brain total RNA are taken from Cullingford et al. (1994).

coordinate increases in the abundance of several of the enzymatic component proteins (Fig. 2). The change in protein content could be attributed to a combination of increases in protein per mitochondrion and the number of mitochondria per cell. This line of inquiry was further advanced by Cullingford et al. (1994) using an RNAse protection assay to determine the steady-state mRNA levels produced by the various subunit genes during brain development. As shown in Figure 2, the message level of E1α begins to increase during the perinatal period until it reaches near-adult levels by 10 days postpartum which precedes the changes in protein abundance and enzymatic activity by several days. The lack of temporal coordination between the changes in mRNA and protein levels suggests control at the translational and/or post-translational level(s).

The distribution and region-specific development of PDC expression in different anatomically defined sections of the brain (cortex, cerebellum, hypothalamus, striatum, midbrain, and medulla) during postnatal development showed similar rates of increase in PDC activity during this period although the timing of when the different regions began to increase varied (Butterworth and Guiguere, 1984). The medulla seemed to attain adult levels earliest (within 15–20 days post partum, whereas the cerebellum seemed most delayed (30 days or longer). The differences in PDC expression patterns in different regions of the brain emphasize that the brain is a metabolically heterogeneous tissue; a fact which is reinforced by studies finding distinct patterns of expression of PDC in the adult rat brain (Milner et al., 1987; Bagely et al., 1989).

To begin to address PDC expression patterns in the brain at earlier developmental stages, Takakubo and Dahl (1994) have recently examined the fetal mouse brain with a three-pronged analysis incorporating enzymatic assays, reverse transcriptase polymerase chain reaction, and immunohistochemistry. The development of total PDC activity in brain homogenates increased from 1% of adult level at embryonic day 9 (E9) to 15% at E18. The samplings at E11 and E15 showed the most dramatic change with an increase of 8 percentage points within the 4-day span. The reverse transcriptase polymerase chain reaction studies did not show a similar trend since the E1α mRNA levels exhibited a U-shaped pattern going from high levels at E9 and then dropping more than 50% by E15 and then regaining the earlier levels by E18. The loose relationship between the apparent protein and mRNA levels seems to suggest involvement of a translational or posttranslational mechanism. The immunohistochemical localization of the E1α protein in the developing embryo showed most prominent expression in the central and peripheral nervous system and the heart at E11. The developmental pattern of expression of E1α in the brain is first detected in the metencephalon, the basal region of the mesencephalon, and the roof of the diencephalon at E11 and eventually becomes expressed strongly in the frontal lobes by E15 and the cerebellum by E18 similar to the pattern of postnatal induction. Interestingly, the areas with high embryonic expression tend to be most susceptible to damage in cases of PDC deficiency documented in humans.

The liver and skeletal muscle of mammals also seem to follow the general trend of steadily increasing levels of total PDC activity from the perinatal period until leveling off when adult levels are reached. Knowles and Ballard (1974) quantified the increase in rat liver to be approximately three-fold in total activity although the activity of the active complex remained relatively constant. Focusing around the immediate postnatal period, Chitra et al. (1985) found a significant induction of active and total PDC within the first 2 and 6 h of extrauterine life, respectively. Subsequent studies using Western blot analysis showed coordinate increases in the three readily detected subunit proteins: E1α, E1β, and E2 (Serrano et al., 1989). Skeletal muscle as measured in the quadriceps appears to undergo a more gradual increase in which PDC activity multiplies three-fold over the first three weeks of postnatal life in the rat (Sperl et al., 1992). This increase in activity is paralleled by increases in abundance of the three detectable subunit proteins. Histologic studies also point to a correlation between PDC induction and skeletal muscle maturation.

The role of PDC in the lipogenic pathway in which glucose carbon is converted to long chain fatty acids has served as a rationale for studies to examine its expression patterns during adipose tissue development. To examine PDC regulation in this developing tissue, the *in vitro* model system in which murine 3T3-L1 preadipocytes are terminally differentiated was exploited. A seven fold increase in PDC activity was registered within 11 days after these cells were induced to differentiate by using standard pharmacologic treatments of the cells with insulin, dexamethasone and isobutylmethylxanthine. By examining the phosphorylation status of the complex combined with immunoprecipitation and pulse-labeling methods, it was shown that the increase in activity was marked by a comparable increase in the synthesis of the E1α and E1β proteins whereas the proportion of active complex and the rate of degradation of these proteins remained unchanged (Hu et al., 1983). In a follow-up study, it was observed that the E3 subunit protein also experienced an approximately three fold increase in activity which could be explained by enhanced protein synthesis (Carothers et al.,1988). More recently, there has been a suggestion that this induction may be regulated at the pretranslational level (Huh et al.,1991). A related study involving the rat mammary gland has shown an increase in PDC activity during lactation which also would support the proposed idea that PDC levels are elevated during periods of increased lipogenesis (Coore and Field, 1974).

Cloning and characterization of the PDC complex genes

The cDNAs of the E1α and the E1β subunits have been isolated from several species. The E1α cDNA (1423 base pairs) encoding the first 29 amino acids of the mitochondrial leader sequence and 361 amino acids of the mature peptide was isolated from a human liver cDNA library (Brown et al., 1989). The human liver E1β cDNA encodes a mature protein of 329 amino acids as well as 26 amino acids of the leader peptide (Koike et al., 1990). The cDNAs for the α and β subunits of rat E1 have been characterized and consist of 2367 and 1410 bp, respectively (Matuda et al., 1991). The predicted amino acid sequences for the α and β subunits of rat PDC are 98 and 90% identical to that of the human α and β subunits.

The E1α subunit of PDC exists as two isozymes encoded by two different genes. One isoform gene, Pdha1, is located on chromosome X (Xp22.1-22.2) in humans (Dahl et al., 1987) and is expressed in somatic tissues. The second variant, Pdha2, maps to an autosome (chromosome 4 in humans and chromosome 19 in mice) and is testis-specific (Dahl et al., 1990). In addition, Pdha2 is intronless suggesting that it may have evolved as a functional processed retroposon. In contrast, a single Pdha homologous gene was isolated in marsupials. The gene is autosomal and maps to a region that contains other human Xp genes (Fitzgerald et al., 1993). This suggests that the translocation in eutherian mammals of Pdha1 to the X chromosome which is inactivated during spermatogenesis, led to the evolution of a second testis-specific, intronless Pdha variant by retroposition to an autosome.

The X-linked gene for the E1α subunit is approximately 17 kb long, containing 11 exons ranging in size from 61 to 174 bp and introns ranging from 600 bp to 5.7 kb (Maragos et al., 1989; Koike et al., 1990). There is a great degree of conservation at the amino acid sequence of the somatic E1α subunits when comparing human to mouse or rat (98 to 99% identity). Similarly, comparison of the testis-specific forms of E1α for rat and mouse shows 92% identity, whereas that between rat and human and between mouse and human are slightly less conserved with 78 and 75% identities respectively (Cullingford et al., 1993).

Several consensus sequences namely, activator protein (AP-1), cAMP-responsive element (CRE), activator proteins (AP-2), CCAAT/enhancer binding protein (C/EBP) and an insulin

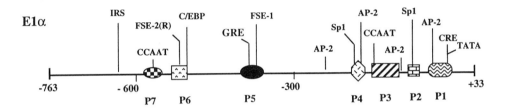

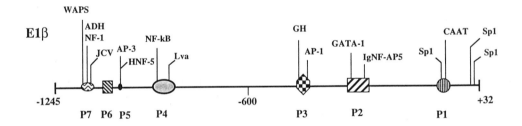

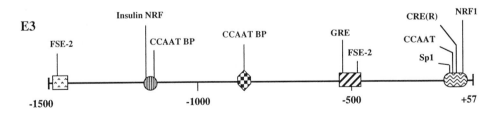

Figure 3. Schematic representation of cis-elements and protein-binding domains determined by DNase I footprinting analysis of the human E1α, E1β and E3 promoters. The information on the E1α, E1β and E3 promoters summarized here is derived from the papers of Chang et al. (1993), Madhusudhan et al. (1995), and Johanning et al. (1993), respectively.

responsive sequence (IRIS) have been identified in a segment (–763 to +33) of the human E1α gene (Maragos et al., 1987; Chang et al., 1993) (Fig. 3). DNase I footprinting analysis using rat liver nuclear extracts has identified seven major protein-binding domains termed P1 through P7 (Chang et al., 1993). Several of these footprint domains show sequence similarities to the known consensus sequences (Fig. 3). It was further shown that purified Fos and Jun bound to the CRE in P1 and AP-2 bound to its site that overlapped P2 and P3 and a distal site within P4 region. Furthermore, purified recombinant C/EBP bound to its corresponding motif within P6 and Sp1 protected two sites within P2 and P4 from nuclease digestion.

Chimeric gene constructs containing a series of nested deletions of the E1α promoter ligated to the reporter chloramphenicol acetyltranferase (CAT) gene were transiently transfected into human hepatoma cells (HepG2) (Fig. 4).The chimeric gene containing the 796 bp (–763 to +33) resulted in a very high CAT expression (Chang et al., 1993). The smallest construct, containing only 102 bp of the E1α 5'-flanking region resulted in a significant decrease in CAT activity relative to the construct containing the 796 bp fragment, although it was still capable of efficiently promoting CAT expression when compared with other standard promoters. It should be noted that this 102 bp region contains the TATA box, CRE, Sp1 and AP-2 sites and was almost completely protected by liver nuclear proteins from DNase I footprint analysis.

Universal expression of PDC suggests constitutive expression of all its gene products. The 5'-flanking region of E1α bears structural similarities to both facultative and housekeeping gene promoters.The former group includes genes that are tissue- or temporal-specific in their expression and are regulated by appropriately positioned TATA and CAAT boxes with upstream control cis-elements.On the other hand, constitutively expressed genes are characterized by the absence of TATA and CAAT boxes and the presence of multiple GC boxes. Therefore the E1α promoter, due to its dual characteristics, may provide an interesting system in elucidating the fundamental differences in the transcriptional regulation of these two types of promoters.

Regulation of the proximal promoter region (–545 to +22 bp) of the intron-less Pdha-2 gene in mouse has been recently reported. The Pdha-2 gene harbors a transcriptionally active core region between nucleotides –187 to +22 which expresses relatively high levels of CAT activity in Hela cells (Iannello et al., 1993). DNase I footprinting analysis using nuclear extracts from mouse testis has identified four regions of protection, namely an Sp1 binding site, an ATF/CREB site, MEP-2 (Mouse E1α Promoter site) and MEP-3 sites. The factor which recognizes MEP-3 appears to be ubiquitous, whereas the MEP-2 complexes were tissue specific involving the formation of a complex between MEP-2 and a putative testis-specific factor (τ-MEP-2BF), first observed in the testis of 2-week old mice (Iannello et al., 1993). A similar sequence was identified in the promoter region of the mouse testis-specific phosphoglycerate kinase gene (Pgk-2) (Goto

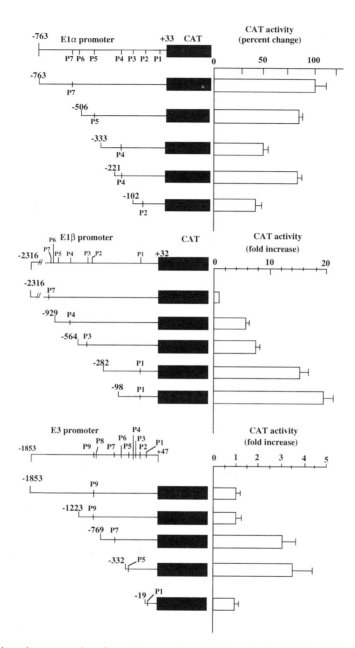

Figure 4. Schematic representation of transient expression of CAT activity in HepG2 cells by various deletion constructs of the promoter-regulatory regions of the human E1α (Pdha-1), E1β and E3 genes ligated to the CAT structural gene. The information on the E1α, E1β and E3 promoters summarized here is derived from the papers of Chang et al. (1993), Madhusudhan et al. (1995), and Johanning et al. (1993), respectively. The information on E3 promoter protein-binding domains identified by DNaseI footprint analyses are unpublished findings of J. Morris, G. Johanning and M.S. Patel.

et al., 1993). The nucleotide sequence that comprises this element bears homology to an Ets-binding motif and appears to interact with an Ets-related factor (TAP-1). Sp1 binding may work in concert with other factors to regulate the testis-specific expression of Pdha-2 and Pgk-2 due to the fact that Sp1 is also differentially regulated during spermatogenesis and is present down-stream from the Ets motif in the promoters of both genes (Saffer et al., 1991).

The human E1β gene located on chromosome 3 is approximately 8 kb long and contains 10 exons ranging in size from 36 to 550 bp (Koike et al., 1990). Analysis of approximately 1200 bp of the 5'-flanking region of the E1β gene shows that the E1β promoter lacks a TATA box but contains two initiator sequences and three Sp1 sites, frequently found in TATA-less promoters (Madhusudhan et al., 1995) (Fig. 3). Initiator elements are believed to mediate the same functions as TATA elements in localization of the start site of transcription by RNA polymerase II (Zawel and Reinberg, 1993). DNase I footprinting using crude rat liver nuclear extracts, revealed at least seven regions of protein binding termed P1 through P7 (Madhusudhan et al., 1995) (Fig. 3). The seven protected regions contain several known consensus sequences, namely, Sp1, GATA-1, NF-kB, HNF-5, ADH, WAP5 and IgNF-A. Using nested deletions of the human E1β promoter region (−2316 to +32 bp), it was observed that CAT expression increased approximately seven-fold and 20-fold for the promoter regions −929/+32 and −98/+32 bp, respectively, compared to the largest fragment (Fig. 4) suggesting the presence of one or more negative elements in the promoter region of the E1β gene. Such a repressor activity has been identified between residues −2316 to −929 using a heterologous thymidine kinase promoter to measure CAT activity (Madhusudhan et al., 1995). The minimal promoter (−98/+32 bp) contains a CAAT box in the P1 region and two potential Sp1 binding sites in the 5'-untranslated region of the E1β gene.

The human E3 gene has been localized to chromosome 7 (7q31.1-32) and has been determined to consist of 14 exons with sizes from 69 to 780 bp that are contained within an approximately 20 kb genomic region (Feigenbaum and Robinson, 1993; Scherer et al., 1991). Northern analysis using the human E3 cDNA as a probe has shown two hybridizing species of 2.2 and 2.4 kb in size in total RNA isolated from various human tissues whereas only a single 2.4 kb band was seen in rat tissues (Pons et al., 1988). Approximately 2 kb of 5' flanking region of this gene has been sequenced and appears to have some of the characteristics of a housekeeping-type promoter as it lacks a canonical TATA box and contains a CpG-rich region near the transcription start site. On the other hand, the presence of several putative cis-acting regulatory motifs including at least one cyclic AMP response element (CRE), fat-specific element 2 (FSE 2) and insulin negative response element (NRE) and nuclear respiratory factor-1 (NRF-1) (Fig. 3) may imply some inducible features to this gene (Johanning et al., 1992; Feigenbaum and Robinson, 1993). To begin to define functional domains in the promoter/regulatory region, transient expression assays were performed using a series of 5'-nested deletions of this promoter ligated to a CAT reporter. Of note, there appears to be a region from –769 to –1223 that suppresses basal transcription and a very small fragment of the 5' region from –19 to +47 that maintains a significant level of expression (Fig. 4). Further work is needed to define the cis and trans elements involved in and hormonal effects on transcriptional regulation of this gene.

The cDNA sequence of the PDC-E2 has been determined independently by three groups with each having used a different tissue source (Coppel et al., 1988; Thekkumkara et al., 1988; Moehario et al., 1990). Differences among these sequences have raised as yet unresolved questions as to whether these discrepancies can be explained by the existence of tissue-specific isoforms, polymorphisms, or inaccuracies. The open reading frame encodes a protein of 561 amino acid residues with a leader sequence of 54 residues. Northern blot analysis shows at least three different sized messages (2.3, 2.9, and 4 kb) which hybridize to the cloned human liver cDNA when either human heart or rat kidney was used as an RNA source (Thekkumkara et al., 1988). Recently, the E2 cDNA sequence was determined from rat heart and appears to be most closely related to this human liver sequence (Matuda et al., 1992).

At present, neither the cDNA nor the gene has been cloned for the mammalian protein X, or more recently termed the E3 binding protein, E3BP. The sequence of the E3BP cDNA from *Saccharomyces cerevisiae* seems to indicate a common ancestry with the E2 gene as determined by similarity of the amino terminal lipoyl domains. The open reading frame within the 1.5 kb mRNA indicates a mature protein of 380 amino acids with a 30 residue leader peptide (Behal et al., 1989).

Conclusions

This overview of the dietary and developmental regulation of PDC in various tissues brings out several important points. Firstly, there appears to be a general trend in which the total PDC activity increases with or without a corresponding increase in the active PDC as cells differentiate. The need for an added reserve of PDC activity may be important in adult mammals during certain physiologic conditions. Secondly, certain dietary and hormonal manipulations result in changes in total PDC activity over a period of days and weeks. In the few cases examined, the increase in PDC activity is a result of increased component proteins with the major regulatory step in this synthetic process being positioned at the translational and/or posttranslational level. The parallel expression of several of the subunit proteins implies some form of coordinate regulation although little is known presently. Thirdly, the initial characterization of the promoter-regulatory regions of several PDC genes have shed light on unique features of these promoters. Additional studies on these promoters expressing chimeric gene products in transfected cells and transgenic animals may provide better insight into gene expression and possible coordinate regulation of the PDC genes.

Acknowledgement
The work referred to in this article and performed in the authors' laboratory was supported by U.S. Public Health Service Grants DK20478 and DK 42885.

References

Bagley, P.R., Tucker, S.P., Nolan, C., Lindsey, J.G., Davies, A., Baldwin, S.A., Cremer, J.E. and Cunningham, V.J. (1989) Anatomical mapping of glucose transporter protein and pyruvate dehydrogenase in rat brain: an immunogold study. *Brain Res.* 409: 214–224.

Benelli, C., Caron, M., de Galle, B., Fouque, F., Cherqui, G. and Clot, J.-P. (1994) Evidence for a role of protein kinase C in the activation of the pyruvate dehydrogenase complex by insulin in Zajdela hepatoma cells. *Metabolism* 43: 1030–1034.

Behal, R.H., Browning, K.S., Hall, B.T. and Reed, L.J. (1989) Cloning and nucleotide sequence of the gene for protein X from Saccharomyces cerevisiae. *Proc. Natl. Acad. Sci. USA* 86: 8732–8736.

Behal, R.H., Buxton, D.B., Robertson, J.G. and Olson, M.S. (1993) Regulation of the pyruvate dehydrogenase multienzyme complex. *Ann. Rev. Nutr.* 13: 497–520.

Begum, N., Tepperman, H.M. and Tepperman, J. (1983) Effects of high fat and high carbohydrate diets on liver pyruvate dehydrogenase and its activation by a chemical mediator released from insulin-treated liver particulate fraction: Effect of neuraminidase treatment on the chemical mediator activity. *Endocrinology* 112: 50–59.

Brown, R.M., Dahl, H.-H. M. and Brown, G.K. (1989) X chromosome location of the functional gene for the E1α subunit of the human pyruvate dehydrogenase complex. *Genomics* 4: 174–181.

Butterworth, R.F. and Giguere, J.F. (1984) Pyruvate dehydrogenase activity in regions of the rat brain during postnatal development. *J. Neurochem.* 43: 280–282.

Carothers, D.J., Pons, G. and Patel, M.S. (1988) Induction of dihydrolipoamide dehydrogenase in 3T3-L1 cells during differentiation. *Biochem. J.* 214: 177–181.

Chang, M., Naik, S., Johanning, G.L., Ho, L. and Patel, M.S. (1993) Multiple protein-binding domains and functional cis-elements in the 5'-flanking region of the human pyruvate dehydrogenase α-subunit gene. *Biochemistry* 32: 4263–4269.

Chicco, A., Gutman, R., Basilico, M. and Lombardo, Y. (1986) Pyruvate dehydrogenase (PDH) activity in heart, liver and adipose tissue of rats fed a sucrose-rich diet. *Nutr. Rep. Intl.* 33: 465–475.

Chitra, C.I., Cuezva, J.M. and Patel, M.S. (1985) Changes in the activity of active pyruvate dehydrogenase complex in the newborn of normal and diabetic rats. *Diabetologia* 28: 148–152.

Coore, H.G. and Field, B. (1974) Properties of pyruvate dehydrogenase of rat mammary tissue and its changes during pregnancy, lactation, and weaning. *Biochem. J.* 142: 87–95.

Coppel, R.L., McNeilage, L.J., Surh, C.D., Van De Water, Spithill, T., Whittingham, S. and Gershwin, M.E. (1988) Primary structure of the human M2 mitochondrial autoantigen of primary biliary cirrhosis: dihydrolipoamide acetyltransferase. *Proc. Natl. Acad. Sci. USA* 85: 7317–7321.

Cremer, J.E. and Teal, H.M. (1974) The activity of pyruvate dehydrogenase in rat brain during postnatal development. *FEBS Lett.* 39: 17–20.

Cross, J.H., Connelly, A., Gadian, D.G., Kendall, B.E., Brown, G.K., Brown, R.M. and Leonard, J.V. (1994) Clinical diversity of pyruvate dehydrogenase deficiency. *Pediat. Neurol.* 10: 276–283.

Cullingford, T.E., Clark, J.B. and Philips, I.R. (1993) Characterization of cDNAs encoding the rat testis-specific E1α subunit of the pyruvate dehydrogenase complex: comparison of expression of the corresponding mRNA with that of the somatic E1α subunit. *Biochim. Biophys. Acta* 1216: 149–153.

Cullingford, T.E., Clark, J.B. and Phillips, I.R. (1994) The pyruvate dehydrogenase complex: cloning of the rat somatic E1α subunit and its coordinate expression with the mRNA for the E1β, and E2 and E3 catalytic subunits in developing rat brain. *J. Neurochem.* 62: 1682–1690.

Dahl, H.-H.M., Hunt, S.M., Hutchison, W.M. and Brown, G.K. (1987) The human pyruvate dehydrogenase complex: Isolation of cDNA clones for the E1α subunit, sequence analysis and characterization of the mRNA. *J. Biol. Chem.* 262: 7398–7405.

Dahl, H.-H.M., Brown, R.M., Hutchinson, W.M., Maragos, C. and Brown, G.K. (1990) A testis-specific form of the human pyruvate dehydrogenase E1α subunit is coded for by an intronless gene on chromosome 4. *Genomics* 8: 225–232.

Da Silva, L.A., De Marcucci, O.L. and Carmona, A. (1992) Adaptive changes in total pyruvate dehydrogenase activity in lipogenic tissues of rats fed high-sucrose or high-fat diets. *Comp. Biochem. Physiol.* 103A: 407–411.

Da Silva, L.A., De Marcucci, O.L. and Kuhnle, Z.R. (1993) Dietary polyunsaturated fats suppress the high-sucrose induced increase of rat liver pyruvate dehydrogenase levels. *Biochim. Biophys. Acta* 1169: 126–134.

Farese, R.V., Standaert, M.L., Barnes, D.E., Davis, J.S. and Pollet, R.J. (1985) Phorbol ester provokes insulin-like effects on glucose transport, amino acid uptake and pyruvate dehydrogenase activity in BC3H-1 cultured myocytes. *Endocrinology* 116: 2650–2655.

Feigenbaum, A.S. and Robinson, B.H. (1993) The structure of the human dihydrolipoamide dehydrogenase gene (DLD) and its upstream elements. *Genomics* 17: 376–381.

Fitzgerald, J., Wilcox, S.A., Marshall Graves, J.A. and Dahl, H.-H.M. (1993) A eutherian X-linked gene, PDHA1, is autosomal in marsupials: A model for the evolution of a second, testis-specific variant in eutherian mammals. *Genomics* 18: 636–642.

Goodison, S., Kenna, S. and Ashcroft, S.J.H. (1992) Control of insulin gene expression by glucose. *Biochem. J.* 285: 563–568.

Goto, M., Masamune, Y. and Nakanishi, Y. (1993) A factor stimulating transcription of the testis-specific Pgk-2 gene recognizes a sequence similar to the binding site for a transcription inhibitor of the somatic-type Pgk-1 gene. *Nucleic Acids Res.* 21: 209–214.

Gottschalk, W.K. (1991) The pathway mediating insulin's effect on pyruvate dehydrogenase bypasses the insulin receptor tyrosine kinase. *J. Biol. Chem.* 266: 8814–8819.

Hu, C.-W.C., Utter, M.F. and Patel, M.S. (1983) Induction of pyruvate dehydrogenase in 3T3-L1 cells during differentiation. *J. Biol. Chem.* 258: 2315–2320.

Huh, T.L., Huh, J.W., Veech, R.L. and Song, B.J. (1991) Coordinate, pretranslational expression of pyruvate dehydrogenase complex during adipocyte differentiation. *FASEB J.* 5: A1200.

Iannello, R.C., Kola, I. and Dahl, H.-H. M. (1993) Temporal and tissue-specific interactions involving novel transcriptional factors and the proximal promoter of the mouse Pdha-2 gene. *J. Biol. Chem.* 268: 22581–22590.

Johanning, G.L., Morris, J.I., Madhusudhan, K.T., Samols, D. and Patel, M.S. (1992) Characterization of the transcriptional regulatory region of the human dihydrolipoamide dehydrogenase gene. *Proc. Natl. Acad. Sci. USA* 89: 10964–10968.

Kerbey, A.L. and Randle, P.J. (1982) Pyruvate dehydrogenase kinase/activator in rat heart mitochondria. Assay, effect of starvation and effect of protein synthesis inhibitors in starvation. *Biochem. J.* 206: 103–111.

Knowles, S.E. and Ballard, F.J. (1974) Pyruvate dehydrogenase activity in rat liver during development. *Biol. Neonate* 24: 41–48.

Koike, K., Urata, Y. and Koike, M. (1990) Molecular cloning and characterization of human pyruvate dehydrogenase β-subunit gene. *Proc. Natl. Acad. Sci. USA* 87: 5594–5597.

Lilley, K., Zhang, C., Villar-Palasi, C., Larner, J. and Huang, L. (1992) Insulin mediator stimulation of pyruvate dehydrogenase phosphatases. *Arch. Biochem. Biophys.* 296: 170–174.

Malloch, G.D.A., Munday, L.A., Olson, M.S. and Clark, J.B. (1986) Comparative development of pyruvate dehydrogenase complex and citrate synthase in rat brain mitochondria. *Biochem. J.* 238: 729–736.

MacDonald, M.J., Kaysen, J.H., Moran, S.M. and Pomije, C.E. (1991) Pyruvate dehydrogenase and pyruvate carboxylase: Sites of pretranslational regulation by glucose of glucose-induced insulin release in pancreatic islets. *J. Biol. Chem.* 266: 22392–22397.

Madhusudhan, K.T., Naik, S. and Patel, M.S. (1995) Characterization of the promoter regulatory region of the human pyruvate dehydrogenase β gene. *Biochemistry* 34: 1288–1294.

Maragos, C., Hutchison, W.M., Hayasaka, K., Brown, G.K. and Dahl, H.-H.M. (1989) Structural organization of the gene for the E1α subunit of the human pyruvate dehydrogenase complex. *J. Biol. Chem.* 264: 12294–12298.

Matuda, S., Nakano, K., Ohta, S., Saheki, T., Kawanishi, Y. and Miyata, T. (1991) The α-ketoacid dehydrogenase complexes. Sequence similarities of rat pyruvate dehydrogenase with *Escherichia coli* and *Azotobacter vinelandii* α-ketoglutarate dehydrogenase. *Biochim. Biophys. Acta* 1089: 1–7.

Matuda, S., Nakano, K., Ohta, S., Shimura, M., Yamanaka, T., Nakagawa, S., Titani, K. and Miyata, T. (1992) Molecular cloning of the dihydrolipoamide acetyltransferase of the rat pyruvate dehydrogenase complex: sequence comparison and evolutionary relationship to other dihydrolipoamide acyltransferases. *Biochim. Biophys. Acta* 1131: 114–118.

Milner, T.A., Aoki, C. Rex Sheu, K.-F., Blass, J.P. and Pickel, V.M. (1987) Light microscopic immunocytochemical localization of pyruvate dehydrogenase complex in rat brain: topographical distribution and relation to cholinergic and catecholaminergic nuclei. *J. Neurochem.* 7: 3171–3190.

Moehario, L.H., Smooker, P.M., Devenish, R.J., Mackay, I.R., Gershwin, M.E.and Marzuki, S. (1990) Nucleotide sequence of a cDNA encoding the lipoate acetyltransferase (E2) of human heart pyruvate dehydrogenase complex differs from that of human placenta. *Biochem. Int.* 20: 417–422.

Orfali, K.A., Fryer, L.G.D., Holness, M.J. and Sugden, M.C. (1993) Long-term regulation of pyruvate dehydrogenase kinase by high-fat feeding. Experiments *in vivo* and in cultured cardiomyocytes. *FEBS Lett.* 336: 501–505.

Patel, M.S. and Roche, T.E. (1990) Molecular biology and biochemistry of pyruvate dehydrogenase complexes. *FASEB J.* 4: 3224–3233.

Patel, M.S. and Harris, R.A. (1995) α-Keto acid dehydrogenase complexes: gene regulation and genetic defects. *FASEB J.* 9: 1164–1172.

Pons, G., Raefsky-Estrin, C., Carothers, D.J., Pepin, R.A., Javed, A.A., Jesse, B.W., Ganapathi, M.K., Samols, D. and Patel, M.S. (1988) Cloning and cDNA sequence of the dihydrolipoamide dehydrogenase component of human α-ketoacid dehydrogenase complexes. *Proc. Natl. Acad. Sci. USA* 85: 1422–1426.

Reed, L.J. (1974) Multienzyme complexes. *Acc. Chem. Res.* 7: 40–46.

Robinson, B.H., MacMillan, H., Petrova-Benedict, R. and Sherwood, W.G. (1987) Variable clinical presentation in patients with defective E1 component of pyruvate dehydrogenase complex. *J. Pediatr.* 111: 525–533.

Roche, T.E. and Patel, M.S. (1989) Alpha-keto acid dehydrogenase complexes: organization, regulation and biomedical ramifications. *Ann. N.Y. Acad. Sci.* 573: 1–474.

Romero, G. and Larner, J. (1993) Insulin mediators and the mechanism of insulin action. *Adv. Pharm.* 24: 21–50.

Saffer, J.D., Jackson, S.P. and Annarella, M.B. (1991) Developmental expression of Sp1 in the mouse. *Mol. Cell. Biol.* 11: 2189–2199.

Scherer, S.W., Otulakowski, G., Robinson, B.H. and Tsui, L.-C. (1991) Localization of the human dihydrolipoamide dehydrogenase gene (DLD) to 7q31-q32. *Cytogenet. Cell Genet.* 56: 176–177.

Serrano, E., Luis, A.M., Encabo, P., Alconada, A., Ho, L., Patel, M.S. and Cuezva, J.M. (1989) Rapid postnatal induction of the pyruvate dehydrogenase complex in the rat liver mitochondria. *Ann. N.Y. Acad. Sci.* 573: 412–415.

Sperl, W., Sengers, R.C.A., Trijbels, J.M.F., Ruitenbeek, W., De Graaf, R., Terlaak, H., Van Lith, T., Kerkhoff, C. and Janssen, A. (1992) Postnatal development of pyruvate dehydrogenase oxidation in quadriceps muscle of rat. *Biol. Neonate* 61: 188–200.

Sugden, M.C. and Holness, M.J. (1994) Interactive regulation of the pyruvate dehydrogenase complex and the carnitine palmitoyltransferase system. *FASEB J.* 8: 54–61.

Takakubo, F. and Dahl, H.-H.M. (1994) Analysis of pyruvate dehydrogenase expression in embryonic mouse brain: localization and developmental regulation. *Dev. Brain Res.* 7: 63–75.

Thekkumkara, T.J., Ho, L., Wexler, I.D., Pons, G., Leu, T.C. and Patel, M.S. (1988) Nucleotide sequence of a cDNA for the dihydrolipoamide acetyltransferase component of human pyruvate dehydrogenase complex. *FEBS Lett.* 240: 45–48.

Thompson, K.S. and Towle, H.C. (1991) Localization of the carbohydrate response element of the rat L-type pyruvate kinase gene. *J. Biol. Chem.* 266: 8679–8682.

Weinberg, M.B. and Utter, M.F. (1979) Effect of thyroid hormone on the turnover of rat liver pyruvate carboxylase and pyruvate dehydrogenase. *J. Biol. Chem.* 254: 9492–9499.

Wieland, O.H. (1983) The mammalian pyruvate dehydrogenase complex: structure and regulation. *Rev. Physiol. Biochem. Pharmacol.* 96: 123–169.

Wieland, O.H., Urumow, T. and Drexler, P. (1989) Insulin, phospholipase, and the activation of the pyruvate dehydrogenase complex: an enigma. *Ann. N.Y. Acad. Sci.* 573: 274–284.

Wilbur, D.O. and Patel, M.S. (1974) Development of mitochondrial pyruvate metabolism in rat brain. *J. Neurochem.* 22: 709–715.

Zawel, L. and Reinberg, D. (1993) Initiation of transcription by RNA polymerase II: a multi-step process. *Prog. Nucl. Acid Res. Mol. Biol.* 44: 67–108.

Alpha-Keto Acid Dehydrogenase Complexes
M.S. Patel, T.E. Roche and R.A. Harris (eds)
© 1996 Birkhäuser Verlag Basel/Switzerland

The sperm-specific pyruvate dehydrogenase E1α genes

H.-H.M. Dahl, J. Fitzgerald[1] and R. Iannello[2]

The Murdoch Institute, The Royal Children's Hospital, Flemington Road, Parkville, Melbourne, Victoria 3052, Australia
[1]*Present address: Department of Anesthesia, University of California, San Francisco (UCSF), San Francisco, CA 94143, USA;* [2]*Present address: Centre for Early Human Development, Institute of Reproduction and Development, Monash Medical Centre, Clayton, Victoria 3168, Australia*

Introduction

All cells require energy for function. This is generated partly through catabolic processes in which carbohydrates, fatty acids and amino acids are converted to metabolites which then can be used in energy-producing reactions. Although some energy can be, and is, generated by anaerobic metabolism (e.g., in glycolysis), most energy in mammalian cells is generated through aerobic energy production in the mitochondria. Nearly all tissues can use carbohydrates, fatty acids and amino acids for energy production. Despite this, they do have certain individual preferences and this leads to cooperation between various cells and tissues in the body. For example, under normal circumstances the liver uses mainly fatty acids and exports glucose. Muscles mainly use glucose for energy production, but can, if necessary, use some of the alternative energy sources. However, this is not true of all tissues. Brain and sperm cells are unique in that they generate nearly all their energy from carbohydrates or derivatives thereof (Fig. 1) and are therefore likely to be more severely affected by deficiencies in their cells' ability to convert carbohydrates to energy.

Pyruvate dehydrogenase (PDH) is an essential and rate-limiting enzyme complex connecting glycolysis with the TCA cycle (Reed, 1974). The PDH complex contains multiple copies of at least seven subunits: E1α, E1β, E2, E3, protein X, a PDH-specific kinase and a PDH-specific phosphatase. Apart from E1α, it has been shown that the genes coding for the PDH complex subunits are all autosomally located (Patel and Roche, 1990). However, it was found that the PDH E1α in somatic cells is coded for by a gene, named PDHA1, on the short arm of the X chromosome, band 22.13 (Brown et al., 1989a). Although at the time this was an unexpected finding, it has provided answers to some of the puzzling features noted in patients with PDH deficiency,

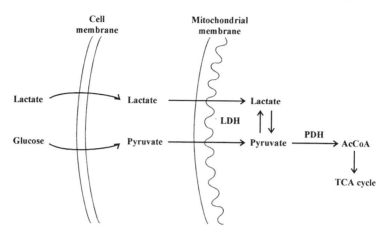

Figure 1. Energy sources in sperm. Carbohydrates such as lactate and glucose are taken up by the sperm and imported into the mitochondria. Lactate dehydrogenase (LDH) and pyruvate dehydrogenase convert these substrates to AcetylCoA (AcCoA) which then enters the tricarboxylic acid (TCA) cycle.

most of whom have PDH E1α deficiency (Brown et al., 1989b). Because PDH deficiency affects a cell's ability to generate ATP from carbohydrates one would expect that the brain would be more severely affected than most other tissues. This is indeed the case, as most or all clinically affected PDH E1α deficient patients have neurological abnormalities (Robinson, 1989; Shevell et al., 1994). However, the nature and severity of abnormalities are dependent on the mutation and, in females, on the X chromosome inactivation pattern (Brown et al., 1989b; Brown et al., 1994; Dahl, 1995).

The X chromosome location of the normally expressed PDH E1α gene has implications for energy production in sperm. Sperm maturation takes approximately 64 days, with the meiotic divisions occurring half-way during the spermatogenesis (Fig. 2). After the spermatogenic cells have gone through the meiotic divisions and become haploid, they retain only one of the sex chromosomes. Half the sperm cells will therefore not contain an X chromosome and the X chromosome located PDH E1α gene. In addition, the X chromosome appears to be inactivated early in spermatogenesis, so even those spermatogenic cells containing an X chromosome appear not to express genes located on this chromosome (Solari, 1989). If sperm either lack the X chromosomal PDH E1α gene or do not use it, how can they make this essential enzyme subunit? There are several possibilities. At a time when the X chromosome is still fully functional, PDH E1α or a precursor, could be made and stored for later use. Another possibility is that spermatogenic cells generate PDH E1α from an autosomally located gene. Autosomes, unlike the sex chromosomes,

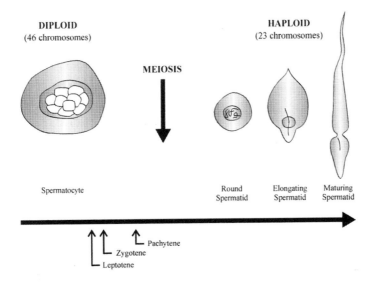

Figure 2. Schematic illustration of mammalian spermatogenesis.

are still functional in the haploid sperm cells. As it turns out, both mechanisms appear to co-operate in mammalian cells.

The PDHA2 gene

When the human PDHA1 gene was mapped to the X chromosome by *in situ* hybridisation, it was noted that the probe also detected a locus on chromosome 4 (Brown et al., 1989a). This second human locus was designated PDHA2 and the gene cloned. It was shown that this gene codes for a sperm-specific PDH E1α subunit (Dahl et al., 1990). Since then, similar PDH E1α genes have been described in mouse (Fitzgerald et al., 1992) and rat (Cullingford et al., 1993). In mouse the two genes, named Pdha-1 and Pdha-2, have been mapped to chromosomes X and 19 (Brown et al., 1990). It was shown that, as in humans, the X chromosome located gene is expressed in somatic tissues and the autosomal gene in spermatogenic cells (Iannello and Dahl, 1992; Iannello et al., 1993a). Two PDH E1α genes have also been characterised in the nematode *Ascaris suum* (Johnson et al., 1992). One gene is expressed mainly in adult muscle and the other in third-stage larvae as well as in adult muscle.

The amino acid sequences of the somatic PDH E1α subunits are highly homologous with fewer than 5% changes. Less homology is seen when the testis-specific forms are compared. For example, the amino acid homology between PDHA2 and Pdha-2 is 75% and between PDHA1 and PDHA2 86%. Most of the changes are conservative changes and located towards the C-terminus. Sequences near the 3 phosphorylation sites and the thiamine pyrophosphate binding site are highly conserved.

No mutations have been detected and characterised in the human PDHA2 gene. This is to be expected. A severe mutation in one of the chromosome 4 genes might lead to non-viable sperm, but only half of the sperm will be affected. This should not lead to infertility and the mutant PDHA2 gene will be strongly selected against.

Gene structure

The X chromosome located human and mouse PDH E1α genes both contain introns. The human PDHA1 gene contains 11 exons and is approximately 17 kb long (Maragos et al., 1989). Our studies suggest that the mouse Pdha-1 gene has a very similar structure to that of the human PDHA1. The human and mouse sperm-specific autosomal genes, on the other hand, are intron-less. This is equivalent to the situation with the human and mouse phosphoglycerate kinase (PGK) genes (McCarrey and Thomas, 1987). The reason for having intronless sperm-specific genes is not known. It has been suggested that these genes are functional retroposons, based on the presence of direct repeats flanking the reverse transcriptase-processed mRNA (McCarrey and Thomas, 1987). Two possible direct repeats have also been detected in the human (Dahl et al., 1990) and mouse sperm-specific PDH E1α genes (Fitzgerald et al., 1992). It is unlikely that such repeats would be conserved during evolution if they have no function. It is not clear if the PDHA2 and Pdha-2 genes arose in a common ancestor or as a result of two independent events (see below). If they arose in a common ancestor and have been conserved due to strong selection pressure, one would also expect homology between the repeats found near the human and mouse sperm-specific PDH E1α genes. However, no obvious similarities were found. The relevance and possible function of these repeats are therefore still obscure. Another explanation for the intron-less genes is that the RNA splicing machinery in the spermatogenic cells is not fully expressed and that the cells therefore are not able to efficiently process the hnRNA. In support of this is the fact that a number of other proteins expressed in sperm are coded for by intronless genes, e.g., PGK (McCarrey and Thomas, 1987), glutamate dehydrogenase (Shashidharan et al., 1994), SRY (Su and Lau, 1993), MYCL2 (Robertson et al., 1991). However, other genes expressed in sperm contain introns, e.g., the mammalian sperm-specific protamine genes (Johnson et al., 1988) and the lactate dehydrogenase-C gene (Takano and Li, 1989) which indicates that the spermatogenic cells do have some capacity to process intron-containing RNAs.

Expression pattern

Messenger RNA was isolated from human tissues, including liver, heart, fibroblast, brain, kidney and adult testis and analysed for expression of the sperm-specific PDH E1α mRNA. RNA blots were probed with a sperm-specific PDH E1α probe, or the mRNAs were used in RT-PCR reactions using PDHA2 specific oligonucleotide primers. These experiments proved that PDHA2 is expressed in a tissue-specific manner, but they did not allow us to determine in which spermatogenic cells PDHA2 is expressed (Dahl et al., 1990).

Mammalian spermatogenesis is a process of germ cell differentiation and can be divided into several stages. Firstly, primitive type A spermatogonia proliferate into type A and type B cells. Type B cells then sequentially differentiate into leptotene, zygotene and pachytene primary spermatocytes (Fig. 2). Secondly, the diploid pachytene spermatocytes pass through two stages of meiosis to become haploid round spermatids. Thirdly, round spermatids differentiate into elongated spermatids and eventually mature sperm. Spermiogenesis culminates with the excision of the residual body and the release of the spermatozoon into the tubule lumen. During mouse development, germ cell differentiation is synchronised as the mouse proceeds towards sexual maturity. Therefore, information about the timing of expression of developmentally regulated sperm-specific genes can relatively easily be obtained in mice. It is also easy to obtain mouse tissues and, as a consequence, the expression pattern of the mouse Pdha-2 gene has been studied in detail. RNA blots probed with a Pdha-2 specific fragment show that Pdha-2 is expressed as a 2.0 kb mRNA from day 16 post-natal in the mouse (Fitzgerald et al., 1994; Iannello and Dahl, 1992). This corresponds to the emergence of leptotene and zygotene spermatocytes (Fitzgerald et al., 1994; Iannello and Dahl, 1992; Takakubo and Dahl, 1992). Thus, Pdha-2 is expressed at the same time as another testis-specific gene, lactate dehydrogenase C (LDH-C) (Thomas et al., 1990), and prior to Pgka-2. Pgka-2 mRNA is first detected at low levels in pachytene spermatocytes during the prophase of meiosis and then in increasing amount following meiosis (Erickson et al., 1980; Gold et al., 1983; McCarrey et al., 1992). The presence of Pdha-2 mRNA in premeiotic spermatocytes was also confirmed by *in situ* hybridisation (Takakubo and Dahl, 1992).

A second, smaller 1.7 kb transcript is observed from day 25 (Fitzgerald et al., 1994; Iannello and Dahl, 1992). Sedimentation on a BSA gradient (Romrell et al., 1976) can be used to fractionate spermatogenic cells, and we isolated highly enriched primary pachytene, round spermatid and elongated spermatid sub-populations. Analysis of RNA from these cell-fractions showed that the 2.0 kb Pdha-2 mRNA was present in pachytene spermatocytes and round spermatids. The 1.7 kb mRNA was present in round spermatids and none of the Pdha-2 mRNAs were detected in elongated spermatids. The 1.7 kb mRNA was present in haploid cells. Sequence analysis of cDNA clones has revealed that the 1.7 kb mRNA only differs from the 2.0 kb mRNA in the length of the 3' untranslated region (3'-UTR) (Fitzgerald et al., 1992). Two typical polyadenylation signals are present in the 3-UTR of the Pdha-2 gene approximately 250 bp apart, and the two mRNAs are due to alternative use of these signals. A similar situation exists in humans (Dahl et al., 1990). It is not known if the 1.7 kb mRNA is generated by post-transcriptional modification of the 2.0 kb mRNA or if it is a result of active transcription in round spermatids. It has been suggested that the 1.7 kb mRNA is more stable than the longer transcript, due to the absence of destabilising AUUU tetranucleotide motifs (Shaw and Kamen, 1986).

To determine if the Pdha-2 mRNAs were being translated, polysomal and non-polysomal fractions of purified populations of pachytene spermatocytes and round spermatids were fractionated on sucrose gradients. RNA from the fractions was Northern blotted and probed for the presence of Pdha-2 mRNA (Fitzgerald et al., 1994). This polysomal analysis showed that the 2.0 kb transcript is translated in pachytene spermatocytes. Because Pdha-2 mRNA is not detected in non-polysomal fractions it was concluded that all 2.0 kb Pdha-2 mRNA is translated immediately following transcription and that no translational control mechanism affects PDH E1α synthesis at this stage. The polysomal analysis indicated that although some of the 1.7 kb Pdha-2 mRNA is translated in spermatids, the majority of the transcript is found in non-polysomal fractions. The inferred stability of the 1.7 kb mRNA combined with the observation that this transcript is largely present in non-polysomal spermatid fractions, suggests that a translational delay mechanism exists. Such translational control may be necessary in order to provide the required levels of E1α protein in the late stages of spermatogenesis where transcription is not generally detected.

Structural analysis of the mouse Pdha-2 promoter

Through transfection studies a Pdha-2 core promoter has been identified (Iannello et al., 1993a; Iannello et al., 1993b; Iannello et al., 1994). This core promoter region spans nucleotide positions –187 to +22 (Fig. 3). When the core promoter was used to direct expression of chloramphenicol transacetylase (CAT) in transgenic mice it was demonstrated that CAT transcription was both tissue- and temporal specific (Iannello et al., manuscript in preparation). The Pdha-2 core promoter therefore contains the necessary sequences for correct expression of the gene *in vivo*.

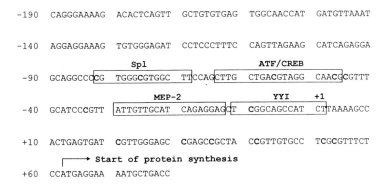

Figure 3. The mouse PDHA-2 promoter. DNA regions protected by DNA binding factors are shown as boxes. Cytosines that are part of a CpG dinucleotide are shown in bold. The initiator ATG codon at nucleotide +62 is indicated.

These studies also identified an enhancer element further upstream of the core promoter. DNaseI footprinting assays and gel shift experiments have revealed that nuclear factors bind to at least four regions in the Pdha-2 core promoter. These four regions contain binding sites for the transcription factors Sp1, ATF/CREB, YY1 and an undefined factor named MEP-2. Because the promoter does not contain classical TATA or CAAT boxes, transcription may be initiated *via* an Sp1-dependent mechanism. Of particular interest is MEP-2. Gel shift studies revealed a difference in mobility, depending on the source of the nuclear extract used (Iannello et al., 1993b). Nuclear factors from either somatic tissues or sexually immature mouse testis formed a large, slow migrating complex with the MEP-2 binding region. A faster migrating complex was observed when factors isolated from adult mouse testis was used. Formation of the faster migrating complex coincided with the appearance of pachytene spermatocytes and Pdha-2 expression in the developing mouse, suggesting that MEP-2 may be directly involved in the tissue and cell specific regulation of this gene.

Methylation analysis of the mouse Pdha-2 promoter

The Pdha-2 core promoter-CAT construct (named pt[-187/+22]E1α-CAT) was cloned in *E. coli* and therefore did not contain methylated CpG dinucleotides often seen in mammalian genomic DNA. This construct was introduced into NIH 3T3 mouse cells and CAT activity measured in transiently and stably transfected cells, and it was shown that the introduced Pdha-2 promoter was able to direct CAT expression (Iannello et al., 1993a; Iannello et al., 1993b). However, the endogenous Pdha-2 gene was not transcribed. These results suggest that this transfected core promoter, whether integrated or not, does not contain the full information for tissue-specific expression. However, as mentioned above, the core promoter will direct tissue and cell-specific expression in transgenic mice. A difference is that the transgene could have been methylated *de novo* during early embryo development. Sequence analysis revealed 8 CpG dinucleotides in the core promoter region.

To establish whether methylation of these sites affected the transcriptional activity of the Pdha-2 promoter, the pt[-187/+22]E1α-CAT construct was treated with Sss1 methylase and transiently

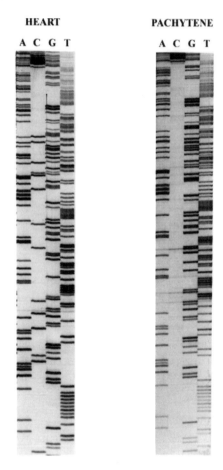

Figure 4. Pdha-2 methylation pattern in DNA from heart and pachytene spermatocytes. Bisulfite treated DNA was cloned and sequenced. In heart the CpG dinucleotides are not converted and therefore methylated.

transfected into an F9 carcinoma cell line. These experiments demonstrate that the activity of the Sss1 methylated promoter is significantly reduced (less than 25%) compared to that of appropriate control transfections (Iannello et al., manuscript in preparation). To further investigate the correlation between methylation status and Pdha-2 promoter activity, genomic DNA from stably transfected NIH 3T3 cells or mouse tissues were analysed for cytosine methylation using the method developed by Frommer et al., 1992. Treatment of the genomic DNA with bisulfite and alkali results in deamination and conversion to uracil of non-methylated cytosine residues. However, 5-methylcytosine do not react. Following PCR amplification and sequencing, the methylation pattern of the region can be determined. When the stably transfected NIH 3T3 cells were analysed it was found that all CpG dinucleotides in the endogenous and inactive Pdha-2 core promoter were methylated. In contrast, none of the CpG dinucleotides in the transfected, and active, Pdha-2 core promoter were methylated (Fig. 4). The *in vivo* methylation status of the mouse Pdha-2 promoter was also determined in somatic tissues and fractionated spermatogenic cells (Dahl et al., manuscript in preparation). This showed that the promoter region is methylated in somatic tissues, such as liver heart and brain, but that it is not methylated in pachytene spermatocytes. In conclusion, these experiments suggest that there is a strong correlation between transcriptional activity and CpG methylation status of the Pdha-2 promoter, and in particular that methylation of the promoter might be associated with repression of Pdha-2 expression in somatic tissues.

Evolution of the sperm-specific PDH E1α genes

Comparative gene mapping has increased our understanding of sex chromosome evolution in mammals (Graves and Watson, 1991). Although the gene content of the X chromosome is conserved amongst eutherian (placental) mammals, at least one major addition, to the short arm of the X chromosome, has occurred since the divergence of eutherian and marsupial infraclasses about 130 MYA. PDHA1 maps to Xp22.13, a region of the human X chromosome that is autosomal in marsupials. It is therefore of interest to determine the number, location and expression pattern of PDHA homologues in marsupials in order to understand the evolution of the sperm-specific E1α gene in eutherians. Genomic and cDNA clones were isolated for a marsupial (*S. macroura*) homologue to human PDHA1 (Fitzgerald et al., 1993). Importantly, there appears to be only one copy of the PDHA gene in the marsupial genome, as the testis cDNA and genomic clones all appeared to originate from the same gene, and only one significant signal was detected by *in situ* hybridisation. Southern blot analysis of genomic DNA from males and females of the two marsupial species *S. macroura* and *M. eugenii*, and from a somatic cell hybrid that retains a marsupial X chromosome, show that PDHA does not map to the X chromosome in marsupials. *In situ* hybridisation indicated that there is a single, autosomally located PDHA gene in marsupials, mapping to chromosome 5p in *M. eugenii*. Other genes that map to 5p in this species include DMD, MOAA, CYBB and ZFX. Hence, the marsupial PDHA gene seems to be part of this block of genes that are autosomal in marsupials and monotremes, and X-linked in eutherians. Further-more, the marsupial PDHA gene contains introns and is expressed not only in testis, but also in tissues such as liver.

The absence of a second intronless, sperm-specific PDHA gene in marsupials, and the fact that modern marsupials possess a single, autosomal PDHA gene, led us to the conclusion that a common ancestor of marsupials and eutherians also had a single autosomal PDHA gene and that the PDHA2 gene arose in eutherian mammals after the marsupial/eutherial divergence (Fitzgerald et al., 1993). Presumably the PDHA gene was part of a group of autosomal genes that were translocated to the short arm of what was to become the eutherian X chromosome. As the X chromosome evolved, an autosomal PDHA gene was selected for in spermatogenic cells, to compensate for either lack of an X chromosome or for the apparent inactivation of this chromo-some in haploid sperm. In eutherians, the genes for PGK are analogous to the two PDHA genes in organisation, and expression pattern. PGK-1 is located on the long arm of the X chromosome. PGK-2 is autosomal, testis-specific and intronless (McCarrey and Thomas, 1987; Michelson et al., 1985). Alloenzyme and *in situ* hybridisation studies have detected two functional PGK genes in eutherians and marsupials, but only a single autosomal locus in monotremes. PDHA and PGK seem to have followed a similar evolutionary course, except that the testis-specific PGK-2 gene arose prior to the marsupial/eutherial divergence (McCarrey, 1990; VandeBerg, 1985).

In an effort to establish events that may have occurred during the evolution of the E1α genes, all PDH E1α sequences were aligned by the CLUSTALV computer program (Higgins, 1988; Higgins, 1989). A neighbour-joining tree drawn using the method of Saitou, 1987 was drawn based on the sequence alignment (Fig. 5). The tree indicates that there is considerable homology between a wide variety of organisms (including eutherians, marsupial, yeast and eubacteria). This is presumably due to the relatively inflexible functional constraint on the subunit.

Further examination of the neighbour-joining tree suggest that the sperm-specific human and mouse PDHA2 genes arose independently following the eutherian/marsupial divergence. Figure 5 shows that PDHA2 and pdha-2 are placed on different branches of the tree. If the genes shared a common ancestor, one would expect them to have a higher degree of homology with each other and therefore be placed closer together perhaps on the same branch. This is not the case. The sequences are placed on separate branches and furthermore the human PDHA2 sequence is more

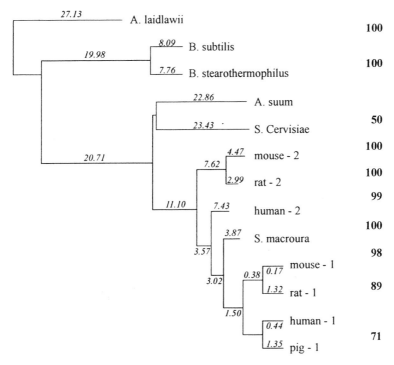

Figure 5. Phylogenetic relationship of 13 PDH E1α sequences using the CLUSTALV alignment programme. The bold numbers to the right are bootstrap values for branchings directly to the left of each number.

closely related to the marsupial sequence. To explore this further, a direct comparison of the somatic and sperm-specific isoforms DNA sequences using the Relative Rate Test was made. The Relative Rate Test is a powerful analytic tool for examining evolutionary relationships (Li and Tanimura, 1987; Li and Wu, 1987; Sarich and Wilson, 1967; Wu and Li, 1985). With this technique related sequences are compared one at a time to a reference or outgroup sequence, usually one that is phylogenetically the most distant of all the sequences. The PDHA sequence cloned from the marsupial species *Sminthopsis macroura* was used as the outgroup in this analysis. Corrections for multiple nucleotide substitutions at single sites were made by the method of Kimura, 1980. The synonymous substitution rate of 41% for PDHA1/PDHA2 and 124% for Pdha-1/Pdha-2 comparisons indicate that the sperm-specific variants evolved independently, with mouse Pdha-2 arising before human PDHA2. The difference in rate between species is too large to be accounted for by the general increase in substitution rate reported for mouse genes (Martin and Palumbi, 1993). In addition, the human and mouse PDHA2 genes do not map to syntenic chromosome regions. PDHA2 and Pdha-2 map to human 4q22-q23 and mouse chromosome 19 respectively. At least three genes that flank PDHA2 on chromosome 4 map to a different location in mice, namely chromosome 3. Thus, these regions are not syntenic between human and mouse. This suggests that the sperm-specific isoforms of PDHA were created at least twice since the eutherian/ marsupial divergence. Both genes arose independently *via* retroposition to different autosomes indicating that retroposition, as a mechanism for gene duplication, may be common.

Conclusion

Mammals have developed a sperm-specific isoform of the PDH E1α. The gene for this subunit is intronless in human and mouse. In mice this gene is initially expressed in the leptotene stage of spermatogenesis, with maximal expression in the pachytene spermatocytes. Induction of expression seems to be associated with changes in promoter binding factors and methylation status. Only a single PDH E1α gene was found in marsupials and monotremes, suggesting that the mammalian sperm-specific genes arose after the eutherian/marsupial divergence 130 MYA. Furthermore, sequence analysis suggests that the human PDHA2 and mouse Pdha-2 genes are functional retroposons, but the result of independent events.

Acknowledgements
H.-H.M.D. is a NH&MRC Senior Research Fellow. Our work is partly supported by a NH&MRC block grant to the Murdoch Institute.

References

Brown, R.M., Dahl, H.-H.M. and Brown, G.K. (1989a) X-Chromosome location of the functional gene for the E1α subunit of the human pyruvate dehydrogenase complex. *Genomics* 4: 174–181.

Brown, G.K., Brown, R.M., Scholem, R.D., Kirby, D.M. and Dahl, H.-H.M. (1989b) The clinical and biochemical spectrum of human pyruvate dehydrogenase deficiency. *Ann. N.Y. Acad. Sci.* 573: 360–368.

Brown, R.M., Dahl, H.-H.M. and Brown, G.K. (1990) Pyruvate dehydrogenase E1α subunit genes in the mouse: mapping and comparison with human homologues. *Som. Cell. Mol. Gen.* 16: 487–492.

Brown, G.K., Otero, L.J., LeGris, M. and Brown, R.M. (1994) Pyruvate dehydrogenase deficiency. *J. Med. Genet.* 31: 875–879.

Cullingford, T.E., Clark, J.B. and Phillips, I.R. (1993) Characterization of cDNAs encoding the rat testis-specific E1α subunit of the pyruvate dehydrogenase complex: comparison of expression of the corresponding mRNA with that of the somatic E1α subunit. *Biochim. Biophys. Acta* 1216: 149–153.

Dahl, H.-H.M., Brown, R.M., Hutchison, W.M., Maragos, C. and Brown, G.K. (1990) A testis-specific form of the human pyruvate dehydrogenase E1α subunit is coded for by an intronless gene on chromosome 4. *Genomics* 8: 225–232.

Dahl, H.-H.M. (1995) Pyruvate dehydrogenase E1α deficiency: Males and females differ yet again. *Am. J. Hum. Genet.* 56: 553–557.

Erickson, R.P., Kramer, J.M., Rittenhouse, J. and Salkeld, A. (1980) Quantitation of mRNAs during spermatogenesis: protamine-like histone and phosphoglyceratekinase-2 mRNAs increase after meiosis. *Proc. Natl. Acad. Sci. USA* 77: 6086–6090.

Fitzgerald, J., Hutchison, W.M. and Dahl, H.H. (1992) Isolation and characterisation of the mouse pyruvate dehydrogenase E1 alpha genes. *Biochim. Biophys. Acta* 1131: 83–90.

Fitzgerald, J., Wilcox, S.A., Graves, J.A.M. and Dahl, H.-H.M. (1993) A eutherian X-linked gene, PDHA1, is autosomal in marsupials: A model for the evolution of a second, testis-specific variant in eutherian mammals. *Gemonics* 18: 636–642.

Fitzgerald, J., Dahl, H.-H.M. and Iannello, R.C. (1994) Differential expression of two testis-specific transcripts of the mouse pdha-2 gene during spermatogenesis. *DNA & Cell Biol.* 13: 531–537.

Frommer, M., McDonald, L.E., Millar, D.S., Collis, C.M., Watt, F., Grigg, G.W., Molloy, P.L. and Paul, C.L. (1992) A genomic sequencing protocol that yields a positive display of 5-methylcytosine residues in individual DNA strands. *Proc. Natl. Acad. Sci. USA* 89: 1827–1831.

Gold, B., Fujimoto, H., Kramer, J.M., Erickson, R.P. and B., H.N. (1983) Haploid accumulation and translational control of phosphoglycerate kinase-2 mRNA during mouse spermatogenesis. *Dev. Biol.* 8: 392–399.

Graves, J.A.M. and Watson, J.M. (1991) Mammalian sex chromosomes: Evolution of organisation and function. *Chromosoma* 101: 63–68.

Higgins, D.G. and Sharp, P.M. (1988) CLUSTAL: a package for performing multiple sequence alignments on a microcomputer. *Gene* 73: 237–244.

Higgins, D.G. and Sharp, P.M (1989) Fast and sensitive multiple sequence alignments on a microcomputer. *CABIOS* 5: 151–153.

Iannello, R.C. and Dahl, H.-H.M. (1992) Transcriptional expression of a testis-specific variant of the mouse pyruvate dehydrogenase E1α subunit. *Biol. Reprod.* 47: 48–58.

Iannello, R.C., Fitzgerald, J. and Dahl, H.-H.M. (1993a) The mouse testis-specific PDH E1α subunit: Transcriptional expression and interaction of a proximal promoter with putative transcription factors. *In*: K. Reed and J. Graves (eds): *Mammalian Sex Chromosomes and Sex-determining Genes.* Harwood Academic Publishers, Switzerland, pp 87–104.

Iannello, R.C., Kola, I. and Dahl, H.H. (1993b) Temporal and tissue-specific interactions involving novel transcription factors and the proximal promoter of the mouse pdha-2 gene. *J. Biol. Chem.* 268: 22581–22590.

Iannello, R.C., Young, J.C. and Kola, I. (1994) Pdha-2: a model for studying transcriptional regulation in early spermatocytes. *Mol. Reprod. Dev.* 39: 194–199.

Johnson, P.A., Peschon, J.J., Yelick, P.C., Palmiter, R.D. and Hecht, N.B. (1988) Sequence homologies in the mouse protamine 1 and 2 genes. *Biochim. Biophys. Acta* 950: 45–53.

Johnson, K.R., Komuniecki, R., Sun, Y. and Wheelock, M.J. (1992) Characterization of cDNA clones for the alpha subunit of pyruvate dehydrogenase from *Ascaris suum*. *Mol. Biochem. Parasitol.* 51: 37–47.

Kimura, M. (1980) A simple method for estimating evolutionary rates of base substitutions through comparative studies of nucleotide sequences. *J. Mol. Evol.* 16: 111–120.

Li, W.H. and Tamura, M. (1987) The molecular clock runs more slowly in man than in apes and monkeys. *Nature* 326: 93–96.

Li, W.H. and Wu, C.I. (1987) Rates of nucleotide substitution are evidently higher in rodents than in man. *Mol. Biol. Evol.* 4: 74–82.

Maragos, C., Hutchison, W.M., Hayasaka, K., Brown, G.K. and Dahl, H.-H.M. (1989) Structural organization of the gene for the E1α subunit of the human pyruvate dehydrogenase complex. *J. Biol. Chem.* 264: 12294–12298.

Martin, A.P. and Palumbi, S.R. (1993) Body size, metabolic rate, generation time, and the molecular clock. *Proc. Natl. Acad. Sci. USA* 90: 4087–4091.

McCarrey, J.R. and Thomas, K. (1987) Human testis-specific PGK gene lacks introns and possesses characteristics of a processed gene. *Nature* 326: 501–505.

McCarrey, J.R. (1990) Molecular evolution of the human Pgk-2 retroposon. *Nucleic Acids Res.* 18: 949–955.

McCarrey, J.R., Berg, W.M., Paragioudakis, S.J., Zhang, P.L., Dilworth, D.D., Arnold, B.L. and Rossi, J.J. (1992) Differential transcription of Pgk genes during spermatogenesis in mouse. *Dev. Biol.* 154: 160–168.

Michelson, A.M., Blake, C.C., Evans, S.T. and Orkin, S.H. (1985) Structure of the human phosphoglycerate kinase gene and the intron-mediated evolution and dispersal of the nucleotide-binding domain. *Proc. Natl. Acad. Sci. USA* 82: 6965–6969.

Patel, M.S. and Roche, T.E. (1990) Molecular biology and biochemistry of pyruvate dehydrogenase complexes. *FASEB J.* 4: 3224–3233.

Reed, L.J. (1974) Multienzyme complexes. *Acc. Chem. Res.* 7: 40–46.

Robertson, N.G., Pomponio, R.J., Mutter, G.L. and Morton, C.C. (1991) Testis-specific expression of the human MYCL2 gene. *Nucleic Acids Res.* 19: 3129–3137.

Robinson, B.H. (1989) Lactic acidemia. *In*: C.R. Schriver, A.L. Beaudet, W.S. Sly and D. Valle (eds): *The Metabolic Basis of Inherited Disease,* Sixth Edition. McGraw Hill, New York, pp 869–888.

Romrell, L.J., Bellve, A.R. and Fawcett, D.W. (1976) Separation of mouse spermatogenic cells by sedimentation velocity. *Dev. Biol.* 49: 119–131.

Saitou, N. and Nei, M. (1987) The neighbor-joining method: a new method for reconstructing phylogenetic trees. *Mol. Biol. Evol.* 4: 406–425.

Sarich, V.M. and Wilson, A.C. (1967) Immunological time scale for hominid evolution. *Science* 158: 1200–1203.

Shashidharan, P., Michaelidis, T.M., Robakis, N.K., Kresovali, A., Papamatheakis, J. and Plaitakis, A. (1994) Novel human glutamate dehydrogenase expressed in neural and testicular tissues and encoded by an X-linked intronless gene. *J. Biol. Chem.* 269: 16971–6.

Shaw, G. and Kamen, R. (1986) A conserved AU sequence from the 3' untranslated region of GM-CSF mRNA mediates selective mRNA degradation. *Cell* 46: 659–67.

Shevell, M.I., Matthews, P.M., Scriver, C.R., Brown, R.M., Otero, L.J., Legris, M., Brown, G.K. and Arnold, D.L. (1994) Cerebral dysgenesis and lactic acidemia: An MRI/MRS phenotype associated with pyruvate dehydrogenase deficiency. *Pediatr. Neurol.* 11: 224–229.

Solari, A.J. (1989) Sex chromosome pairing and fertility in the heterogametic sex of mammals and birds. *In*: C.B. Gillies (ed.): *Fertility and Chromosome Pairing: Recent Studies in Plants and Animals.* CRC Press, Boca Raton, pp 77–107.

Su, H. and Lau, Y.F. (1993) Identification of the transcriptional unit, structural organization, and promoter sequence of the human sex-determining region Y (SRY) gene, using a reverse genetic approach. *Am. J. Hum. Genet.* 52: 24–38.

Takakubo, F. and Dahl, H.-H.M. (1992) The expression pattern of the pyruvate dehydrogenase E1α subunit genes during spermatogenesis in adult mouse. *Exp. Cell Res.* 199: 39–49.

Takano, T. and Li, S.S. (1989) Human testicular lactate dehydrogenase-C gene is interrupted by six introns at positions homologous to those of LDH-A (muscle) and LDH-B (heart) genes. *Biochem. Biophys. Res. Comm.* 159: 579–83.

Thomas, K., Del, M.J., Eversole, P., Bellve, A., Hiraoka, Y., Li, S.S. and Simon, M. (1990) Developmental regulation of expression of the lactate dehydrogenase (LDH) multigene family during mouse spermatogenesis. *Development* 109: 483–493.

VandeBerg, J.L. (1985) The phosphoglycerate kinase isozyme system in mammals: biochemical, genetic, developmental, and evolutionary aspects. *Curr. Top. Biol. Med. Res.* 12: 133–187.

Wu, C.I. and Li, W.H. (1985) Evidence for higher rates of nucleotide substitution in rodents than in man. *Proc. Natl. Acad. Sci. USA* 82: 1741–1745.

Alpha-Keto Acid Dehydrogenase Complexes
M.S. Patel, T.E. Roche and R.A. Harris (eds)
© 1996 Birkhäuser Verlag Basel/Switzerland

Molecular defects of the branched-chain α-keto acid dehydrogenase complex: Maple syrup urine disease due to mutations of the E1α or E1β subunit gene

Y. Indo and I. Matsuda

Department of Pediatrics, Kumamoto University, School of Medicine, Kumamoto 860, Japan

Introduction

Leucine, isoleucine, and valine, the three branched-chain amino acids (BCAA), are essential components of the human diet. Following uptake by the cell, BCAAs are either incorporated into proteins or are catabolized for energy (Danner and Elsas II, 1989). The BCAA are catabolized through analogous mechanisms for the first three steps: transamination, oxidative decarboxylation of the branched-chain α-keto acid (BCKA), and dehydrogenation of the resulting branched-chain acyl-coenzyme A (CoA) to enoyl CoA (Tanaka and Rosenberg, 1983). The oxidative decarboxylation of the BCKAs is performed by the branched-chain α-keto acid dehydrogenase complex (BCKDH), a constituent of all mammalian tissues which is associated with the mitochondrial inner membrane.

Maple syrup urine disease (MSUD) is an autosomal recessive inborn error of metabolism caused by dysfunction of any of the subunits (E1α, E1β and E2) of the BCKDH complex.

First we will review historical studies on MSUD. These have been not only useful for analyses of BCAA metabolism but also complemented biochemical studies of mitochondrial BCKDH.

Subsequently, we refer to analysis of genetic defects in the E1 component of BCKDH which have been identified in patients with MSUD. Analysis of these defects provides unique insights with respect to the structure-function and regulatory aspects of the BCKDH.

Maple syrup urine disease

Clinical phenotypes of MSUD

MSUD is characterized by ketoacidosis and in some cases by developmental retardation. The distinctive odor of the urine (or the patient) is characteristic in this disease, although the substance responsible for the maple syrup odor is unknown; the odor is not due to BCKAs. Several phenotypes have been distinguished on the basis of clinical features: classical (Menkes et al., 1954; Dancis et al., 1959), intermittent (Morris et al., 1961; Dancis et al., 1967), intermediate (Schulman et al., 1970), thiamine responsive (Scriver et al., 1971) types. The classification is based on the rapidity of onset, severity of the disease, tolerance for dietary protein, response to thiamine supplements, and results of enzyme assay.

Disease incidence varies with the population studied from 1/290 000 in a New England newborn screening experience to 1/760 from a selective screening in an inbred Mennonite group (Danner and Elsas II, 1989). Another group reported that this incidence could rise to 1/176 in the Mennonite community (DiGeorge et al., 1982). The incidence was estimated to be 1/620 000 in Japan, based on data of newborns.

Identification of the metabolic step which is blocked in MSUD

In 1954, Menkes and associates described a family in which four of six infants died during the first weeks of life (Menkes et al., 1954). Accumulation of the three BCAAs in blood and urine of affected patients was first reported by Westall et al. (1957). Subsequently, Menkes identified accumulation of the corresponding BCKAs (Menkes, 1959). These data suggested that the metabolic pathway of the three BCAAs was blocked at the α-keto acid level. This impairment produces increased concentrations of BCAAs and BCKAs in body cells and fluids. The accumulation of the BCAAs was considered to be due to a reversible reaction of BCAA transaminase. These results are consistent with the hypothesis that a single common enzyme which catalyzes the oxidative decarboxylation of the three BCKAs is deficient in patients with MSUD.

Biochemical abnormalities of cells derived from patients with MSUD

Oxidation of BCAAs by intact cells

Dancis et al. (1960) demonstrated that the ability of isolated leukocytes from patients with MSUD to produce $^{14}CO_2$ from [1-^{14}C]Leu, [1-^{14}C]Ile, and [1-^{14}C]Val was drastically reduced, while the ability of the cells to produce the corresponding keto acids was normal. This supports the hypothesis that the three amino acids share one enzyme in the degradative pathway which is deficient in maple syrup urine disease. The concept was further supported by similar experiments *in vitro* in which [1-^{14}C]BCKAs were used as substrates (Seegmiller and Westall, 1967; Goedde et al., 1967).

Dancis and co-workers (1963) also measured the ability of skin fibroblast cells from patients to produce $^{14}CO_2$ from [1-^{14}C] labeled BCAAs. They also noted that the level of activity in the cells reflects the ability to degrade the amino acids, thus providing an index of the severity of the disease (Dancis et al., 1972). In general, cells from patients with the classical phenotype showed lower residual activity than did cells from patients with a milder phenotype. However, they pointed out that there was not always a consistent relationship between residual activity and clinical course, attributing this discrepancy primarily to diet. They proposed a more logical approach to classify MSUD patients according to protein tolerance, which more directly reflects the magnitude of the enzyme defect. Thus, enzyme level in the cell was related to the ability of the individual to tolerate protein.

Similar studies were shown to be feasible, using lymphoblastoid cells established from MSUD patients (Skaper et al., 1976; Jinno et al., 1984b).

Genetic heterogeneity of MSUD

The nature of the genetic heterogeneity observed in MSUD has been explored in several studies in genetic complementation analyses (Lyons et al., 1973; Singh et al., 1977; Jinno et al., 1984a). These studies provided evidence for the existence of two complementation groups among patients with classical MSUD. It was not clear at that time whether such complementation is intergenic (implying defects at two different loci) or interallelic (implying two different defects at the same locus).

Activities and kinetics of the BCKDH complex from MSUD subjects

Radio-labeled BCAAs were used to determine the residual activity of the BCKDH in intact cultured cells and were found to be useful for analysis of MSUD patients. Several steps, including uptake of BCAAs by cells, transamination of these amino acids to corresponding to BCKAs and transport of these keto acids to mitochondria are necessary before BCKDH reacts with these substrates. Thus, it is necessary to use BCKAs as substrates to more directly characterize BCKDH properties. Danner and Elsas II (1975) noted BCKDH activity in broken but not intact cell suspensions of normal and mutant cultured human fibroblasts. They reconstituted BCKDH activity in a freeze-thaw disrupted preparation of broken fibroblast cells and found that this system required several cofactors including thiamine pyrophosphate (TPP), Mg^{2+}, CoA and NAD^+. They also showed a genetic dissociation between pyruvate dehydrogenase and BCKDH activity as well as mitochondrial localization of BCKDH function to the inner membrane-matrix.

Wendel et al. (1975) studied kinetics of BCKDH activity in intact cultured cells with BCKAs as substrates. They showed biphasic degradation kinetics for each BCKA in normal control subjects and suggested that the component with a higher substrate affinity was affected in MSUD. Although this study was not a direct assay of BCKDH activity, the results paved the way for further analysis of BCKDH in MSUD subjects.

Chuang et al. (1981) studied activities of BCKDH using $[1-^{14}C]BCKA$ as a substrate in disrupted preparations of cultured cells from MSUD patients. They compared the activity and kinetics of the BCKDH in cells from normal and classical MSUD subjects. In overall reactions of the multienzyme complex, normal controls showed typical hyperbolic kinetics as a function of substrate concentration. In contrast, classical MSUD patients exhibited sigmoidal kinetics. By separately measuring E1, E2 and E3 activity, they suggested that there was a defect in the E1 step of BCKDH in classical MSUD patients. However, GM-612, one of two cell lines they studied, was shown to be due to mutations of the E2 gene of the BCKDH, by the same group (Fisher et al., 1993). They speculated that the deficiency of E1 activity might be due to absence of E1 and E2 interactions which were necessary for the high catalytic efficiency of E1.

Structural analyses of the BCKDH protein in MSUD

Isolation and purification of bovine BCKDH were reported by two groups (Pettit et al., 1978; Danner et al., 1979). The mammalian BCKDH is located in the inner membrane of the mitochondrial matrix, and is both structurally and mechanistically analogous to pyruvate and α-ketoglutarate dehydrogenase complexes. Unlike these dehydrogenases, E3 was readily dissociated from the rest of complex and it was necessary to add E3 from pyruvate dehydrogenase to the assay mix to obtain maximal activity. This observation indicated that BCKDH shares a common E3 with pyruvate dehydrogenase. This was supported by genetic evidence from a study on a patient with E3 deficiency (Robinson et al., 1977). The BCKDH was also purified from human liver (Ono et al., 1987). Each protein of the BCKDH is encoded by nuclear genes, synthesized in the cytosol and imported into mitochondria where assembly occurs (Litwer and Danner, 1988; Danner et al., 1989; Lindsay, 1989). These studies advanced analysis of the BCKDH in patients with MSUD.

Components (subunit) of the BCKDH

This complex is composed of three catalytic components, i.e., branched-chain α-keto acid decarboxylase (E1), dihydrolipoyl transacylase (E2) and dihydrolipoyl dehydrogenase (E3). The E1 component is further composed of α (E1α) and β (E1β) subunits. This mitochondrial multienzyme complex is organized around a 24-mer E2 core, to which multiple copies of a heterotetrameric ($\alpha_2\beta_2$) E1, a dimeric E3, a specific kinase and a specific phosphatase are attached by noncovalent interactions (Reed and Hackert, 1990). E1 catalyzes both the decarboxylation of α-keto acid and the subsequent reductive acylation of the lipoyl moiety that is covalently bound to E2. The TPP is an essential cofactor for E1 reaction. E2 catalyzes a transfer of the acyl group from the lipoyl moiety to coenzyme A (Pettit et al., 1978; Heffelfinger et al., 1983). E3, a flavoprotein, is a homodimer and catalyzes the reoxidation of the dihydrolipoyl moiety with NAD^+ as the ultimate electron acceptor. The E1 and E2 components are specific for the BCKDH complex, whereas the E3 component is identical to that associated with the pyruvate and α-ketoglutarate dehydrogenase complexes (Reed and Hackert, 1990). These enzyme components catalyze a coordinate sequence of reactions constituting the overall reaction as follows (Pettit et al., 1978):

$$R-CO-COOH + CoA-SH + NAD^+ \xrightarrow{\text{TPP, Mg}^{2+}} R-CO-S-CoA + CO_2 + NADH + H^+$$

It was shown that BCKDH also decarboxylates the α-keto acids from methionine and threonine (Pettit et al., 1978; Jones and Yeaman, 1986; Paxton et al., 1986). Interestingly, these amino acids are not elevated in plasma from MSUD patients. The BCKDH complex interacts with two other proteins, a kinase (Odessey, 1982; Paxton and Harris, 1982; Shimomura et al., 1990; Popov et al., 1992) and a phosphatase (Damuni et al., 1984; Damuni and Reed, 1987), which regulate activity of BCKDH. The E1 component is the substrate for the kinase. Phosphate incorporation into E1α inactivates the complex. Structure and function of the catalytic components as well as regulatory components of the BCKDH will be described by Wynn and Chuang and by Shimomura (this volume).

Immunological analyses of BCKDH in cultured cell lines derived from patients with MSUD

Isolation and purification of BCKDH enabled investigators to analyze MSUD at the protein level. Danner et al. (1985) reported data on a patient with MSUD and an E2 component deficiency, as determined by the immunoblot method. The patient from whom the cell line was derived died in infancy despite dietary restriction. This was the first demonstration of the structural basis of an impaired multienzyme complex of mitochondria in man. Indo et al. (1987) studied lymphoblastoid cell lines derived from MSUD patients. These included two (GM1655 and GM1366) and four from the United States and Japan, respectively. First, the BCKDH complexes derived from patients with two different MSUD phenotypes were studied in terms of their catalytic functions (substrate dependent kinetics). One group showed hyperbolic kinetics, whereas the other group exhibited sigmoidal or near sigmoidal kinetics. The former cell lines were derived from two intermittent MSUD patients and GM 1366. The latter cell lines were from classical MSUD and GM 1655. Both cell lines showed significantly reduced enzyme activities measured at 2.0 mM or lower substrate concentration, which corresponds to the apparent Km value of the highly purified bovine BCKDH complex. The enzyme activities in the former group were significantly higher than in the latter group. The immunoblot study revealed a markedly decreased amount of the E1β subunit accompanied by weak staining of the E1α subunit in all three cell lines from the group which exhibited sigmoidal or near sigmoidal kinetics. The E2 components exhibited a cross-reactive peptide in these cells. In contrast, the E1α and E1β subunits were identified on immunoblots in three cell lines from the former group which showed hyperbolic kinetics of BCKDH. In this group, the E2 component was moderately reduced in a cell line derived from an intermittent type patient, and the protein was absent in GM 1366. Another cell line contained three proteins, normally. The E3 components exhibited a cross-reactive peptide in all the cells examined.

These results indicated that mutations of the E1β subunit might provide an explanation for the altered kinetic properties of the BCKDH complex in at least some patients with MSUD. No adequate explanation for the weak staining of the E1α subunit was available at this stage. It was only speculated that a defect in one of the two peptides of E1 might inhibit transport of the mutated precursor peptide into the mitochondria, or assembly of the heterotetramer, leading to rapid degradation of one or both subunits. As described later, mutations of E1β subunit gene were identified in two cell lines among them (Nobukuni et al., 1993; Hayashida et al., 1994), and another cell line (GM1655) derived from a Mennonite kindred proved to be due to missense mutation of the E1α subunit gene but not of E1β (Matsuda et al., 1990).

In combination with immunoblot analysis, a kinetic study of the BCKDH from MSUD subjects revealed that classical, intermediate and intermittent types corresponded to the enzyme properties of sigmoidal kinetics with E1β subunit deficiency, near-sigmoidal kinetics with E1β subunit deficiency and hyperbolic kinetics with E2 subunit deficiency of the BCKDH, respectively (Indo et al., 1988). However, they admitted that there were a few exceptional cases in classifying clinical phenotype. As described later, some of their cell lines were reported to be due to mutations of the E1α subunit gene but not of E1β. Although these studies indicated that kinetic and immunoblot analyses could not always identify subunits with mutations, they supported the thesis that abnormality of E1 component (E1α or E1β subunit) reduced activity of the BCKDH complex by altering stability of the E1 component or affinity of this component against substrates.

Function and primary structure of E1α, E1β and E2 proteins and chromosomal localizations of their genes

The primary structure of each component (subunit) of the BCKDH has been determined, thus enabling investigators to analyze the molecular basis of MSUD at the protein, mRNA and DNA levels.

E1 component consists of E1α and E1β. The function of E1 component is to remove CO_2 from the BCKAs and transfer the acyl group to E2, utilizing TPP as a cofactor. The location of the active site has not been identified; it may be on E1α, E1β, or a site created by domains of both subunits.

A full-length human E1α cDNA encodes 445 amino acids including a leader peptide for mitochondrial targeting. Mature enzyme and leader peptide are composed of 400 and 45 amino acids, respectively. The calculated molecular mass of mature enzyme is 45 552 daltons (Fisher et al., 1989; McKean et al., 1992). The E1α subunit contains phosphorylation sites (serine 292 and 302 of the mature protein) that are responsible for interconversion of the active, non-phosphorylated form and the inactive, phosphorylated form of BCKDH (Cook et al., 1983; Cook et al., 1984). The human E1α gene is located on chromosome 19q13.1-13.2 (Fekete et al., 1989) and contains 9 exons and spans at least 55 kilobases (Chuang et al., 1993).

E1β has no defined function, although it is commonly thought that it has a role in transfer of the acyl group to E2-lipoate (Yeaman, 1989). It was reported that the E1β subunit bound to E2, suggesting an apparent binding order of E2-E1β-E1α (Wynn et al., 1992).

A full-length human E1β cDNA encodes 392 amino acids, including a leader peptide for mito-chondrial targeting. Mature enzyme and leader peptide are composed of 342 and 50 amino acids, respectively. A calculated molecular mass of mature enzyme is 37 585 daltons (Nobukuni et al., 1990). The human E1β gene is located on chromosome 6p21-22 (Zneimer et al., 1991) and contains 10 exons and is over 100 kilobase long (Mitsubuchi et al., 1991a).

E2 catalyzes transfer of the acyl group from the lipoyl moiety to coenzyme A (Pettit et al., 1978; Heffelfinger et al., 1983). A full-length human E2 cDNA encodes 477 amino acids, including a leader peptide for mitochondrial targeting. Mature enzyme and leader peptide are composed of 421 and 56 amino acids, respectively (Nobukuni et al., 1989; Danner et al., 1989). Another group reported that the leader peptide of this component has 61amino acids (Lau et al., 1992; Lau et al., 1992a; Wynn et al., 1994). A lipoic acid is covalently linked at lysine 44 of the mature protein (Hummel et al., 1988). The calculated molecular mass of the mature enzyme is 46 322 daltons. The human E2 gene is located on chromosome 1p31 (Zneimer et al., 1991), contains 11 exons and spans approximately 68 kilobases of genomic DNA (Lau et al., 1992b). The intron-less E2 pseudogene was isolated, mapped to chromosome 3q24 and corresponds to the complete mito-chondrial leader peptide sequence and the lipoyl-bearing domain that are encoded by exons I through IV of the functional gene (Lau et al., 1992b).

E3 is a flavoprotein and reoxidizes the reduced lipoyl sulfur residues of E2. A full-length human E3 cDNA encodes 509 amino acids, including a leader peptide for mitochondrial targeting. Mature enzyme and leader peptide are composed of 474 and 35 amino acids, respec-tively (Otulakowski and Robinson, 1987; Pons et al., 1988). The calculated molecular mass of the mature enzyme is 50 216 daltons. The human E3 gene is located on chromosome 7q31-32 (Scherer et al., 1991).

Molecular analysis of BCKDH genes in MSUD

Here, we will focus on molecular analysis of E1α and E1β genes. An analysis of the E2 gene will be discussed by Wynn and Chuang (this volume).

Mutations of the E1α gene in patients with MSUD

GM 649 cell

A case of MSUD involving the E1α subunit was reported for the first time by Zhang et al. (1989). They studied fibroblasts from a classical MSUD patient (GM649) and his parents (GM

Table 1. Mutations affecting the E1α subunit of the branched-chain α-keto acid dehydrogenase in maple syrup urine disease

Subjects	Phenotype	Genotype	Nucleotide change	Exon	Intron	Functional change	Reference
GM 649	classical	compound-heterozygote	T-1312 → A / 8 bp deletion	9 / 7	– / –	Y393N; / early stop	Zhang et al. (1989) / Chuang et al. (1994)
Mennonite	classical	homozygote	T-1312 → A	9	–	Y393N	Matsuda et al. (1990) / Fisher et al. (1991a) / Mitsubuchi et al. (1992)
KM 06	intermediate	compound-heterozygote	C-568 → A / T-977 → C	5 / 7	– / –	Q145K / I281T	Nobukuni et al. (1993)
KM 09	classical	compound-heterozygote	C-475 → T / G-757 → A	4 / 6	– / –	R114W / A208T	Nobukuni et al. (1993)
KM 22	classical	homozygote	G-757 → A	6	–	A208T	Nobukuni et al. (1993)
Lo	intermediate	compound-heterozygote	A-1238 → G / 1 bp (C) insertion	9 / 2	– / –	Y368C; / early stop	Chuang et al. (1994)

Nucleotide and deduced amino acid sequences of the human E1α cDNA were from references (Fisher et al., 1989; McKean et al., 1992). Numbering system differs from that in the original report because nucleotide number and amino acid number start from the codon corresponding to the initiation methionine and the N-terminal of the mature protein, respectively.

650 and GM 651) and identified a T to A transversion that altered a Tyr to an Asn at residue 393 (this number is 394 according to their numbering system) of E1α subunit. After analyzing cells from both parents, they concluded that the patient was a compound heterozygote, inheriting an allele encoding an abnormal E1α from the father, and an allele from the mother containing a cis-acting defect in regulation which abolished expression of one of the E1α alleles. Later, the same cell (GM 649) was further studied by Chuang et al. (1994) who used an alternate approach, i.e., amplification of exonic DNAs and which led to detection of a heterozygous 8-bp (G-860 ~ C-867) deletion in exon 7 of the E1α gene of the patient (GM 649) and the mother (GM 651). They showed that nonsense codon 255 of the mature E1α subunit were generated by an 8-bp deletion of the E1α gene. Occurrence of this premature stop codon appeared to be associated with a low abundance of the mature E1α mRNA, as determined by allele-specific oligonucleotide probing (Tab. 1).

Mennonite kindred

Matsuda et al. (1990) identified mutation in the E1α subunit in two lymphoblastoid cell lines (GM 1655, GM 1099) from Mennonite patients. One of the cells (GM 1655) had been previously characterized at the protein level by the immunoblot analysis (Indo et al., 1987) (Fig. 1). In the case of this cell line, they thought that the E1β subunit was primarily involved, as deduced from kinetic and immunoblot analysis of BCKDH. After sequencing of the entire region of E1β cDNA, no mutation was found in this subunit. They then analyzed the E1α subunit from the same patient and identified a T-to-A substitution which generated an Asn in place of a Tyr at amino acid 393 (Y393N) of the mature E1α subunit. This substitution was present in both alleles in these two cell lines. They concluded that this was the mutation in the Mennonite patient, since they had sequenced the entire coding region of both E1α and E1β cDNAs. This mutation is in the last exon (exon 9) of the E1α gene and the substitution of amino acid is located in the vicinity of the C-terminal end of E1α subunit protein. Interestingly, this was the same as one allele of two mutations previously identified in a MSUD subject (GM 649) described above (Zhang et al., 1989).

Fisher et al. (1991a) identified the same mutation in a MSUD patient of a Mennonite family. The functional significance of this missense mutation was assessed by transfection studies using E1α-deficient MSUD lymphoblastoid cell (Lo) as a host (Fisher et al., 1991b). The level of E1β subunit was also greatly reduced in Lo cells. Mutations of E1α subunit gene in Lo cells were reported later (Chuang et al., 1994), as described below. They transfected chimeric bovine-human cDNAs which encode mitochondrial import competent E1α subunit precursors. Transfection with normal E1α cDNA restored decarboxylation activity of intact cells and both E1α and E1β subunits were markedly increased. Transfection of Y393N (the Tyr to Asn substitution at position 393) mutant E1α cDNA failed to produce any measurable decarboxylation activity. The mutant E1α subunit was expressed at a normal level, however, the E1β subunit was undetectable. These results suggest that this mutation impedes the assembly of E1α with E1β into a stable $\alpha_2\beta_2$ structure, resulting in degradation of the free E1β subunit. Therefore, immunoblot analysis may identify an affected subunit but does not always reveal the subunit primarily involving the E1 component.

Mitsubuchi et al. (1992) examined genomes from 70 members, including 12 patients belonging to eight different Mennonite MSUD pedigrees, by primer-specified restriction map modification. The same mutation previously identified was present in all the patients and in a single allele in all obligate carriers and several siblings. These family studies provided additional evidence that Mennonite MSUD was caused by a missense mutation of the E1α gene of BCKDH (Tab. 1).

KM 06, 09 and 22

These patients were diagnosed at screening of newborns in Japan. To characterize the mutations present in these patients with MSUD, three-step analyses were made: identification of the involved subunit by complementation analysis using three different cell lines derived from homozygotes having E1α, E1β or E2 mutant gene; screening for a mutation site in cDNA of the corresponding subunit by RT-PCR-SSCP; mutant analysis by sequencing the amplified cDNA fragment (Nobukuni et al., 1993).

The clinical phenotype of KM 06 was intermediate. There were no clinical signs or symptoms in the early neonatal period. Tolerance of KM 06 for leucine was relatively high compared with that of classical case. The BCKDH showed near-sigmoidal kinetics and its activity obtained at low substrate concentrations was higher than that of GM 1655 derived from the Mennonite patient (Indo et al., 1988). The immunoblot study revealed a markedly decreased amount of the E1β subunit accompanied by moderately decreased staining of the E1α subunit in this cell line. In other experimental conditions, however, the band of the E1β subunit was faintly stained. Later, two mutations were identified in this cell line (Nobukuni et al., 1993). The patient was a compound heterozygote with different paternal and maternal alleles of E1α subunit. These were a C-to-A and a T-to-C substitutions located in exons 5 and 7, generating a Lys in place of a Gln at amino acid 155 (Q155K) and a Thr in place of a Ile at amino acid 281(I281T) in the mature E1α subunit, respectively. The amino acid numbers are according to the published sequences (Fisher et al., 1989; McKean et al., 1992). The numbering system differs from that in original reports because nucleotide number and amino acid number start from the codon corresponding to the initiation Met and the N-terminal of the mature protein, respectively.

Clinical phenotype of both KM 09 and 22 was classical. Mutations in these two cell lines were identified in the E1α gene. KM 09 was a compound heterozygote with two missense mutations, a C-to-T and G-to-A transitions located in exons 4 and 6, generating R114W and A208T, respectively. KM 22 was a homozygote of the mutant allele A208T of the E1α subunit which was shared by one of the mutant alleles in KM 09 (Tab. 1).

Lo

This case was reported as a distinct variant of intermediate maple syrup urine disease (Gonzalez-Rios et al., 1985). The defect in the catalytic activity of the E1 component was suggested by measuring component activities of the BCKDH. The levels of both E1α and E1β subunit proteins and the E1α mRNA level were reduced in cells from this patient (Fisher et al., 1989), whereas the E1β mRNA level was normal (Chuang et al., 1990). The E1α cDNAs prepared from this cell were amplified and subclones were sequenced (Chuang et al., 1994). The patient Lo was a compound heterozygote with two mutations, a A-to-G transition in exon 9 and a single C nucleotide insertion (between A-110 and A-118) in exon 2 which resulted in the Y368C mutation and a frameshift at codon 4 of the mature E1α, respectively. The frameshift generated a premature stop codon at base 174 of the E1α mRNA (Tab. 1).

The Y368C mutation was characterized in an expression study (Chuang et al., 1994). The plasmids carrying the normal and the mutant E1α cDNAs were transfected into E1α-deficient lymphoblastoid cells derived from a Mennonite patient homozygous for the Y393N substitution in the E1α subunit. It was shown that the Y368C mutant subunit was a defective polypeptide which failed to restore the BCKDH activity. It had been demonstrated in immunoblot analysis that the Mennonite cells had reduced amounts of both E1α and E1β subunits (Indo et al., 1987; Fisher et al., 1989). Immunoblot study of the transfected cells revealed that a similar increase in the E1α level occurred in both cell lines transfected with the Y368C and normal E1α cDNA. The E1β subunit in transfected cells with the normal cDNA was restored to ~50% of that in normal cells, but the E1β subunit transfected with the Y368C retained a trace amount. These results suggested that this mutation impaired proper folding and subsequent assembly of E1α with E1β. The failure of Y368C E1α to rescue the E1β subunit appears to be similar to the effect of the Y393N mutation in Mennonite patients. These results suggested that the carboxy-terminal region is critical for proper assembly of E1α with E1β.

The mutant mRNA containing the single C insertion was in low abundance, as determined by allele-specific oligonucleotide (ASO) probing.

Mutations of E1β gene in patients with MSUD

KM 08 (E.K.)
The first case of MSUD involving the E1β subunit was reported by Nobukuni et al. (1991). The proband (E.K.) was the progeny of first cousin Japanese parents. Elevated leucine levels were detected at age 10 days, at the time of newborn screening for MSUD. Specific treatment with peritoneal dialysis was started immediately, however, opisthotonos and hypertonia became evident at age 26 days. At this stage, the blood leucine level was over 10 times the normal. The phenotype was categorized as a classical type of MSUD. The lymphoblastoid cells from this patient had been previously characterized at the protein level, by immunoblot analysis (Fig. 1) (Indo et al., 1988). The BCKDH showed near-sigmoidal kinetics and activity was low. The immunoblot study revealed that the E1β subunit of BCKDH was absent, accompanied by severely decreased staining of the E1α subunit in this cell line. The cDNAs of both E1α and E1β subunits from the patient's cell were amplified and sequenced. The deduced amino acid sequence for the E1α subunit was normal. An 11-bp deletion was identified in the region encoding the mitochondrial targeting leader peptide in the E1β cDNA. This 11-bp sequence locates in the first exon of the E1β gene, as a direct tandem repeat. This deletion removed nucleotides 80–90 (81–91) or 91–101 (92–102) of the cDNA, resulting in a frameshift and early stop codon downstream (Tab. 2).

Table 2. Mutations affecting the E1β subunit of the branched-chain α-keto acid dehydrogenase in maple syrup urine disease

Subjects	Phenotype	Genotype	Nucleotide change	Exon	Intron	Functional change	Reference
KM 08	classical	homozygote	11 bp deletion from 80 to 102	1	–	early stop	Nobukuni et al.(1991)
KM 04	classical	homozygote	single base change: 5' splice donor site (GT → TT)	–	5	splice error (deletion of exon 5 and both exons 5 and 6)	Hayashida et al. (1994)
KM 10	classical	compound-heterozygote	1 bp (G) insertion between 51 and 56;	1	–	early stop	Nobukuni et al. (1993)
			1 bp (T) deletion at 954	9	–		
KM 14	classical	compound-heterozygote	A-617 → G;	5	–	H156R;	Nobukuni et al. (1993)
			11 bp deletion from 80 to 102	1	–	early stop	

Nucleotide and deduced amino acid sequences of the human E1β cDNA were from Nobukuni et al. (1990). Nucleotide number and amino acid number start from the codon corresponding to the initiation methionine and the N-terminal of the matute protein, respectively.

Amplification of genomic DNA revealed that the consanguineous parents were heterozygous for this mutant allele, and the sister and the brother of the patient with the disease were each homozygous for this mutant allele. These observations suggested that the absence of the E1β subunit resulted in instability of the E1α subunit, affecting normal function of the BCKDH activity.

Interestingly, the same 11-bp deletion in exon 1 of the E1β gene was identified in two unrelated Italian patients (Parrella et al., 1994). Two of their parents and one sibling were carriers of this mutation.

KM 04 (K.Y.)

Hayashida et al. (1994) identified a unique splice mutation of the E1β subunit gene in a patient with typical classical MSUD. This patient was diagnosed at the time of newborn screening in Japan. The lymphoblastoid cells from this patient had been previously characterized at the protein level by the immunoblot analysis (Fig. 1) (Indo et al., 1987). The BCKDH showed sigmoidal kinetics similar to that of GM 1655 from the Mennonite patient and activity obtained at low substrate concentrations was 0.2% of the normal. The immunoblot study revealed that the E1β subunit of BCKDH was absent, accompanied by severely decreased staining of the E1α subunit in this cell line. The defect responsible for the deficiency of the E1β subunit protein was identified by analysis of cDNA and genomic DNA by PCR. A single base substitution from G to T of

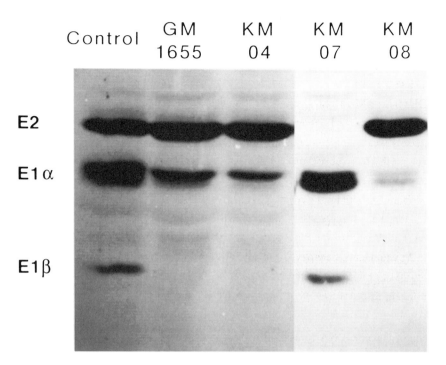

Figure 1. Immunoblot analysis of the BCKDH (E1+E2) complex in lymphoblastoid cell lines from patients with maple syrup urine disease. The BCKDH complex peptides are, in decreasing size: E2, E1α and E1β. Mutations of E1α (GM 1655), E1β (KM 04 and 08) and E2 (KM 07) were demonstrated in each cell line. Mutations of E1 component cause reduction of both E1α and E1β subunits. In contrast, the mutation of the E2 component causes no change in E1α and E1β subunits. A splice mutation in the E2 gene of KM 07 was reported (Mitsubuchi et al., 1991b). (See text for details).

the invariant GT dinucleotides at 5' splice site of the intron 5 was identified in the proband's E1β gene. Analysis of family members using primer-specified restriction map modification showed that the patient was homozygous for this mutation. This mutation led to skipping of the preceding exon 5 as well as both exons 5 and 6, thus producing two shortened E1β mRNA. This was the first documented example of exon skipping in the E1β gene as the cause of MSUD and the novel mutation of the invariant G at the 5' splice site which resulted in two alternatively spliced mRNA due to the skipping of the preceding exon as well as both preceding and following exon (Tab. 2).

KM 10 (R.D.) and KM 14 (Y.T.)

These patients were diagnosed at the time of newborn screening in Japan. In the case of KM 10, clinical course during early-neonatal period was not especially eventful and medical intervention was started when the female infant was 1 month of age. Restriction of BCAA intake was not strictly followed and mild metabolic acidosis and constantly elevated leucine levels were observed. The clinical phenotype was initially assigned as "so-called intermediate" (Indo et al., 1988), but she later died of an acute episode. Thus, it might be appropriate to assign her phenotype as classical rather than intermediate.

The BCKDHs from KM 14 and 10 showed sigmoidal and near-sigmoidal kinetics, respectively (Indo et al., 1987; Indo et al., 1988). Compared with intermittent phenotype patients, their BCKDH activities were significantly low at the low substrate concentration. The immunoblot study revealed that the E1β subunit of BCKDH was absent, accompanied by decreased staining of the E1α subunit in these cell lines.

Mutations of these two cell lines were identified in the E1β gene (Nobukuni et al., 1993). KM 10 was a compound heterozygote with two mutations, one base G insertion between 51 and 56 in exon 1 and one base T deletion at 954 in exon 9, generating termination codons at the next codon. KM 14 was also a compound heterozygote, a A-to-G transition in exon 5 and the 11-bp deletion in exon 1 of the E1β gene which resulted in the H156R mutation and a frameshift and early stop codon downstream, respectively. The 11-bp deletion was the same mutation noted in KM 08 (Tab. 2).

Significance of molecular analysis of the BCKDH in cell lines derived from patients with MSUD

Molecular basis of genetic heterogeneity

As described above, evidence was obtained for the existence of two complementation groups among patients with MSUD. Such complementation is intergenic in at least some combinations since mutations of the three genes encoding E1α, E1β and E2 in MSUD cells were identified and complementation of each pair of these cells was demonstrated (Nobukuni et al., 1993).

Activities of BCKDH in MSUD subjects measured using intact or disrupted cells

BCKAs are direct substrates for BCKDH. BCAAs are utilized for protein synthesis or converted to the corresponding BCKAs by transaminase(s) in metabolic pathway of the cell.

Dancis et al. (1972) reported that the enzyme level in MSUD subjects determined using intact cells and BCAA as a substrate was related to the potential to tolerate protein. They pointed out that BCAA was a more suitable substrate for intact cell assay than was BCKA, to evaluate protein tolerance of MSUD patients (Dancis et al., 1977). In contrast, it was shown that tolerance for dietary protein in patients correlated with levels of enzyme activity at lower BCKA concentrations, determined, using disrupted cells (Indo et al., 1988).

The availability of BCKAs for the BCKDH depends not only on the rate of formation of BCKAs but also on the rate of entry of BCKAs into the mitochondria and the rate of release of BCKAs into the extracellular fluid. These processes seem to maintain concentrations of BCKAs below those required to saturate the BCKDH when BCAA concentrations are little above the usual physiological range. Therefore, true V_{max} activities may not be attainable and the BCKDH may function at a relatively low concentration of substrate. Wendel and Langenbeck (1984) suggested that when leucine was used as the substrate for BCKDH in intact cells, the intracellular α-keto acid concentrations were very low. The observations of BCKDH activities at a low substrate concentration might be comparable with findings in assays using intact cells and BCAAs as substrates.

Activities and kinetic properties of the BCKDH

Activities and kinetic properties of the BCKDH were compared with mutations of specific subunits identified by gene or immunoblot analyses (Indo et al., 1987; Indo et al., 1988; Matsuda et al., 1990; Nobukuni et al., 1991; Mitsubuchi et al., 1991b; Nobukuni et al., 1993; Hayashida et al., 1994). Cells with E2 mutation showed hyperbolic kinetics and some were derived from patients with the intermittent phenotype. It seems that affinity of the BCKDH decarboxylation activity was not so affected in E2 deficient cells. Cells with E1α or E1β mutations exhibited sigmoidal or near-sigmoidal kinetics for BCKDH over the substrate concentration; these included cells from classical MSUD subjects (KM 04, 08, 10 and 14) and GM 1655. GM 1655 is derived from a Mennonite patient. Kinetic differences could give significantly different enzyme activities among MSUD subjects, especially at lower substrate concentrations. Differences of activities were obvious when measured at lower substrate concentrations, which corresponded to the apparent Km value of a normal control. These differences are obscured when measured at high substrate concentration. Activities measured at high substrate concentration were approximately 5% and 10% of normal control in the E1 and E2 deficient cells, respectively and Km value of the E2 deficient cell was similar to that of the normal control. The overall reaction of the BCKDH complex is similar to that of the pyruvate dehydrogenase complex and these two complexes share the same component (E3). Hence, the possibility of cross-reaction between the components (subunits) of these two complexes would need to be considered. Cook et al. (1985) investigated this possibility by monitoring the NADH produced by the reconstituted system. They found that the enzyme components of the two complexes do not interact to any significant extent. However, the low but significant BCKDH activity in patients with E2 mutation suggests that interaction of E1 and E2 components of other α-keto acid dehydrogenase might occur in the presence of a pathology, the result being the reconstitution of decarboxylation.

In case of E1 deficient cells, it should be noted that the E1α subunit protein decreased in amount and was only faintly detected even in the cell without the E1β subunit. However, some residual activity of BCKDH was observed at high substrate concentration. This low affinity component which acts at high substrate concentration might be due to activities of other α-keto acid dehydrogenase(s) such as pyruvate dehydrogenase and α-ketoglutarate dehydrogenase, as speculated by Danner and Elsas II (1989). Another possibility is that the E1α subunit not associated with E1β subunit could catalyze decarboxylation reaction, but this is only speculation. Mechanisms of the BCKDH reaction will be explored, using cells derived from patients with MSUD.

Stability of each subunit of BCKDH

Based on analyses of the mutations described above, the following was elucidated. Generally, mutations of the E1 component lead to a reduction in both E1α and E1β subunits. Mutations of each subunit of the E1 component could affect the stability of the counterpart in the same component, however, the E2 protein remains intact. E1α and E1β must be intact to maintain stability of the E1 component. In contrast, mutations of the E2 component cause no apparent change in the E1α and E1β subunits (Fig. 1). Thus, stability of E1 or E2 might be maintained independently by an intrinsic quality control system in the mitochondria. This notion is consistent with the speculation that dissociation and reconstitution of E1 with the E2 core may occur naturally in the mitochondria (Cook and Yeaman, 1988).

It will be of interest to examine how the E1 component ($\alpha_2\beta_2$ tetramer) stably exists in mitochondria of the cell lacking E2 protein or how E1α exists in the cell lacking the E1β subunit.

Clinical phenotypes and mutations of BCKDH complex in MSUD patients

Historically, MSUD classification was based on clinical presentation and outcome. Typical clinical phenotypes could be observed in many but not all patients. Clinical phenotypes could be determined by many variables, including mutations of BCKDH genes, and environmental differences such as quantity of protein ingested, intercurrent insults such as infections and the proficiency of continued dietary control. When BCKDH activity was measured at a low substrate concentration, the residual activity of each patient correlated with the patient's tolerance for leucine intake. Biochemical and mutation analyses of BCKDH revealed that, for all but a few exceptions, sigmoidal and hyperbolic kinetics correspond to the abnormalities of E1 and E2 component, respectively.

Mutations causing MSUD are heterogeneous, including missense, deletion, insertion and splicing abnormalities of any subunit of BCKDH complex. Different mutations in the same gene could lead to different degrees of impaired BCKDH function. Mutations in entirely different genes could lead to similar degrees of impaired BCKDH function. Thus, genotype and phenotype of MSUD are heterogeneous.

Acknowledgements
We are indebted to physicians in our laboratory for continuous collaboration: Drs. H. Mitsubuchi, Y. Hayashida, M. Tsuruta, K. Ohta, Y. Nobukuni and F. Endo. M. Ohara helped us prepare this report. This work was supported by a grant-in-aid for scientific research (C) from The Ministry of Education, Science and Culture of Japan, the research grant (5B-4) for nervous and mental disorders from The Ministry of Health and Welfare of Japan, a grant from The Okukubo Memorial Fund for Medical Research in the Kumamoto University School of Medicine and a grant from The Vehicle Racing Commemorative Foundation.

References

Chuang, D.T., Niu, W.-L. and Cox, R.P. (1981) Activities of branched-chain 2-oxo acid dehydrogenase and its components in skin fibroblasts from normal and classical-maple-syrup-urine-disease subjects. *Biochem. J.* 200: 59–67.

Chuang, J.L., Cox, R.P. and Chuang, D.T. (1990) Molecular cloning of the mature E1b-β subunit of human branched-chain α-keto acid dehydrogenase complex. *FEBS Lett.* 262: 305–309.

Chuang, J.L., Cox, R.P. and Chuang, D.T. (1993) Characterization of the promoter-regulatory region and structural organization of E1α gene (BCKDHA) of human branched-chain α-keto acid dehydrogenase complex. *J. Biol. Chem.* 268: 8309–8316.

Chuang, J.L., Fisher, C.R., Cox, R.P. and Chuang, D.T. (1994) Molecular basis of maple syrup urine disease: novel mutations at the E1α locus that impair E1($\alpha_2\beta_2$) assembly or decrease steady-state E1α mRNA levels of branched-chain α-keto acid dehydrogenase complex. *Am. J. Hum. Genet.* 55: 297–304.

Cook, K.G., Lawson, R., Yeaman, S.J. and Aitken, A. (1983) Amino acid sequence at the major phosphorylation site on bovine kidney branched-chain 2-oxoacid dehydrogenase complex. *FEBS Lett.* 164: 47–50.

Cook, K.G., Bradford, A.P., Yeaman, S.J., Aitken, A., Fearnley, I.M. and Walker, J.E. (1984) Regulation of bovine kidney branched-chain 2-oxoacid dehydrogenase complex by reversible phosphorylation. *Eur. J. Biochem.* 145: 587–591.

Cook, K.G., Bradford, A.P. and Yeaman, S.J. (1985) Resolution and reconstitution of bovine kidney branched-chain 2-oxo acid dehydrogenase complex. *Biochem. J.* 225: 731–735.

Cook, K.G. and Yeaman, S.J. (1988) Purification, resolution, and reconstitution of branched-chain 2-keto acid dehydrogenase complex from bovine kidney. *Meth. Enzymol.* 166: 303–308.

Damuni, Z., Merryfield, M.L., Humphreys, J.S. and Reed, L.J. (1984) Purification and properties of branched-chain α-keto acid dehydrogenase phosphatase from bovine kidney. *Proc. Natl. Acad. Sci. USA* 81: 4335–4338.

Damuni, Z. and Reed, L.J. (1987) Purification and properties of the catalytic subunit of the branched-chain α-keto acid dehydrogenase phosphatase from bovine kidney mitochondria. *J. Biol. Chem.* 262: 5129–5132.

Dancis, J., Levitz, M., Miller, S. and Westall, R.G. (1959) Maple syrup urine disease. *Br. Med. J.* 1: 91–93.

Dancis, J., Hutzler, J. and Levitz, M. (1960) Metabolism of the white blood cells in maple-syrup-urine disease. *Biochim. Biophys. Acta* 43: 342–343.

Dancis, J., Jansen, V., Hutzler, J. and Levitz, M. (1963) The metabolism of leucine in tissue culture of skin fibroblasts of maple-syrup-urine disease. *Biochim. Biophys. Acta* 77: 523–524.

Dancis, J., Hutzler, J. and Rokkones, T. (1967) Intermittent branched-chain ketonuria – variant of maple-syrup-urine disease. *N. Eng. J. Med.* 276: 84–89.

Dancis, J., Hutzler, J., Snyderman, S.E. and Cox, R.P. (1972) Enzyme activity in classical and variant forms of maple syrup urine disease. *J. Pediatr.* 81: 312–320.

Dancis, J., Hutzler, J. and Cox, R.P. (1977) Maple syrup urine disease: branched-chain keto acid decarboxylation in fibroblasts as measured with amino acids and keto acids. *Am. J. Hum. Genet.* 29: 272–279.

Danner, D.J. and Elsas II, L.J. (1975) Subcellular distribution and cofactor function of human branched chain α-ketoacid dehydrogenase in normal and mutant cultured skin fibroblasts. *Biochem. Med.* 13: 7–22.

Danner, D.J., Lemmon, S.K., Besharse, J.C. and Elsas II, L.J. (1979) Purification and characterization of branched chain α-ketoacid dehydrogenase from bovine liver mitochondria. *J. Biol. Chem.* 254: 5522–5526.

Danner, D.J., Armstrong, N., Heffelfinger, S.C., Sewell, E.T., Priest, J.H. and Elsas, L.J. (1985) Absence of branched chain acyl-transferase as a cause of maple syrup urine disease. *J. Clin. Invest.* 75: 858–860.

Danner, D.J. and Elsas II, L.J. (1989) Disorders of branched chain amino acid and keto acid metabolism. *In*: C.R. Scriver, A.L. Beaudet, W.S. Sly and D. Valle (eds): *The Metabolic Basis of Inherited Disease*, 6th Edition. McGraw-Hill Information Services Company, New York, pp 671–692.

Danner, D.J., Litwer, S., Herring, W.J. and Pruckler, J. (1989) Construction and nucleotide sequence of a cDNA encoding the full-length preprotein for human branched chain acyltransferase. *J. Biol. Chem.* 264: 7742–7746.

DiGeorge, A.M., Rezvani, I., Garibaldi, L.R. and Schwartz, M. (1982) Prospective study of maple-syrup-urine disease for the first four days of life. *N. Eng. J. Med.* 307: 1492–1495.

Fekete, G., Plattner, R., Crabb, D.W., Zhang, B., Harris, R.A., Heerema, N. and Palmer, C.G. (1989) Localization of the human gene for the E1α subunit of branched chain keto acid dehydrogenase (BCKDHA) to chromosome 19q13.1----q13.2. *Cytogenet. Cell Genet.* 50: 236–7.

Fisher, C.R., Fisher, C.W., Chuang, D.T. and Cox, R.P. (1991a) Occurrence of a Tyr393→Asn (Y393N) mutation in the E1α gene of the branched-chain α-keto acid dehydrogenase complex in maple syrup urine disease patients from a Mennonite population. *Am. J. Hum. Genet.* 49: 429–434.

Fisher, C.R., Chuang, J.L., Cox, R.P., Fisher, C.W., Star, R.A. and Chuang, D.T. (1991b) Maple syrup urine disease in Mennonites. Evidence that the Y393N mutation in E1α impedes assembly of the E1 component of branched-chain α-keto acid dehydrogenase complex. *J. Clin. Invest.* 88: 1034–1037.

Fisher, C.W., Chuang, J.L., Griffin, T.A., Lau, K.S., Cox, R.P. and Chuang, D.T. (1989) Molecular phenotypes in cultured maple syrup urine disease cells. Complete E1α cDNA sequence and mRNA and subunit contents of the human branched chain α-keto acid dehydrogenase complex. *J. Biol. Chem.* 264: 3448–3453.

Fisher, C.W., Fisher, C.R., Chuang, J.L., Lau, K.S., Chuang, D.T. and Cox, R.P. (1993) Occurrence of a 2-bp (AT) deletion allele and a nonsense (G-to-T) mutant allele at the E2 (DBT) locus of six patients with maple syrup urine disease: multiple-exon skipping as a secondary effect of the mutations. *Am. J. Hum. Genet.* 52: 414–424.

Goedde, H.W., Hüfner, M., Möhlenbeck, F. and Blume, K.G. (1967) Biochemical studies on branched-chain oxoacid oxidases. *Biochim. Biophys. Acta* 132: 524–525.

Gonzalez-Rios, M.C., Chuang, D.T., Cox, R.P., Schmidt, K., Knopf, K. and Packman, S. (1985) A distinct variant of intermediate maple syrup urine disease. *Clin. Genet* 27: 153–159.

Hayashida, Y., Mitsubuchi, H., Indo, Y., Ohta, K., Endo, F., Wada, Y. and Matsuda, I. (1994) Deficiency of the E1β subunit in the branched-chain α-keto acid dehydrogenase complex due to a single base substitution of the intron 5, resulting in two alternatively spliced mRNAs in a patient with maple syrup urine disease. *Biochim. Biophys. Acta* 1225: 317–325.

Heffelfinger, S.C., Sewell, E.T. and Danner, D.J. (1983) Identification of specific subunits of highly purified bovine liver branched-chain ketoacid dehydrogenase. *Biochemistry* 22: 5519–5522.

Hummel, K.B., Litwer, S., Bradford, A.P., Aitken, A., Danner, D.J. and Yeaman, S.J. (1988) Nucleotide sequence of a cDNA for branched chain acyltransferase with analysis of the deduced protein structure. *J. Biol. Chem.* 263: 6165–6168.

Indo, Y., Kitano, A., Endo, F., Akaboshi, I. and Matsuda, I. (1987) Altered kinetic properties of the branched-chain α-keto acid dehydrogenase complex due to mutation of the β-subunit of the branched-chain α-keto acid decarboxylase (E1) component in lymphoblastoid cells derived from patients with maple syrup urine disease. *J. Clin. Invest.* 80: 63–70.

Indo, Y., Akaboshi, I., Nobukuni, Y., Endo, F. and Matsuda, I. (1988) Maple syrup urine disease: a possible biochemical basis for the clinical heterogeneity. *Hum. Genet.* 80: 6–10.

Jinno, Y., Akaboshi, I. and Matsuda, I. (1984a) Complementation analysis in lymphoid cells from five patients with different forms of maple syrup urine disease. *Hum. Genet.* 68: 54–56.

Jinno, Y., Akaboshi, I., Katsuki, T. and Matsuda, I. (1984b) Study on established lymphoid cells in maple syrup urine disease. Correlation with clinical heterogeneity. *Hum. Genet.* 65: 358–361.

Jones, S.M.A. and Yeaman, S.J. (1986) Oxidative decarboxylation of 4-methylthio-2-oxobutyrate by branched-chain 2-oxo acid dehydrogenase complex. *Biochem. J* 237: 621–623.

Lau, K.S., Chuang, J.L., Herring, W.J., Danner, D.J., Cox, R.P. and Chuang, D.T. (1992a) The complete cDNA sequence for dihydrolipoyl transacylase (E2) of human branched-chain α-keto acid dehydrogenase complex. *Biochim. Biophys. Acta* 1132: 319–321.

Lau, K.S., Herring, W.J., Chuang, J.L., McKean, M., Danner, D.J., Cox, R.P. and Chuang, D.T. (1992b) Structure of the gene encoding dihydrolipoyl transacylase (E2) component of human branched chain α-keto acid dehydrogenase complex and characterization of an E2 pseudogene. *J. Biol. Chem.* 267: 24090–24096.

Lindsay, J.G. (1989) Targeting of 2-oxo acid dehydrogenase complexes to the mitochondrion. *Ann. N.Y. Acad. Sci.* 573: 254–266.

Litwer, S. and Danner, D.J. (1988) Mitochondrial import and processing of an *in vitro* synthesized human prebranched chain acyltransferase fragment. *Am. J. Hum. Genet.* 43: 764–769.

Lyons, L.B., Cox, R.P. and Dancis, J. (1973) Complementation analysis of maple syrup urine disease in heterokaryons derived from cultured human fibroblast. *Nature* 243: 533–535.

Matsuda, I., Nobukuni, Y., Mitsubuchi, H., Indo, Y., Endo, F., Asaka, J. and Harada, A. (1990) A T-to-A substitution in the E1α subunit gene of the branched-chain α-ketoacid dehydrogenase complex in two cell lines derived from Menonite maple syrup urine disease patients. *Biochem. Biophys. Res. Comm.* 172: 646–651.

McKean, M.C., Winkeler, K.A. and Danner, D.J. (1992) Nucleotide sequence of the 5' end including the initiation codon of cDNA for the E1α subunit of the human branched chain α-ketoacid dehydrogenase complex. *Biochim. Biophys. Acta* 1171: 109–112.

Menkes, J.H., Hurst, P.L. and Craig, J.M. (1954) A new syndrome: progressive familial infantile cerebral dysfunction associated with an unusual urinary substance. *Pediatrics* 14: 462–466.

Menkes, J.H. (1959) Maple syrup disease – Isolation and identification of organic acids in the urine. *Pediatrics* 23: 348–353.

Mitsubuchi, H., Nobukuni, Y., Endo, F. and Matsuda, I. (1991a) Structural organization and chromosomal localization of the gene for the E1β subunit of human branched chain α-keto acid dehydrogenase. *J. Biol. Chem.* 266: 14686–14691.

Mitsubuchi, H., Nobukuni, Y., Akaboshi, I., Indo, Y., Endo, F. and Matsuda, I. (1991b) Maple syrup urine disease caused by a partial deletion in the inner E2 core domain of the branched chain α-keto acid dehydrogenase complex due to aberrant splicing. A single base deletion at a 5'-splice donor site of an intron of the E2 gene disrupts the consensus sequence in this region. *J. Clin. Invest.* 87: 1207–1211.

Mitsubuchi, H., Matsuda, I., Nobukuni, Y., Heidenreich, R., Indo, Y., Endo, F., Mallee, J. and Segal, S. (1992) Gene analysis of Mennonite maple syrup urine disease kindred using primer-specified restriction map modification. *J. Inherit. Metab. Dis.* 15: 181–187.

Morris, M.D., Lewis, B.D., Doolan, P.D. and Harper, H.A. (1961) Clinical and biochemical observations on an apparently nonfatal variant of branched-chain ketoaciduria (Maple syrup urine disease). *Pediatrics* 28: 918–923.

Nobukuni, Y., Mitsubuchi, H., Endo, F. and Matsuda, I. (1989) Complete primary structure of the transacylase (E2b) subunit of the human branched chain α-keto acid dehydrogenase complex. *Biochem. Biophys. Res. Comm.* 161: 1035–1041.

Nobukuni, Y., Mitsubuchi, H., Endo, F., Akaboshi, I., Asaka, J. and Matsuda, I. (1990) Maple syrup urine disease. Complete primary structure of the E1β subunit of human branched chain α-ketoacid dehydrogenase complex deduced from the nucleotide sequence and a gene analysis of patients with this disease. *J. Clin. Invest.* 86: 242–247.

Nobukuni, Y., Mitsubuchi, H., Akaboshi, I., Indo, Y., Endo, F., Yoshioka, A. and Matsuda, I. (1991) Maple syrup urine disease. Complete defect of the E1β subunit of the branched chain α-ketoacid dehydrogenase complex due to a deletion of an 11-bp repeat sequence which encodes a mitochondrial targeting leader peptide in a family with the disease. *J. Clin. Invest.* 87: 1862–1866.

Nobukuni, Y., Mitsubuchi, H., Hayashida, Y., Ohta, K., Indo, Y., Ichiba, Y., Endo, F. and Matsuda, I. (1993) Heterogeneity of mutations in maple syrup urine disease (MSUD): screening and identification of affected E1α and E1β subunits of the branched-chain α-keto-acid dehydrogenase multienzyme complex. *Biochim. Biophys. Acta* 1225: 64–70.

Odessey, R. (1982) Purification of rat kidney branched-chain oxo acid dehydrogenase complex with endogenous kinase activity. *Biochem. J.* 204: 353–356.

Ono, K., Hakozaki, M., Nishimaki, H. and Kochi, H. (1987) Purification and characterization of human liver branched-chain α-keto acid dehydrogenase complex. *Biochem. Med. Metab. Biol.* 37: 133–141.

Otulakowski, G. and Robinson, B.H. (1987) Isolation and sequence determination of cDNA clones for porcine and human lipoamide dehydrogenase – homology to other disulfide oxidoreductases. *J. Biol. Chem.* 262: 17313–17318.

Parrella, T., Surrey, S., Iolascon, A., Sartore, M., Heidenreich, R., Diamond, G., Ponzone, A., Guardamagna, O., Burlina, A.B., Cerone, R., Parini, R., Dionisi-Vici, C., Rappaport, E. and Fortina, P. (1994) Maple syrup urine disease (MSUD): screening for known mutations in Italian patients. *J. Inherit. Metab. Dis.* 17: 652–660.

Paxton, R. and Harris, R.A. (1982) Isolation of rabbit liver branched chain α-ketoacid dehydrogenase and regulation by phosphorylation. *J. Biol. Chem.* 257: 14433–14439.

Paxton, R., Scislowski, P.W.D., Davis, E.J. and Harris, R.A. (1986) Role of branched-chain 2-oxo acid dehydrogenase and pyruvate dehydrogenase in 2-oxobutyrate metabolism. *Biochem. J.* 234: 295–303.

Pettit, F.H., Yeaman, S.J. and Reed, L.J. (1978) Purification and characterization of branched chain α-keto acid dehydrogenase complex of bovine kidney. *Proc. Natl. Acad. Sci. USA* 75: 4881–4885.

Pons, G., Raefsky-Estrin, C., Carothers, D.J., Pepin, R.A., Javed, A.A., Jesse, B.W., Ganapathi, M.K., Samols, D. and Patel, M.S. (1988) Cloning and cDNA sequence of the dihydrolipoamide dehydrogenase component of human α-ketoacid dehydrogenase complexes. *Proc. Natl. Acad. Sci. USA* 85: 1422–1426.

Popov, K.M., Zhao, Y., Shimomura, Y., Kuntz, M.J. and Harris, R.A. (1992) Branched-chain α-ketoacid dehydrogenase kinase – molecular cloning, expression, and sequence similarity with histidine protein kinases. *J. Biol. Chem.* 267: 13127–13130.

Reed, L.J. and Hackert, M.L. (1990) Structure-function relationships in dihydrolipoamide acyltransferases. *J. Biol. Chem.* 265: 8971–8974.

Robinson, B.H., Taylor, J. and Sherwood, W.G. (1977) Deficiency of dihydrolipoyl dehydrogenase (a component of the pyruvate and α-ketoglutarate dehydrogenase complexes): a cause of congenital chronic lactic acidosis in infancy. *Pediat. Res.* 11: 1198–1202.

Scherer, S.W., Otulakowski, G., Robinson, B.H. and Tsui, L.-C. (1991) Localization of the human dihydrolipoamide dehydrogenase gene (DLD) to 7q31-q32. *Cytogenet. Cell Genet.* 1991: 176–177.

Schulman, J.D., Lustberg, T.J., Kennedy, J.L., Museles, M. and Seegmiller, J.E. (1970) A new variant of maple syrup urine disease (branched chain ketoaciduria). *Am. J. Med.* 49: 118–124.

Scriver, C.R., Mackenzie, S., Clow, C.L. and Delvin, E. (1971) Thiamine-responsive maple-syrup-urine disease. *Lancet* 1: 310–312.

Seegmiller, J.E. and Westall, R.G. (1967) The enzyme defect in maple syrup urine disease. *J. Ment. Defic. Res.* 11: 288–294.

Shimomura, Y., Nanaumi, N., Suzuki, M., Popov, K.M. and Harris, R.A. (1990) Purification and partial characterization of branched-chain α-ketoacid dehydrogenase kinase from rat liver and rat heart. *Arch. Biochem. Biophys.* 283: 293–299.

Singh, S., Willers, I. and Goedde, H.W. (1977) Heterogeneity in maple syrup urine disease: aspects of cofactor requirement and complementation in cultured fibroblasts. *Clin. Genet* 11: 277–284.

Skaper, S.D., Molden, D.P. and Seegmiller, J.E. (1976) Maple syrup urine disease: branched-chain amino acid concentrations and metabolism in cultured human lymphoblasts. *Biochem. Genet.* 14: 527–539.

Tanaka, K. and Rosenberg, L.E. (1983) Disorders of branched chain amino acid and organic acid metabolism. *In*: J.B. Stanbury, J.B. Wyngaaden, D.S. Fredrickson, J.L. Goldstein and M.S. Brown (eds): *The Metabolic Basis of Inherited Disease*, 5th Edition. McGraw-Hill, New York, pp 440–473.

Wendel, U. and Langenbeck, U. (1984) Intracellular levels and metabolism of leucine and α-ketoisocaproate in normal and maple syrup urine disease fibroblasts. *Biochem. Med.* 31: 294–302.

Wendel, U., Wentrup, H. and Rudiger, H.W. (1975) Maple syrup urine disease: analysis of branched chain ketoacid decarboxylation in cultured fibroblasts. *Pediat. Res.* 9: 709–717.

Westall, R.G., Dancis, J. and Miller, S. (1957) Maple sugar urine disease. *A. M. A. J. Dis. Child.* 94: 571–572.

Wynn, R.M., Chuang, J.L., Davie, J.R., Fisher, C.W., Hale, M.A., Cox, R.P. and Chuang, D.T. (1992) Cloning and expression in *Escherichia coli* of mature E1β subunit of bovine mitochondrial branched-chain α-keto acid dehydrogenase complex. Mapping of the E1β-binding region on E2. *J. Biol. Chem.* 267: 1881–1887.

Wynn, R.M., Kochi, H., Cox, R.P. and Chuang, D.T. (1994) Differential processing of human and rat E1α precursors of the branched-chain α-keto acid dehydrogenase complex caused by an N-terminal proline in the rat sequence. *Biochim. Biophys. Acta* 1201: 125–128.

Yeaman, S.J. (1989) The 2-oxo acid dehydrogenase complexes: recent advances. *Biochem. J.* 257: 625–632.

Zhang, B., Edenberg, H.J., Crabb, D.W. and Harris, R.A. (1989) Evidence for both a regulatory mutation and a structural mutation in a family with maple syrup urine disease. *J. Clin. Invest.* 83: 1425–1429.

Zneimer, S.M., Lau, K.S., Eddy, R.L., Shows, T.B., Chuang, J.L., Chuang, D.T. and Cox, R.P. (1991) Regional assignment of two genes of the human branched-chain α-keto acid dehydrogenase complex: the E1β gene (BCKDHB) to chromosome 6p21–22 and the E2 gene (DBT) to chromosome 1p31. *Genomics* 10: 740–7.

Alpha-Keto Acid Dehydrogenase Complexes
M.S. Patel, T.E. Roche and R.A. Harris (eds)
© 1996 Birkhäuser Verlag Basel/Switzerland

Human defects of the pyruvate dehydrogenase complex

D.S. Kerr[1], I.D. Wexler[1], A. Tripatara[2] and M.S. Patel[3]

Departments of [1]Pediatrics, [1]Biochemistry, and [2]Nutrition, School of Medicine, Case Western Reserve, University, and [3]Department of Biochemistry, State University of New York at Buffalo, NY 14214, USA

Introduction

Deficiency of the pyruvate dehydrogenase complex (PDC), a critical enzyme of energy metabolism, would appear incompatible with survival, particularly in the intrauterine environment where energy production depends primarily on glucose oxidation. The now well-established occurrence of these defects presents a challenge to provide physiological explanations of what factors determine survival and adaptation of affected individuals. Unfortunately, for most affected individuals, deleterious effects on the central nervous system are severe. Biochemical and genetic explanations for why this group of disorders is so heterogeneous are beginning to emerge. Several reviews concerning clinical, biochemical, and genetic features of PDC deficiency and related disorders have been published recently (Robinson, 1995; Kerr and Zinn, 1995; Dahl, 1995; Patel and Harris, 1995).

History and prevalence

At least 150 cases of various sorts of PDC deficiency have been reported. The first report of an affected male with lactic acidosis and intermittent ataxia was based on measurement of pyruvate decarboxylation activity in cultured skin fibroblasts (Blass et al., 1970). Further characterization of various types of PDC deficiency emerged as specific assays for the catalytic components were applied to cells from patients (Robinson et al., 1980), indicating that pyruvate dehydrogenase (E1) deficiency is the most common type of defect. Application of specific immunoassays did not resolve which E1 subunit was involved, because either both were present or both were missing (Wexler et al., 1988). A key development was the recognition that the gene for E1α is located on chromosome X (Brown et al., 1989a), which explained the apparent prevalence of males with severe PDC deficiency (Ho et al., 1989a) and facilitated searching for E1α mutations in hemizygous males (Endo et al., 1989). As more and more mutations have been characterized, the importance and greater frequency of E1α deficiency in heterozygous females has become evident (Brown et al., 1989b; Dahl, 1995). Variable patterns of inactivation of the mutant chromosome X in different cells and tissues results in unpredictable manifestations in affected females.

Less is known as yet about autosomally transmitted forms of PDC deficiency. There have been no confirmed reports of E1β deficiency, and several cases of E2 (Robinson et al., 1990), protein-X (Marsac et al., 1993), and E1-phosphatase (Ito et al., 1992b) deficiencies have not been characterized as thoroughly as has E1α deficiency. One case of mixed heterozygosity for two mutations of the E3 gene has been described (Liu et al., 1993) (see below). While PDC deficiency remains one of the most frequently recognized causes of congenital lactic acidosis, it is definitively diagnosed in not more than 10% of undifferentiated cases of lactic acidosis. Among characterized defects of PDC deficiency, E1α defects are the most common; E3 defects are less than 5% as common. Currently, it is not possible to give reliable estimates of the frequency of PDC deficiency in the general population, since there have been no general screening programs.

Clinical manifestations

There is great variation of severity of clinical manifestations of PDC deficiency, with a common theme that the primary manifestations are impairment of neurological function and/or development. The clinical characteristics associated with E1 and E3 deficiencies are listed in Table 1, indicating their relative frequency amongst these heterogeneous groups. The clinical characteristics of E3 deficiency are similar to E1 deficiency, and there are too few reported cases to make useful comparisons of possible differences; so the distinction of these subgroups of PDC deficiency depends on analysis of their metabolic parameters and enzymatic activity (see below).

Some affected infants expire in the newborn period, or perhaps *in utero*, and some have congenital malformations of the brain, while at least one adult male is functioning normally with intermittent ataxia as the only manifestation. Paradoxically, the latter is the first reported case of PDC deficiency mentioned above with virtually no detectable PDC activity in his cultured skin fibroblasts (Sheu et al., 1981). The more common age of presentation is within the first few months of life, and is associated with hypotonia, developmental delay, and frequently, seizures. Less commonly, an infant with PDC deficiency may initially seem normal and then show progressive or degenerative neurological symptoms with an eventual fatal outcome. In some cases, this degenerative progression has been associated with neuropathological changes initially described by Leigh (commonly called "Leigh disease"), which includes focal necrosis of mid-

Table 1. Clinical characteristics of patients with PDC deficiency

Clinical Features	E1-deficiency	E3-deficiency
Delayed growth and development	+++	+++
Hypotonia	+++	+++
Seizures	++	+
CNS Degeneration (including Leigh disease)	++	+
CNS Malformations	+	–
Ataxia, choreoathetosis	+	+
Apnea, hypoventilation	+	+
Sudden death	+	–
Dysmorphic features	+	–
Peripheral neuropathy	+	–
Skeletal/cardiac myopathy	–	–

(+++) Very common (> 75%), (++) common (25–75%), (+) uncommon (< 25%), (–) not noted; (E1) pyruvate dehydrogenase, (E3) dihydrolipoamide dehydrogenase.

and lower brain nuclei with surrounding capillary proliferation (Leigh, 1951). Such lesions have been described in only a minority of cases (Kretzschmar et al., 1987; Sorbi and Blass, 1982; DeVivo et al., 1979; Miyabayashi et al., 1985), and are not specific to PDC deficiency as they also occur in certain nuclear and mitochondrial DNA encoded defects of the electron transport chain (Miranda et al., 1989; Tatuch et al., 1992; Wijburg et al., 1991).

The hypotonia of PDC deficiency appears to be mainly secondary to changes in the central nervous system, although peripheral neuropathy can occur. There is usually no evidence of pathological lesions in skeletal or cardiac muscle, even though normally the highest level of PDC expression is in cardiac muscle and postmortem studies have shown negligible residual PDC activity in muscle and heart (Kerr et al., 1987). Presumably, certain regions within the central nervous system ordinarily depend on glucose oxidation as a source of energy and acetyl-CoA, while muscle and heart may oxidize fatty acids as alternative substrates; possibly, certain brain cells are more susceptible to local lactic acidosis.

Virtually all identified cases of PDC deficiency have some degree of lactic acidemia, which is usually the factor leading to more specific testing and diagnosis. A few cases have been identified on the basis of isolated elevation of CSF lactate (Brown et al., 1988). The finding of a normal ratio of blood lactate/pyruvate is useful in distinguishing PDC deficiency from disorders of impaired tissue oxygenation or defects of electron transport, since PDC deficiency does not alter the NAD/NADH ratio. Lactate and pyruvate are frequently increased in urine, without an increase of other organic acids. Quantitative plasma amino acid analysis shows a specific increase in alanine. These parameters are distinctive from defects of other α-ketoacid dehydrogenase complexes or E3, in which the branched-chain amino acids and/or glutamate typically are increased in plasma and the corresponding α-keto- or α-hydroxyacids are increased in urine.

Biochemical characterization

Detection of PDC deficiency usually depends on assay of PDC in cells or tissues. Due to variable expression of PDC in different cell/tissue samples from some patients (Kerr et al., 1988), testing two or more different cell types (e.g., blood lymphocytes and cultured skin fibroblasts) is advisable. Females have been identified with neurological manifestations of E1α deficiency who have normal PDC activity in fibroblasts, due to variable skewing of chromosome X inactivation (Matthews et al., 1994; Dahl, 1995). Males with profound PDC deficiency usually have some measurable level of residual PDC activity, especially in cultured fibroblasts. Although there may be some correlation of clinical severity with levels of fibroblast expression (Robinson et al., 1987), this has not proven to be consistent. Severe fibroblast PDC deficiency has been associated with favorable outcome and normal fibroblast PDC activity has been found in males who have died in childhood. Since the clinical manifestations are largely related to dysfunction of focal areas in the central nervous system, it would be ideal to be able to determine exactly how much residual PDC activity was initially present in cells within these specific loci, before local damage occurred.

Enzymatic assays

Assay of total activated (dephosphorylated) PDC by decarboxylation of $1\text{-}^{14}C$-pyruvate to $^{14}CO_2$ has been the most useful method for detection of PDC deficiency in cell/tissue homogenates or isolated mitochondria (Robinson et al., 1980; Sheu et al., 1981; Kerr et al., 1987). Various methods have been employed to activate PDC; the most convenient and widely used is pre-incubation of whole cells or intact mitochondria with dichloroacetate to inhibit E1-kinase (Sheu et al., 1981). Alternatively, phospho-E1-phosphatase can be used with disrupted cells/mitochondria, including frozen specimens (Kerr et al., 1987).

Protein and mRNA analyses

Immunoblotting methods have been used to detect defects of E1 subunits (Wicking et al., 1986; Ho et al., 1986; Wexler et al., 1988), E3 (Matuda et al., 1984; Liu et al., 1993), as well as E2 and component-X (Robinson et al., 1990; Marsac et al., 1993). Since there is no simple functional assay for the non-catalytic component-X, immunoblotting is currently the only method for detection of such defects. The majority of E1 defects do not have reduced levels of either subunit. When the amount of immunoreactivity of either E1 subunit is reduced, both subunits are almost always reduced (Ho et al., 1986; Kitano et al., 1988; Kerr et al., 1988; Wexler et al., 1988). A few cases of E1 deficiency have been shown to have reduced E1α mRNA (Ho et al., 1989a; Wexler et al., 1988); low levels of mRNA may be present in these cases. The molecular basis of mutations resulting in reduced mRNA has not yet been established. Global defects in the E1α gene (e.g., large DNA deletions) have not been described.

Specific mutations

In the past few years there has been a proliferation of reports of specific mutations associated with PDC deficiency, almost all of which are within the E1α coding region. At the time of writing this review, at least 43 different E1α mutations have been reported in 72 individuals (34 male and 38 female). Two mutations of E3 have been reported in a case with compound heterozygosity. No human mutations have been reported as yet which specifically affect the E1β subunit, E2, component-X, or phospho-E1-phosphatase.

E1α Mutations

Reported mutations of E1α are listed in Table 2 and shown in a linear map of the coding region in Figure 1. The tremendous heterogeneity of E1α mutations is consistent with the hypothesis that there is a relatively high frequency of appearance of new mutations in this X-linked gene; assuming a steady-state prevalence, there must be a correspondingly high rate of disappearance of mutant alleles. Failure to date to discover complete deletions or other null mutations of E1α in males is consistent with the hypothesis that total absence of PDC is lethal to cells lacking a functional E1α allele (Dahl, 1995).

The 43 mutations shown in Table 2 all involve the coding region of the mature E1α protein. These include 22 single base-pair mutations resulting in 21 missense codon changes, 10 inser-

Table 2. Reported mutations of the E1α subunit of human PDC

Exon	Nucleotide[1]	Amino acid[2]	Type	Comments[3]	Sex M	Sex F	Fibroblast PDC %	Immuno-blot	References
III	A174G	H15R	Missense	Thiamine responsive; increased Km for TPP	1		0–17	Normal	(Naito et al., 1994)
III	C257T	R43C	Missense	Conserved in mammal + yeast PDC	2		20–30		(Dahl et al., 1992a; Chun et al., 1995)
III	G308A	G60S	Missense	Conserved in all PDCs; increased Km for TPP		1	100		(Matsuda et al., 1994)
IV	C422T	R98W	Missense	Highly conserved in PDC + BKDC	2	1	7–15	Trace	(Fujii et al., 1994)
V	G542A	V138M	Missense	Conserved as V or L in PDC + BKDC		2	1–7		(Chun et al., 1993)
VI	G553(-93)	V142del	Exon skip	Partial VI skipping, 31 aa deleted (part of TPP motif)	1		40		(Dahl et al., 1992a)
VI	G566A	V142del	Exon skip	Total VI skipping, 31 aa deleted (part of TPP motif)	1		4	Absent	(Chun et al., 1995)
VI	G566C	A146P	Missense	Conserved in all PDCs	1		51	Normal	(Takakubo et al., 1993b)
VI	G638A	A170T	Missense	TPP binding loop	1		17		(Chun et al., 1993)
VII	C658A	F176L	Missense	TPP binding loop	2		25–36	Reduced	(Dahl and Brown, 1994; Chun et al., 1995)
VII	A671G	M181V	Missense	Conserved mammal + yeast PDC; TPP binding loop	1		6	Normal	(Tripatara et al., 1995)
VII	C693T	P188L	Missense	Highly conserved; TPP binding loop hinge	1		12	Reduced	(Hemalatha et al., 1995)
VII	A734G	T202A	Missense	Highly conserved; end of TPP motif, adjacent to P3	1		2		(Chun et al., 1993)
VII	T770A	Y214N	Missense	Loss of E1 stability	1		32	Reduced	(Matthews et al., 1994)
VIII	A816C	D229A	Missense	Highly conserved; subunit binding region	1		10	Reduced	(Matthews et al., 1993b)
VIII	C830G	R234G	Missense	Subunit binding region; unstable E1	5	3	16–168	Normal?	(Wexler et al., 1992; Chun et al., 1993; Chun et al., 1995)
VIII	G831A	R234Q	Missense	Subunit binding region; defect dephosphorylation?	2	1	38	Altered	(Awata et al., 1994)
IX	A887C	M253L	Missense	Subunit binding region	1		115		(Matthews et al., 1994)
IX	C904(+1)	R259fs	Insertion	Truncated 283 aa protein	1		88		(Matthews et al., 1994)
IX	G914AorC	G262R	Missense	Near P1; conserved in mammal + yeast PDC	1		100		(Matsuda et al., 1994)
IX	A918T	H263L	Missense	Highly conserved, next to P1; interaction with TPP?	1		2		(Chun et al., 1993)
X	G942(-20)	S271fs	Deletion	Just following P2; truncated 276 aa, unstable E1	1		16	Absent	(Matthews et al., 1993a)
X	C947T	R273C	Missense	Near P2; highly conserved	4		47–89	Normal	(Dahl et al., 1992a; Dahl et al., 1992b)
X	C946(+5)	R273fs	Insertion	Truncated 297 aa protein		1	5	Absent	(Hansen et al., 1993)
X	G959(+21)	E277ins	Insertion	7 aa inserted, elongated protein	1		10	Normal	(De Meirleir et al., 1992)
X	T963(+33)	I278ins	Insertion	11 aa inserted, elongated protein	1		15		(Hansen et al., 1994)

Table 2. (continued)

Exon	Nucleotide[1]	Amino Acid[2]	Type	Comments[3]	Sex M	Sex F	Fibroblast PDC %	Immunoblot	References
X	T972(-3)	R282del	Deletion	Arginine 282 deleted	1		11	Normal	(Chun et al., 1995)
X	A973(-3)	R282del	Deletion	Arginine 282 deleted		1	0	Normal	(Tripatara et al., 1995)
X	G975(-4)	S283fs	Deletion	Truncated 294 aa protein	1		31	Normal	(Lissens et al., 1995)
X	A976(-7)	S283fs	Deletion	Truncated 293 aa protein	4		8–25	Reduced?	(Dahl et al., 1990; Chun et al., 1993; Chun et al., 1995)
X	T979(-3)	K284del	Deletion	Lysine 284 deleted	1		40	Altered	(Hansen et al., 1991)
X	G986A	D286N	Missense	Conserved in all PDCs		1	26	Normal	(Matthews et al., 1994)
X	C990(-2)	P287fs	Deletion	Truncated 308 aa protein	1		1		(Chun et al., 1993)
X	G1023(+13)	S298fs	Insertion	Truncated 313 aa protein	1		61	Sl reduced	(Chun et al., 1995)
XI	T1114(-20)	E329fs	Deletion	Truncated 339 aa protein	1		26	Smaller	(Chun et al., 1991)
XI	A1120(+46)	P331fs	Insertion	Truncated 332 aa protein	1		100	Smaller	(Chun et al., 1995)
XI	T1168(+2)	E347fs	Insertion	Elongated 394 aa protein	1		20		(Chun et al., 1993)
XI	G1176A	R349H	Missense	Highly conserved; possible subunit interaction region	5	1	13–50	Normal?	(Hansen et al., 1991; Matthews et al., 1994; Wexler et al., 1995; Chun et al., 1995)
XI	A1188(+4)	Q353fs	Insertion	Truncated 358 aa protein	4		6–46	Reduced	(Endo et al., 1991; Ito et al., 1992a; Chun et al., 1993; Takakubo al., 1993a)
XI	A1202(+16)	K358fs	Insertion	Elongated 362 aa protein	1		41	Reduced	(Tripatara et al., 1995)
XI	T1201(-2)	K358fs	Deletion	Elongated 400 aa protein	1		25	Trace	(Hansen et al., 1991)
XI	T1205(+4)	S359fs	Insertion	Truncated 358 aa protein	2		8–14	Reduced	(Chun et al., 1995)
XI	T1209(-4)	V360fs	Deletion	Elongated 392 aa protein, thiamine responsive	1		25	Reduced	(Endo et al., 1989)

[1] Nucleotide changed or last nucleotide preceding insert or deletion (±number of bp); numbering according to Ho et al. (1989b).
[2] Amino acid changed or first residue changed by insert (ins), deletion (del), or frameshift (fs); numbering for mature protein according to Ho et al. (1989b).
[3] The normal mature E1α protein includes 361 amino acids. (BKDC) branched-chain α-ketoacid dehydrogenase complex. For sequence comparisons see Wexler et al. (1992).

Human E$_1\alpha$ (X) Protein

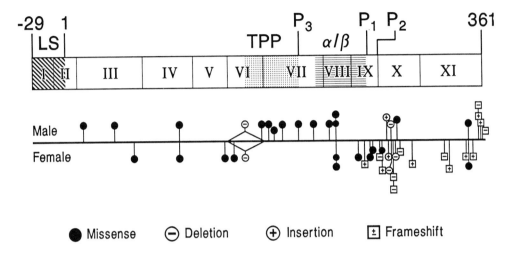

Figure 1. A linear map of the primary structure of the E1α protein showing functional regions and location of identified mutations. The numbering of residues is based on the mature form of the protein (361 amino acids) (Ho et al., 1989b). LS = leader sequence (29 amino acids); TPP = conserved TPP binding motif; α/β = proposed subunit interaction region; P$_1$, P$_2$, P$_3$ = phosphorylated serines. The sections encoded by exons I-XI are delineated with their respective roman numerals. Missense mutations are indicated by solid circles; mutations causing amino acid insertions or deletions are indicated by open circles containing + or −; insertions or deletions causing frameshifts are indicated by open squares containing + or −. Mutations occurring in males and females are shown separately and connected with the same line if identical or within the same codon. Different mutations within the same codon and the same sex are also connected with a single line. See Table 2 for additional details concerning each of these mutations.

tions, and 11 deletions. Almost all of the mutations occurring in exons I-IX are missense mutations which preserve the primary structure and some residual function of the protein. The exceptions are two mutations which result in skipping exon VI. These exceptions maintain the principle that some residual PDC activity is preserved, because one results in partial skipping, preserving 40% of residual PDC activity (Dahl et al., 1992a; De Meirlier et al., 1994), whereas the other mutation which results in complete skipping of exon VI occurred in a heterozygous female (Chun et al., 1995). The insertions and deletions are clustered in exons IX-XI, in most cases resulting in shifting of the reading frame and synthesis of truncated or elongated proteins; most of these modified proteins are presumably inactive although some may have residual activity.

E1α Structure-function correlations

The abundance of human E1α missense mutations provides opportunities to interpret the functional significance of specific amino acid residues or regions within the protein. All of these mutations have been identified because they are associated with reduced PDC activity or indirect evidence of PDC dysfunction. It is possible that some of the reported missense mutations will turn out to be fully functional polymorphisms, but it is improbable that many are, since no alternative candidate mutations were identified in these individuals and all of these mutations have been consistently associated with lack of normal PDC function. Many of these mutations affect conserved residues and/or occur in regions of the molecule which have been previously implicated as having critical functions, as indicated in Figure 1.

Thiamine pyrophosphate interactions

Five missense mutations have been identified which modify conserved residues within the TPP motif common to α-keto acid dehydrogenases (Hawkins et al., 1989). The three dimensional structure of this region in *S. cerevisiae* transketolase has been analyzed by X-ray crystallography (Sundström et al., 1992; Lindqvist et al., 1992), and the corresponding residues in human E1α and E1β which surround the Ca^{++}-pyrophosphate and thiazolium ring ends of TPP, respectively, have been identified by extrapolation of sequence homology (Robinson and Chun, 1993). One of these mutations substitutes leucine for proline-188, which lies in a loop between two regions that surround the Ca^{++}-pyrophosphate complex of TPP. It appears that the structural constraint of this proline, which is highly conserved, may limit mobility of the "hinge" region of the loop, thereby facilitating TPP binding (Hemalatha et al., 1995). The major consequence of this mutation is to decrease activity of PDC by decreasing stability of E1, with only a minor effect on the apparent K_m for TPP activation of residual PDC (Hemalatha et al., 1995). These findings are consistent with evidence that the TPP binding is integral to α and β-subunit interaction (Zhao et al., 1992).

Three other adjacent missense mutations lie on the same side of this loop which extends between residues at either end that directly interact with the Ca^{++}-pyrophosphate complex of TPP (Robinson and Chun, 1993); none of these three residues are highly conserved (Wexler et al., 1991). Modification of alanine-170 to threonine reduces but does not eliminate PDC activity (17% in fibroblasts from an affected male) (Chun et al., 1993); this alanine is only three residues away from aspartate 167, which appears to interact covalently with Ca^{++} (Robinson and Chun, 1993). Substitution of leucine for phenylalanine-176, a little further out on the same loop, is associated with somewhat higher residual PDC activity (36% in fibroblasts from an affected male)

(Chun et al., 1995). Substitution of valine for methionine-181 is associated with somewhat lower PDC activity (6% in fibroblasts from an affected male), without loss of immunoreactivity (Tripatara et al., 1995). It is not clear by what mechanism these three adjacent mutations within the TPP-binding loop alter PDC activity, as kinetic data relating the effect of TPP concentration to PDC activity have not been reported.

A fifth mutation found within the TPP binding motif results in substitution of alanine for threonine-202, which is very highly conserved. PDC activity in skin fibroblasts from this male is only 2%, and is unresponsive to dichloroacetate (Chun et al., 1993). Threonine-202 is immediately adjacent to the serine phosphorylation site P_3, and therefore the mutation could also interfere with dephosphorylation (activation) at this site.

Yet another mutation which may alter the function of the TPP-binding region but is located outside the motif is substitution of leucine for histidine-263. By homology with yeast transketolase, this residue appears to be one of two histidines that coordinate with the pyrophosphate moiety *via* hydrogen bonding (the other is histidine-83) (Robinson and Chun, 1993). Histidine-263 is also immediately adjacent to serine 264 (P_1 in Fig. 1), which is the first of three serines on E1α to be phosphorylated and the most critical for inactivation (Korotchkina and Patel, 1995). It is not clear which of these two functional relationships is more affected by the mutation; they may be inseparable. This mutation was found in a severely affected female with remarkably little residual PDC activity in fibroblasts (2%) and no activation of PDC with dichloroacetate (Chun et al., 1993). The severity of these consequences suggests a combination of effects due to an extremely unfavorable pattern of inactivation of the chromosome X carrying the normal E1α allele, and the effect of the mutation itself, which must knock out virtually all PDC activity (Chun et al., 1993).

A well-documented mutation which is associated with a major change in the relationship of PDC activity to the concentration of TPP is substitution of arginine for histidine-15. This is located quite remotely in the primary structure from the recognized TPP motif, and this is not a conserved residue. The activity of PDC at high concentrations of TPP is 17% in fibroblasts, 35% in cultured lymphoblasts, and about 50% in muscle. The apparent K_m for TPP is increased from around 10^{-9} M to 0.5×10^{-4} M in lymphoblastoid cells (Naito et al., 1994). The affected male improved upon administration of very large amounts of thiamine (see below). How this amino acid residue interacts with the TPP binding region has not yet been determined, but discovery of this mutation is a good example of the benefit of identification of unpredicted functional roles for specific residues/regions of the protein through investigation of human mutations.

Phosphorylation sites

Three other mutations have been found which are adjacent to phosphorylation sites, in addition to the two mutations discussed above which interact with TPP. A mutation converting glycine-262 to arginine, one residue proximal to histidine-263 mentioned above, might interfere with P_1, since this would introduce a large basic residue very close to the primary phosphorylation site. A 20 bp deletion immediately following serine 271, the second phosphorylation site (P_2, Fig. 1), encoded at the beginning of exon X, very probably affects this site, but also causes a frameshift leading to an unstable truncated protein. This mutation also was discovered in a severely affected female with lactic acidosis and markedly reduced PDC activity (16% in the fibroblasts), and direct evidence was obtained confirming predominant inactivation of the chromosome X carrying the normal E1α allele, with loss of immunoreactive E1 (Matthews et al., 1993a). Two amino acids downstream from P_2, substitution of cysteine for arginine-273 very probably also interferes with this site, but 47–89% PDC activity was reported in fibroblasts from affected female subjects (Dahl et al., 1992b).

Subunit interaction regions

In addition to the TPP-binding region, another region extending from amino acid residues 217–261 has been identified as a putative α/β subunit interaction region (Wexler et al., 1991). Four missense mutations have been found within this region in 13 individuals. One of these mutations, R234G, has been identified in eight individuals from four families, making it the most common PDC mutation to date. In the first family reported, variable expression of PDC activity was found in different cells and tissues; fibroblast PDC activity was normal in two of the affected males, whereas PDC deficiency was profound in lymphocytes and various tissues, and the outcome was fatal (Kerr et al., 1988). Other affected males have had 16–39% PDC activity in their fibroblasts without fatal outcome (Chun et al., 1993; Chun et al., 1995), and the outcome in three heterozygous mothers has also been variable and not correlated with fibroblast PDC activity (Kerr et al., 1988; Chun et al., 1995). In the first family, a correlation was found between loss of PDC activity and loss of immunoreactive E1α and E1β subunits; when the E1 protein was present, activity was normal. Arginine-234 is conserved in mammalian and nematode PDCs or replaced by another cationic residue seven amino acids downstream in yeast PDC and branched-chain α-ketoacid dehydrogenases with α and β subunits, such that there are always two or three positively charged groups spaced at intervals along the same side of this helical region of E1α, aligned in a manner that may enable ionic interaction of the two subunits (Wexler et al., 1991; Wexler et al., 1992). Exactly how different cell culture conditions or differences in tissue environments may affect this interaction is not known, but could involve differences in pH or efficiency of specific proteases.

There is evidence from a similar situation that the loss of both subunits is due to their mutual degradation by proteolysis (Huq et al., 1991).

Another mutation of the same residue has been reported in two unrelated males and one of their mothers, changing arginine-234 to glutamine (Awata et al., 1994). This mutation is associated with a shift in mobility and diminished amount of E1α on two-dimensional immunoelectrophoresis, consistent with predominance of the phosphorylated form of E1α (Kitano et al., 1990). Since this residue is situated between phosphorylation sites P1 and P3, it was proposed that this mutation may affect interaction of E1-phosphatase with the phospho-E1α substrate. Apparently, the R234Q mutation does not affect α/β subunit interaction in quite the same way as R234G, although reduced PDC activity is associated with some reduction of immunoreactivity. Substitution of the nearby highly conserved anionic residue aspartate-229 with alanine also results in loss of immunoreactive E1 (Matthews et al., 1993b); this additional negative charge may be critical for ionic interaction of the two subunits as well.

Several insertions and deletions have been described within the carboxy terminal residues of E1α, with the common effect of altering up to seven of these amino acids. Several are associated with loss of immunoreactivity, indicating loss of α/β subunit stability (Tripatara et al., 1995). A missense mutation adjacent to this area is substitution of histidine for arginine-349. Five unrelated affected males have been described whose fibroblasts had 13–50% of normal PDC activity; immunoreactivity was normal in four of these (Wexler et al., 1995; Matthews et al., 1994; Chun et al., 1995) and reduced in one case (Hansen et al., 1991). The outcome was uniformly fatal for these affected males, who died between ages 4–41 months. Arginine-349 is conserved in all mammalian and eukaryotic PDCs and branched-chain α-ketoacid dehydrogenases, and lysine-358, nine residues downstream, is conserved in all PDCs. While specific functions for these positively charged residues in E1α have not been assigned, the finding of loss of immunoreactivity in several mutations adjacent to these residues within the carboxy terminal region suggests that they too may play a role in covalent interaction of the α/β subunits, analogous to the putative α/β interaction site of residues 217–261 discussed above. Preservation of immunoreactivity in fibroblasts from two of these cases of the R349H mutation seems inconsistent, but as found with the R234G mutation, skin fibroblasts in culture may not accurately reflect the status of PDC in other tissues.

Mutations of residues with undetermined function

Functional significance has not been assigned to many of the other missense mutations of E1α, which implicate critical undetermined functions for the affected residues. Several are highly conserved, including: glycine 60, arginine-98, alanine-146, and aspartate-286. Details concerning these mutations are included in Table 2.

E1α Genotype/phenotype correlations

There are now enough E1α mutations that one might hope to be able to derive some correlation of phenotypes with the genotypes. There are several impediments to accomplishing this goal. First, the large heterogeneity of these defects makes it hard to collect very many subjects, especially males, with the same mutation. Secondly, there is intrinsic variation expected in affected females, due to variable inactivation of the two X-linked E1α alleles. This variation of severity has been well-documented in three women with the same mutation (R273C), making it very difficult to describe a typical outcome (Dahl et al., 1992b). In principle, more predictable phenotypic patterns could be established for affected males with identical mutations. However, variable effects of other genes as well as the medical, nutritional, and circumstantial environment of individuals also are likely to affect outcome.

These points are well illustrated by the R234G mutation, which would appear to provide one of the best opportunities to make a correlation, since this mutation has been found in five males and three females. As described above, this mutation is associated with variable levels of expression of PDC activity in different cells/tissues, with the highest level in cultured skin fibroblasts. All of the affected males have had developmental delay with hypotonia or ataxia. The first two reported cases were brothers who died at ages three and nine years and had normal PDC activity in their fibroblasts, whereas the other three males were alive at ages 7–9 years and had relatively lower expression of PDC activity (16–48%) in their fibroblasts. The two brothers initially had similar clinical presentations, but differed in that the diagnosis was made sooner in the younger brother. He was started on a strict, nearly carbohydrate-free diet about a year earlier than his older brother and lived longer with significantly greater mental development (Kerr et al., 1988). Details of medical management of the other three surviving males have not been published. The mother of the two brothers who died is heterozygous for the R234G allele and is clinically normal, while two of the mothers of the other three less severely affected boys who carry the same allele are not normal; one has mental retardation and the other has cerebellar ataxia (Chun et al., 1995).

Another mutation that has been found in four unrelated males is R349H. The consequences are severe, in that all died in early childhood at ages 4 to 41 months. Skin fibroblast PDC activity ranged from 13% to 50% of controls, but does not correlate with outcome or with PDC activity in other cells (lymphocytes) or tissues. All had significant developmental delay and three had early onset of episodes of marked lactic acidosis. The difference between the shortest and longest survivors may in part be attributed to vigorous treatment with a carbohydrate-free, high fat diet (Wexler et al., 1995).

E3 Mutations

One report has been published describing two mutations in a previously described case of E3 deficiency (Liu et al., 1993). The affected male had only 6% E3 activity in skin fibroblasts. Two missense point mutations were identified in this mixed heterozygote; one results in substitution of glutamate for lysine-37 and the other in substitution of leucine for proline 453. These two mutations correlate with previously identified functional regions of this flavoprotein which directly affect the binding of FAD and catalytic site, respectively. Cells from the parents of this child were not available for testing. There is evidence from biochemical assays of other cases that parents express reduced amounts of E3 activity; in contrast to hemizygous females with E1α mutations, there is no indication that parents who are heterozygous for E3 deficiency are functionally abnormal.

Genetic transmission and prenatal diagnosis

Patterns of inheritance

It is now clear that the by far most common type of PDC deficiency is due to mutations of the E1α gene of chromosome X. While initial information suggested a prevalence of males among severely affected cases, the frequency and range of severity are greater in females. Use of terms such as "X-linked dominant" or "X-linked lethal" inheritance have been used to describe this situation (Chun et al., 1995). However, there is a variable degree of penetrance due to unpredictable patterns of inactivation of the two X-linked E1α alleles in females, especially within the brain (Brown et al., 1989b). While mutations found in males appear to be associated consistently with some measurable residual PDC activity or predicted preservation of minimal E1α function, heterozygous females can survive with certain mutations that would not be predicted to have any residual function. In these females, cells in which the normal allele is expressed should function normally, while neighboring cells expressing the abnormal allele may die (Dahl, 1995). The high frequency of new mutations is consistent with the hypothesis that there is a reproductive loss of affected females as well as males. It is not known which stage of development may be the most vulnerable to PDC deficiency. During normal male reproduction, the problem of lack of expression of the E1α gene on chromosome X in haploid spermatozoa is resolved by testis-specific expression of the processed E1α gene on chromosome 4 (Iannello et al., 1993). It is possible that haploid ova containing a single defective E1α allele which is not able to produce any PDC activity may not survive in the ovary, accounting for the apparently frequent disappearance of these alleles.

E3 deficiency appears to be inherited in a more conventional autosomal recessive manner, consistent with the localization of the gene on chromosome 7, but no family studies of the pattern of inheritance of defined mutations have been reported.

Recurrence rate and prenatal testing

Most cases of E1α deficiency are sporadic, apparently due to new mutations, since only a small minority of mothers have been found to carry the mutated X-linked allele in their peripheral cells. Our estimate of the risk of recurrence in subsequent pregnancies after an affected male has been identified is less than 10% (Kerr and Lusk, 1992). There have been reports of successful negative prenatal tests in pregnancies which resulted in normal offspring (Kerr and Lusk, 1992; Brown and Brown, 1994), but there have been no reports of successful prenatal diagnosis of a case of PDC deficiency. Due to variable expression of the E1α gene, DNA mutational analysis would be optimal for this purpose, provided the mutation can be identified in the family.

Treatment strategies

Ketogenic diet

Use of ketogenic diets for treatment of PDC deficiency was first introduced some 20 years ago (Falk et al., 1976). The metabolic rationale for this treatment remains convincing, in that provision of β-hydroxybutyrate and acetoacetate should substitute for pyruvate (derived from glucose) as a source of acetyl-CoA for those brain cells which appear to be most dependent on PDC. However, clinical success with application of this dietary treatment has not been overwhelmingly successful. There are several possible explanations for inadequacy of this treatment. First, there is some evidence, especially from the most severe cases of PDC deficiency, that injury to the developing brain starts *in utero,* during a period when the fetal brain is inevitably glucose dependent. Secondly, because new cases of PDC deficiency are almost all sporadic (not recurrent in the same family), the diagnosis is typically made at some delayed interval after birth and treatment not started until several weeks of age, at best. Thirdly, it is very difficult to sustain ketosis by dietary restriction of carbohydrate over long periods of time, because introduction of small amounts of carbohydrate can effectively suppress ketogenesis. Fourthly, quantitative methods for measuring plasma "ketones" (β-hydroxybutyrate and acetoacetate) have not been readily available in most clinical settings or for home use. Finally, there are currently no guidelines as to what degree of ketosis may be necessary to protect the PDC-deficient brain. It appears reasonable, by analogy to the degree of ketosis induced by fasting in normal infants and children, that the goal should be to sustain a molar concentration of β-hydroxybutyrate + acetoacetate around 4–6 mM (Kerr, 1995). In our experience, this requires a virtually carbohydrate-free diet. In reviewing the experience with ketogenic diet therapy of seven male children with the same mutations (including R234G and R349H), we concluded that the earliest initiation of the ketogenic diet and use of the least amount of carbohydrate was associated with longer survival and greater mental development (Wexler et al., 1995).

Thiamine

Administration of large amounts of thiamine to individuals with PDC deficiency (and a variety of other forms of undifferentiated lactic acidemia) has been commonplace, but the benefits are generally undocumented and only rarely have even been claimed to be significant. An important example of thiamine responsiveness is the H15R mutation described above, in which an increase of the K_m for TPP was demonstrated by assay of PDC; in addition, cultured cells were shown to increase pyruvate oxidation in the presence of high concentrations of thiamine (Naito et al., 1994). Lactic acid decreased when this male child was given massive amounts of thiamine (up to 100 mg/kg/d), and there was clinical improvement. Another mutation associated with reported clinical thiamine responsiveness is a 4 bp deletion of the carboxy terminal codons for valine-360 and serine-361, resulting in a frameshift and extension of 32 amino acids (Endo et al., 1989); an increase of the K_m for TPP also was reported in cultured cells (Narisawa et al., 1992).

Dichloroacetate

Since dichloroacetate inhibits E1-kinase, it might be expected to be of benefit for cases with significant residual PDC activity or specifically cases with E1-phosphatase deficiency. There have been several reports of clinical experience with administration of dichloroacetate in PDC deficiency with variable outcomes (Stacpoole, 1989; Uziel et al., 1994). The outcome has not been correlated with molecular characterization of the defect in these reports. A controlled clinical trial of dichloroacetate therapy for PDC deficiency and other forms of congenital lactic acidemia is currently underway, which may help answer the potential benefit of this agent for specific disorders (P.W. Stacpoole, personal communication).

Conclusions

PDC deficiency is one of the major genetic disorders of oxidative metabolism, causing elevation of lactate in blood and/or CSF. The consequences primarily affect the developing central nervous system, but range greatly in severity. The puzzling reality that affected individuals survive at all with such a critical defect appears to be accounted for by residual levels of PDC activity. The most common defects are associated with mutations of the E1α gene located on chromosome X. To date, some 43 mutations within the reading frame of E1α have been reported in around 72 individuals, with less than 10% recurring in the same family. Reliable prenatal diagnosis has not been established. In males, all these mutations are associated with some residual function of PDC. Mutations with more severe disruption of E1α structure and function have been described in heterozygous females, and the clinical manifestations are variably alleviated by the degree of activation of the normal allele on the other chromosome X. Defects of other components of PDC are less common; to date, mutations have been characterized only for the E3 component.

Characterization of a large variety of missense mutations of E1α and consideration of their consequences provides opportunities for confirmation and new insights into the relationship of structure and function of this protein. Defects within the conserved TPP motif appear to affect the conformation of this binding site and in some cases affect the stability of interaction between the α and β subunits. Other mutations which are located remotely in the primary structure are associated with an increase in the K_m for TPP. Several mutations are adjacent to the three serines whose phosphorylation regulates PDC activity. Other mutations within conserved regions whose function has been attributed to α/β subunit interaction result in loss of E1 stability. Finally, several mutations which reduce PDC activity have been found in conserved residues not previously recognized as playing a role in the active site.

Unfortunately, treatment of PDC deficiency has been generally disappointing, including use of ketogenic diets, thiamine, and dichloroacetate. However, there are some promising exceptions, particularly with rare thiamine responsive cases, and the possibility of developing more effective regimens for ketogenic therapy or selection of dichloroacetate responsive cases remains. Evaluation of these therapies depends on early detection, organization of clinical trials, and correlation with the molecular basis of the defect.

References

Awata, H., Endo, F., Tanoue, A., Kitano, A. and Matsuda, I. (1994) Characterization of a point mutation in the pyruvate dehydrogenase E1α gene from two boys with primary lactic acidaemia. *J. Inherit. Metab. Dis.* 17: 189–195.

Blass, J.P., Avigan, J. and Uhlendorf, B.W. (1970) A defect in pyruvate decarboxylase in a child with an intermittent movement disorder. *J. Clin. Invest.* 49: 423–432.

Brown, G.K., Haan, E.A., Kirby, D.M., Scholem, R.D., Wraith, J.E., Rogers, J.G. and Danks, D.M. (1988) "Cerebral" lactic acidosis: defects in pyruvate metabolism with profound brain damage and minimal systemic acidosis. *Eur. J. Pediatr.* 147: 10–14.

Brown, R.M., Dahl, H.H.M. and Brown, G.K. (1989a) X-chromosome localization of the functional gene for the E1α subunit of the human pyruvate dehydrogenase complex. *Genomics* 4: 174–181.

Brown, G.K., Brown, R.M., Scholem, R.D., Kirby, D.M. and Dahl, H.H.M. (1989b) The clinical and biochemical spectrum of human pyruvate dehydrogenase complex deficiency. *Ann. N.Y. Acad. Sci.* 573: 360–368.

Brown, R.M. and Brown, G.K. (1994) Prenatal diagnosis of pyruvate dehydrogenase E1 alpha subunit deficiency. *Prenat. Diagn.* 14: 435–441.

Chun, K., MacKay, N., Petrova-Benedict, R. and Robinson, B.H. (1991) Pyruvate dehydrogenase deficiency due to a 20-bp deletion in exon 11 of the pyruvate dehydrogenase (PDH) E1α gene. *Am. J. Hum. Genet.* 49: 414–420.

Chun, K., MacKay, N., Petrova-Benedict, R. and Robinson, B.H. (1993) Mutations in the X-linked E1α subunit of pyruvate dehydrogenase leading to deficiency of the pyruvate dehydrogenase complex. *Hum. Mol. Genet.* 2: 449–454.

Chun, K., MacKay, N., Petrova-Benedict, R., Federico, A., Fois, A., Cole, D.E.C., Robertson, E. and Robinson, B.H. (1995) Mutations in the X-linked E1α subunit of pyruvate dehydrogenase: Exon skipping, insertion of duplicate sequence, and missense mutations leading to the deficiency of the pyruvate dehydrogenase complex. *Am. J. Hum. Genet.* 56: 558–569.

Dahl, H.H.M., Maragos, C., Brown, R.M., Hansen, L.L. and Brown, G.K. (1990) Pyruvate dehydrogenase deficiency caused by deletion of a 7-bp repeat sequence in the E1α gene. *Am. J. Hum. Genet.* 47: 286–293.

Dahl, H.H.M., Brown, G.K., Brown, R.M., Hansen, L.L., Kerr, D.S., Wexler, I.D., Patel, M.S., De Meirleir, L., Lissens, W., Chun, K., MacKay, N. and Robinson, B.H. (1992a) Mutations and polymorphisms in the pyruvate dehydrogenase E1α gene. *Hum. Mutat.* 1: 97–102.

Dahl, H.H.M., Hansen, L.L., Brown, R.M., Danks, D.M., Rogers, J.G. and Brown, G.K. (1992b) X-linked pyruvate dehydrogenase E1α subunit deficiency in heterozygous females: variable manifestation of the same mutation. *J. Inherit. Metab. Dis.* 15: 835–847.

Dahl, H.H.M. and Brown, G.K. (1994) Pyruvate dehydrogenase deficiency in a male caused by a point mutation (F205L) in the E1α subunit. *Hum. Mutat.* 3: 152–155.

Dahl, H.H.M. (1995) Pyruvate dehydrogenase E1α deficiency: Males and females differ yet again. *Am. J. Hum. Genet.* 56: 553–557.

De Meirleir, L., Lissens, W., Vamos, E. and Liebaers, I. (1992) Pyruvate dehydrogenase (PDH) deficiency caused by a 21-base pair insertion mutation in the E1α subunit. *Hum. Genet.* 88: 649–652.

De Meirleir, L., Lissens, W., Benelli, C., Ponsot, G., Desguerre, I., Marsac, C., Rodriguez, D., Saudubray, J.M., Poggi, F. and Liebaers, I. (1994) Aberrant splicing of exon 6 in the pyruvate dehydrogenase-E1 alpha mRNA linked to a silent mutation in a large family with Leigh's encephalomyelopathy. *Pediatr. Res.* 36: 707–712.

DeVivo, D.C., Haymond, M.W., Obert, K.A., Nelson, J.S. and Pagliara, A.S. (1979) Defective activation of the pyruvate dehydrogenase complex in subacute necrotizing encephalomyelopathy (Leigh disease). *Ann. Neurol.* 6: 483–494.

Endo, H., Hasegawa, K., Narisawa, K., Tada, K., Kagawa, Y. and Ohta, S. (1989) Defective gene in lactic acidosis: abnormal pyruvate dehydrogenase E1 α-subunit caused by a frame shift. *Am. J. Hum. Genet.* 44: 358–364.

Endo, H., Miyabayashi, S., Tada, K. and Narisawa, K. (1991) A four-nucleotide insertion at the E1α gene in a patient with pyruvate dehydrogenase deficiency. *J. Inherit. Metab. Dis.* 14: 793–799.

Falk, R.E., Cederbaum, S.D., Blass, J.P., Gibson, G.E., Pieter Kark, R.A. and Carrel, R.E. (1976) Ketogenic diet in the management of pyruvate dehydrogenase deficiency. *Pediatrics* 58: 713–721.

Fujii, T., Van Coster, R.N., Old, S.E., Medori, R., Winter, S., Gubits, R.M., Matthews, P.M., Brown, R.M., Brown, G.K., Dahl, H.H.M. and De Vivo, D.C. (1994) Pyruvate dehydrogenase deficiency: molecular basis for intrafamilial heterogeneity. *Ann. Neurol.* 36: 83–89.

Hansen, L.L., Brown, G.K., Kirby, D.M. and Dahl, H.H.M. (1991) Characterization of the mutations in three patients with pyruvate dehydrogenase E1α deficiency. *J. Inherit. Metab. Dis.* 14: 140–151.

Hansen, L.L., Brown, G.K., Brown, R.M. and Dahl, H.H.M. (1993) Pyruvate dehydrogenase deficiency caused by a 5 base pair duplication in the E1α subunit. *Hum. Mol. Genet.* 2: 805–807.

Hansen, L.L., Horn, N., Dahl, H.H.M. and Kruse, T.A. (1994) Pyruvate dehydrogenase deficiency caused by a 33 base pair duplication in the PDH E1α subunit. *Hum. Mol. Genet.* 3: 1021–1022.

Hawkins, C.F., Borges, A. and Perham, R.N. (1989) A common structural motif in thiamin pyrophosphate-binding enzymes. *FEBS Lett.* 255: 77–82.

Hemalatha, S.G., Kerr, D.S., Wexler, I.D., Lusk, M.M., Kaung, M., Du, Y., Kolli, M., Schelper, L. and Patel, M.S. (1995) Pyruvate dehydrogenase complex deficiency due to a point mutation (P188L) within the thiamine pyrophosphate binding loop of the E1α subunit. *Hum. Mol. Genet.* 4: 315–318.

Ho, L., Hu, C.W.C., Packman, S. and Patel, M.S. (1986) Deficiency of the pyruvate dehydrogenase component in pyruvate dehydrogenase complex-deficient human fibroblasts. *J. Clin. Invest.* 78: 844–847.

Ho, L., Wexler, I.D., Kerr, D.S. and Patel, M.S. (1989a) Genetic defects in human pyruvate dehydrogenase. *Ann. N.Y. Acad. Sci.* 573: 347–359.

Ho, L., Wexler, I.D., Liu, T.C., Thekkumkara, T.J. and Patel, M.S. (1989b) Characterization of cDNAs encoding human pyruvate dehydrogenase alpha subunit. *Proc. Natl. Acad. Sci. USA* 86: 5330–5334.

Huq, A.H.M.M., Ito, M., Naito, E., Saijo, T., Takeda, E. and Kuroda, Y. (1991) Demonstration of an unstable variant of pyruvate dehydrogenase protein (E1) in cultured fibroblasts from a patient with congenital lactic acidemia. *Pediat. Res.* 30: 11–14.

Iannello, R.C., Kola, I. and Dahl, H.H.M. (1993) Temporal and tissue-specific interactions involving novel transcription factors and the proximal promoter of the mouse Pdha-2 gene. *J. Biol. Chem.* 268: 22581–22590.

Ito, M., Huq, A.H.M.M., Naito, E., Saijo, T., Takeda, E. and Kuroda, Y. (1992a) Mutation of E1α gene in a female patient with pyruvate dehydrogenase deficiency due to rapid degradation of E1 protein. *J. Inherit. Metab. Dis.* 15: 848–856.

Ito, M., Kobashi, H., Naito, E., Saijo, T., Takeda, E., Huq, A.H.M.M. and Kuroda, Y. (1992b) Decrease of pyruvate dehydrogenase phosphatase activity in patients with congenital lactic acidemia. *Clin. Chim. Acta* 209: 1–7.

Kerr, D.S., Ho, L., Berlin, C.M., Lanoue, K.F., Towfighi, J., Hoppel, C.L., Lusk, M.M., Gondek, C.M. and Patel, M.S. (1987) Systemic deficiency of the first component of the pyruvate dehydrogenase complex. *Pediat. Res.* 22: 312–318.

Kerr, D.S., Berry, S.A., Lusk, M.M., Ho, L. and Patel, M.S. (1988) A deficiency of both subunits of pyruvate dehydrogenase which is not expressed in fibroblasts. *Pediat. Res.* 24: 95–100.

Kerr, D.S. and Lusk, M.M. (1992) Infrequent expression of heterozygosity or deficiency of pyruvate dehydrogenase (E1) among parents and sibs of affected patients. *Pediat. Res.* 31: 133A.

Kerr, D.S. (1995) Treatment of lactic acidosis: A review. *Int. Pediatr.* 10: 75–81.

Kerr, D.S. and Zinn, A.B. (1995) The pyruvate dehydrogenase complex and tricarboxylic acid cycle. *In:* J. Fernandes, J.-M. Saudubray and G. van Berghe (eds): *Inherited Metabolic Diseases. Diagnosis and Treatment.* Second Edition. Springer-Verlag, Berlin, pp 109–119.

Kitano, A., Akaboshi, I., Endo, F., Matsuda, I., Okano, Y., Hase, Y., Nagao, Y., Kamoshita, S., Miyabayashi, S. and Narisawa, K. (1988) Immunochemical evidence of pyruvate dehydrogenase (E1) deficiency. *J. Inherit. Metab. Dis.* 11: 329–332.

Kitano, A., Endo, F. and Matsuda, I. (1990) Immunochemical analysis of pyruvate dehydrogenase complex in 2 boys with primary lactic acidemia. *Neurology* 40: 1312–1314.

Korotchkina, L.G. and Patel, M.S. (1995) Mutagenesis studies of the phosphorylation sites of recombinant human pyruvate dehydrogenase. Site-specific regulation. *J. Biol. Chem.* 270: 14297–14304.

Kretzschmar, H.A., DeArmond, S.J., Koch, T.K., Patel, M.S., Newth, C.J.L., Schmidt, K.A. and Packman, S. (1987) Pyruvate dehydrogenase complex deficiency as a cause of subacute necrotizing encephalopathy (Leigh Disease). *Pediatrics* 79: 370–373.

Leigh, D. (1951) Subacute necrotizing encephalomyelopathy in an infant. *J. Neurol. Neurosurg. Psychiatr.* 14: 216–221.

Lindqvist, Y., Schneider, G., Ermler, U. and Sundström, M. (1992) Three dimensional structure of transketolase, a thiamine diphosphate dependent enzyme, at 2.5 Å resolution. *EMBO J.* 11: 2373–2379.

Lissens, W., Desguerre, I., Benelli, C., Marsac, C., Fouque, F., Haenggeli, C., Ponsot, G., Seneca, S., Liebaers, I. and De Meirleir, L. (1995) Pyrvuvate dehydrogenase deficiency in a female due to a 4 base pair deletion in exon 10 of the E1α gene. *Hum. Mol. Genet.* 4: 307–308.

Liu, T.C., Kim, H., Arizmendi, C., Kitano, A. and Patel, M.S. (1993) Identification of two missense mutations in a dihydrolipoamide dehydrogenase-deficient patient. *Proc. Natl. Acad. Sci. USA* 90: 5186–5190.

Marsac, C., Stansbie, D., Bonne, G., Cousin, J., Jehenson, P., Benelli, C., Leroux, J.P. and Lindsay, G. (1993) Defect in the lipoyl-bearing protein X subunit of the pyruvate dehydrogenase complex in two patients with encephalomyelopathy. *J. Pediatr.* 123: 915–920.

Matsuda, J., Ito, M., Naito, E., Yokota, I. and Kuroda, Y. (1994) DNA diagnosis for pyruvate dehydrogenase complex deficiency in female partients with congenital lactic acidemia. *Proc. VIth Int.Cong. Inborn Errors Metab.* Milan, p. 65.

Matthews, P.M., Brown, R.M., Otero, L., Marchington, D., Leonard, J.V. and Brown, G.K. (1993a) Neurodevelopmental abnormalities and lactic acidosis in a girl with a 20-bp deletion in the X-linked pyruvate dehydrogenase E1α subunit gene. *Neurology* 43: 2025–2030.

Matthews, P.M., Marchington, D.R., Squier, M., Land, J., Brown, R.M. and Brown, G.K. (1993b) Molecular genetic characterization of an X-linked form of Leigh's syndrome. *Ann. Neurol.* 33: 652–655.

Matthews, P.M., Brown, R.M., Otero, L.J., Marchington, D.R., LeGris, M., Howes, R., Meadows, L.S., Shevell, M., Scriver, C.R. and Brown, G.K. (1994) Pyruvate dehydrogenase deficiency. Clinical presentation and molecular genetic characterization of five new patients. *Brain* 117: 435–443.

Matuda, S., Kitano, A., Sakaguchi, Y., Yoshino, M. and Saheki, T. (1984) Pyruvate dehydrogenase subcomplex with lipoamide dehydrogenase deficiency in a patient with lactic acidosis and branched chain ketoaciduria. *Clin. Chim. Acta* 140: 59–64.

Miranda, A.F., Ishii, S., DiMauro, S. and Shay, J.W. (1989) Cytochrome c oxidase deficiency in Leigh's syndrome: genetic evidence for a nuclear DNA-encoded mutation. *Neurology* 39: 697–702.

Miyabayashi, S., Ito, T., Narisawa, K., Iinuma, K. and Tada, K. (1985) Biochemical study in 28 children with lactic acidosis, in relation to Leigh's encephalomyelopathy. *Eur. J. Pediatr.* 143: 278–283.

Naito, E., Ito, M., Takeda, E., Yokota, I., Yoshijima, S. and Kuroda, Y. (1994) Molecular analysis of abnormal pyruvate dehydrogenase in a patient with thiamine-responsive congenital lactic acidemia. *Pediat. Res.* 36: 340–346.

Narisawa, K., Endo, H., Miyabayashi, S. and Tada, K. (1992) Thiamine responsive pyruvate dehydrogenase deficiency. *J. Nutrition Sci. Vitaminol.* Spec. No.: 585–588.

Patel, M.S. and Harris, R.A. (1995) α-Keto acid dehydrogenase complexes: Gene regulation and genetic defects. *FASEB J.* 9: 1164–1172.

Robinson, B.H., Taylor, J. and Sherwood, W.G. (1980) The genetic heterogeneity of lactic acidosis: occurrence of recognizable inborn errors of metabolism in a pediatric population with lactic acidosis. *Pediat. Res.* 14: 956–962.

Robinson, B.H., MacMillan, H., Petrova-Benedict, R. and Sherwood, W.G. (1987) Variable clinical presentation in patients with defective E1 component of pyruvate dehydrogenase complex. *J. Pediatr.* 111: 525–533.

Robinson, B.H., MacKay, N., Petrova-Benedict, R., Ozalp, I., Coskun, T. and Stacpoole, P.W. (1990) Defects in the E2 lipoyl transacetylase and the X-lipoyl containing component of the pyruvate dehydrogenase complex in patients with lactic acidemia. *J. Clin. Invest.* 85: 1821–1824.

Robinson, B.H. and Chun, K. (1993) The relationships between transketolase, yeast pyruvate decarboxylase and pyruvate dehydrogenase of the pyruvate dehydrogenase complex. *FEBS Lett.* 328: 99–102.

Robinson, B.H. (1995) Lactic acidemia (Disorders of pyruvate carboxylase, pyruvate dehydrogenase). *In*: C.R. Scriver, A.L. Beaudet, W.S. Sly and D. Valle (eds): *Metabolic and Molecular Basis of Inherited Disease,* Seventh Edition. McGraw-Hill, New York, pp 1479–1499.

Sheu, K.F.R., Hu, C.W.C. and Utter, M.F. (1981) Pyruvate dehydrogenase complex activity in normal and deficient fibroblasts. *J. Clin. Invest.* 67: 1463–1471.

Sorbi, S. and Blass, J.P. (1982) Abnormal activation of pyruvate dehydrogenase in Leigh disease fibroblasts. *Neurology* 32: 555–558.

Stacpoole, P.W. (1989) The pharmacology of dichloroacetate. *Metabolism* 38: 1124–1144.

Sundström, M., Lindqvist, Y. and Schneider, G. (1992) Three-dimensional structure of apotransketolase. Flexible loops at the active site enable cofactor binding. *FEBS Lett.* 313: 229–231.

Takakubo, F., Thorburn, D.R. and Dahl, H.H.M. (1993a) A four-nucleotide insertion hotspot in the X chromosome located pyruvate dehydrogenase E1α gene (PDHA1). *Hum. Mol. Genet.* 2: 473–474.

Takakubo, F., Thorburn, D.R. and Dahl, H.H.M. (1993b) A novel mutation and a polymorphism in the X chromosome located pyruvate dehydrogenase E1α gene (PDHA1). *Hum. Mol. Genet.* 2: 1961–1962.

Tatuch, Y., Christodoulou, J., Feigenbaum, A., Clarke, J.T., Wherret, J., Smith, C., Rudd, N., Petrova-Benedict, R. and Robinson, B.H. (1992) Heteroplasmic mtDNA mutation (T----G) at 8993 can cause Leigh disease when the percentage of abnormal mtDNA is high. *Am. J. Hum. Genet.* 50: 852–858.

Tripatara, A., Kerr, D.S., Lusk, M.M., Kolli, M., Tan, J. and Patel, M.S. (1995) Three new mutations of the pyruvate dehydrogenase alpha subunit: A point mutation (M181V), 3 bp deletion (-R282), and 16 bp insertion/frameshift (K358SVS → TVDQS). *Hum. Mutat.* 6:; *in press.*

Uziel, G., Bardelli, P., Orefice, R., Colomaria, V., Carnevale, F., Garavaglia, B., Rimoldi, M. and Bertagnolio, B. (1994) Dichloroacetate treatment in 7 patients with mitochondrial encephalomyopathies. *Proc. VIth Int. Cong. Inborn Errors Metab.* Milan, p. 123.

Wexler, I.D., Kerr, D.S., Ho, L., Lusk, M.M., Pepin, R.A., Javed, A.A., Mole, J.E., Jesse, B.W., Thekkumkara, T.J., Pons, G. and Patel, M.S. (1988) Heterogeneous expression of protein and mRNA in pyruvate dehydrogenase deficiency. *Proc. Natl. Acad. Sci. USA* 85: 7336–7340.

Wexler, I.D., Hemalatha, S.G. and Patel, M.S. (1991) Sequence conservation in the α and β subunits of pyruvate dehydrogenase and its similarity to branched-chain α-keto acid dehydrogenase. *FEBS Lett.* 282: 209–213.

Wexler, I.D., Hemalatha, S.G., Liu, T.C., Berry, S.A., Kerr, D.S. and Patel, M.S. (1992) A mutation in the E1α subunit of pyruvate dehydrogenase associated with variable expression of pyruvate dehydrogenase complex deficiency. *Pediat. Res.* 32: 169–174.

Wexler, I.D., Hemalatha, S.G., Dahl, H.H.M., Buist, N.R., Berry, S.A., Cederbaum, S.D., Du, Y., Kuang, M., McConnell, J., Patel, M.S. and Kerr, D.S. (1995) Different outcomes of sibs and patients with identical mutations of pyruvate dehydrogenase treated with ketogenic diets. *Proc. Soc. Inher. Metab. Disorders,* Perdido Beach, FL.

Wicking, C.A., Scholem, R.D., Hunt, S.M. and Brown, G.K. (1986) Immunochemical analysis of normal and mutant forms of human pyruvate dehydrogenase. *Biochem. J.* 239: 89–96.

Wijburg, F.A., Wanders, R.J., van Lie Peters, E.M., Vos, G.D., Loggers, H.G., Bolhuis, P.A., Herzberg, N.H., Ruitenbeek, W., van Wilsem, A., ten Houten, R. and Barth, P.G. (1991) NADH:Q1 oxidoreductase deficiency without lactic acidosis in a patient with Leigh syndrome: implications for the diagnosis of inborn errors of the respiratory chain. *J. Inherit. Metab. Dis.* 14: 297–300.

Zhao, Y., Kuntz, M.J., Harris, R.A. and Crabb, D.W. (1992) Molecular cloning of the E1β subunit of the rat branched chain α-ketoacid dehydrogenase. *Biochim. Biophys. Acta* 1132: 207–210.

Multigenic basis for maple syrup urine disease with emphasis on mutations in branched chain dihydrolipoyl acyltransferase

D.J. Danner and M.C. McKean

Department of Genetics and Molecular Medicine, Emory University, School of Medicine, Atlanta, GA 30322, USA

History of maple syrup urine disease

Our understanding of inborn errors of metabolism began with Sir Archibald Garrod's description of alcaptonuria and the fact that this disorder tended to "run in families"(Garrod, 1908, 1923). By the 1950s, this understanding extended to knowing specific enzyme defects for several inborn errors. Thus, the stage was set for describing new disorders of which maple syrup urine disease (MSUD) was one. MSUD was first described clinically in 1954 by Menkes (Menkes et al., 1954). Four of six family members died in infancy, all having the odor of maple syrup or burnt sugar in their body fluids, especially urine (Menkes, 1959; Westall et al., 1957). Analysis showed that the odor was due to elevated concentrations of the branched chain amino acids (BCAA) and their respective α-ketoacids (BCKA). The interpretation, at that time and later proven true, was that a block occurred in the catabolic pathway of these compounds (Dancis et al., 1959). As more individuals were identified with this phenotype, family studies revealed that the disorder was transmitted as an autosomal recessive trait (McKusick 248600) (McKusick et al., 1994).

With the awareness of MSUD in the clinical community, reports appeared from around the world indicating that all ethnic populations express the disease (Chemke and Levin, 1975; Dastur et al., 1966; Kam-pui et al., 1985; Merinero et al., 1983; Popa et al., 1970; Raven, 1969; Rokkones, 1970). When states and countries began testing for elevated plasma leucine as part of newborn screening programs, additional reports appeared (Antonozzi et al., 1980; Fernhoff et al., 1982; Yadav and Reavey, 1988). Current data puts the incidence of MSUD in the general population near 1/150 00 – 1/200 000 live births. Within isolated populations, inbreeding can increase this value; for instance, the Mennonite population where the frequency is 1/176 (Peinemann and Danner, 1994). Along with these surveillance programs and the increased general awareness of the MSUD phenotype, variant clinical phenotypes were described (Dancis et al., 1972; Elsas et al., 1972; Gonzalez-Rios et al., 1985; Langenbeck, 1986; Scriver et al., 1971). These distinctions are

Table 1. Clinical classification of maple syrup urine disase based on presenting phenotype.

Type	Onset	Plasma [leucine] mM	BCKD activity% control*	Phenotype
Classic	Birth – 1 week	2 – 4	0 – 2	Neurologic abnormality, seizures, coma, ketoacidosis
Severe variant	Birth – 1 month	1 – 2	2 – 20	Similar to classic
Mild variant	Infant – childhood	0.5 – 1	2 – 40	Few, ketoacidosis periodically
Thiamin-responsive	Birth – 17 months	0.5 – 2	0 – 40	Similar to classic

* Activity is based on the $^{14}CO_2$ released from 1-[^{14}C]-leucine and % reflects activity compared to α-chloroiso-caproate activated wild-type control cells. (BCKD) Branched chain α-ketoacid dehydrogenase.

based on age of presentation, BCAA concentrations in plasma, and clinical features. Clinical phenotypes are summarized in Table 1. A genetic classification relating phenotype and genotype would be ideal, except now early intervention prevents full phenotype expression. Molecular genetic changes are being described and will be addressed below and in the chapters by Wynn and Chuang, and Indo and Matsuda, this volume.

Once identified, the affected individuals are treated with a protein restricted diet with special attention to the leucine intake (Berry et al., 1991; Elsas and Acosta, 1987; Snyderman et al., 1964; Thomas, 1992). Given the essential nature of the BCAAs, adequate supplies are necessary for proper growth and development. Some programs treat all patients with pharmacologic thiamin (10 mg/kg for newborns to a maximum of 200 mg/day in graded doses in the older children and adults) since thiamin is not known to produce a toxicity in high doses. It is important that these special diets are begun as soon as possible to minimize brain dysfunction and maximize intellec-tual outcome (Hilliges et al., 1993; Kaplan et al., 1991). Individuals identified by newborn screening are started on diet within 10 days after birth. Even with careful monitoring, some patients will experience acute episodes of ketoacidosis. Dialysis is used in these rare instances to rapidly reduce plasma concentrations of the BCAAs and BCKAs and minimize complications (Gouyon et al., 1994; Wendel et al., 1980). Pathologies that result from elevated BCAA/BCKAs are mental and physical dysfunction with changes in brain function seen by several different tech-niques (Kamei et al., 1992; Riviello et al., 1991; Taccone et al., 1992). Brain scans have shown localized edema when patients are out of diet control. Seven to ten days after reestablishing diet control with concomitant lowering of the BCAA plasma values, the pathologic conditions are returned to normal (Felber et al., 1993; Muller et al., 1993). It is interesting that in the only animal model for MSUD, a naturally occurring Poll Herford calf, brain and neurologic patterns similar to the untreated human MSUD patient are found (Harper et al., 1986, 1990).

Most patients in the United States are identified through newborn screening programs with a blood sample taken only after a protein load. The test quantifies plasma leucine concentration and values in excess of 153 μM are indicative of further testing (Fernhoff et al., 1982). Other cases are diagnosed by failure to thrive, lethargy, coma and other complications whereby the patient is brought to the attention of the pediatrician. The dinitrophenylhydroazone test for urinary keto-acids is still reliably used in some physician's offices. Confirmation of the diagnosis is most often made by quantitation of plasma amino acid concentrations by ion-exchange, gas, or high performance liquid chromatography (Sherwood et al., 1990; Tanaka et al., 1980). The most definitive diagnosis is enzyme assay on freshly isolated peripheral white blood cells or cultured cells established from the patient (Peinemann and Danner, 1994). More recently, analysis of whole body leucine oxidation has been established which is a simple and reliable test that can also differentiate the heterozygote state (Elsas et al., 1993). Individuals are given an oral load of 1-[^{13}C]-leucine and breath samples collected over 3 h. Enrichment of expelled $^{13}CO_2$, which can be produced only by activity of the branched chain α-ketoacid dehydrogenase (BCKD) complex, is quantified by gas-isotope-ratio mass spectrometry. Once diagnosed within a family, prenatal monitoring is possible by enzyme assays on cultured cells from amniotic fluid (Elsas et al., 1974) or chorionic villi samples. If the molecular genetic basis of the disorder has been established for the family, molecular biological techniques are used (Ellerine et al., 1993).

When a patient is maintained on the diet restricted in the BCAAs his/her development proceeds normally and many individuals show minimal complications with a near-normal lifestyle (Hilliges et al., 1993; Treacy et al., 1992). As the individual grows, the diet must be monitored and adjusted to maintain the plasma BCAA concentration within normal ranges. Most important is for leucine concentration to remain between 80 and 200 μM with the ranges for valine at 200–425 μM and isoleucine at 40–90 μM (Danner and Elsas, 1989). Special care is needed during usual childhood illnesses, but with this care ketoacidosis, and the complications from this condition, are greatly minimized. In fact, the population of individuals first treated successfully by diet therapy are now young adults and able to reproduce. A successful pregnancy has been reported for one woman with MSUD (Van Calcar et al., 1992). Obviously, new complications must be addressed with the situation known as maternal MSUD. These expectant mothers are benefiting from the long experience of studies on maternal phenylketonuria, a similar amino acid-intolerant inherited disorder (Koch et al., 1993; Thompson et al., 1991). Extreme care is necessary to provide the fetus with adequate nutrition and protein without compromising the MSUD-mother. In most pregnancies of this type the fetus will be heterozygous for the mutant allele, receiving only one allele from the homozygous mother. It is presumed that homozygous males are also fertile but this should present no complications to the pregnancy if the mother is homozygous normal.

Metabolism of branched chain amino acids

Leucine, isoleucine and valine are the three branched chain amino acids which enter cells by the system L transporter (Gazzola et al., 1980). These amino acids are converted to their α-ketoacids by the action of reversible transaminases that function both in the cytosol and mitochondria (Aki et al., 1967; Aki et al., 1968). The ketoacids, if not formed in the mitochondria, must be transported into this organelle and a BCKA-specific mitochondrial transporter has been described (Hutson and Rannels, 1985). Once inside the mitochondria, the BCKAs become committed to catabolism by the action of branched chain α–ketoacid dehydrogenase (BCKD). The reaction releases CO_2, produces the branched chain acyl-CoA ester and reduces NAD^+ (Yeaman, 1989). All three BCKAs are oxidatively decarboxylated by a single enzyme complex (Danner et al., 1979; Pettit et al., 1978) after which they follow separate, enzyme-specific degradative pathways. End products from the three amino acids are succinyl-CoA from valine, acetyl-CoA from leucine and a mix of the two from isoleucine (Fig. 1). Defective BCKD function, as in MSUD, results in

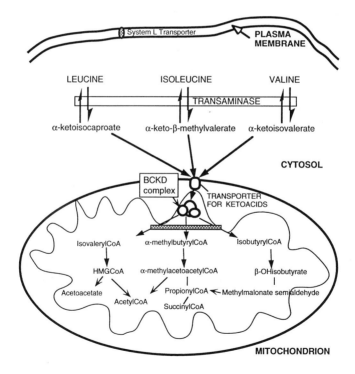

Figure 1. Diagrammatic representation of cellular branched chain amino acid metabolism.

Table 2. Physical properties of human branched chain α–ketoacid dehydrogenase subunits

Subunit	Chromosome locations	Precursor size AA[1]	Precursor size M_r[2]	Mature size AA[1]	Mature size M_r[2]	Cofactor
Lipoamide dehydrogenase	7	54 760	55	49 702	52	FAD NAD
Branched chain acyltransferase	1p21-31	53 487	51	46 369	47	Lipoic acid
Decarboxylase α	19Q13.1-13.2	50 455	47	45 513	43	TPP[3]
Deccarboxylase β	6	42 959	41	37 830	37	TPP[3]

[1]Size based on amino acid content deduced from cDNA; [2]mobility in SDS-PAGE of *in vitro*-made subunits before and after import into mitochondria; [3]thiamin pyrophosphate.

a decreased output of the CoA compounds from all three pathways as well as the accumulation of the precursor substrates. Whether the absence of these CoA esters of the BCKAs plays any role in the pathology associated with the disease is not known. Recently two laboratories have presented evidence that α–ketobutyrate, an intermediate in the catabolism of threonine and methionine, is also decarboxylated by the BCKD complex (Jones and Yeaman, 1986; Paxton et al., 1986). No accumulation of methionine, threonine or α–ketobutyrate in the plasma of MSUD patients has been reported, although complete amino acids profiles have been studied.

The BCKD complex is constitutively expressed in all tissues but the activity state of the complex varies among tissues depending upon the phosphorylation state (see Harris and Popov, this volume; Shimomura, this volume; Harris et al., 1985, 1994). Three catalytic components composed of four different proteins and two regulatory components associate to form the BCKD complex. A summary of the physical properties of the proteins in the complex is found in Table 2. Discussion of the regulatory components appears elsewhere in this book. Suffice it to say, the substrate specific kinase phosphorylates two serine residues in the E1α component of the decarboxylase resulting in inhibition of the oxidative decarboxylation of the BCKAs. A substrate specific phosphatase removes the phosphate to reactivate the complex (Damuni et al., 1984; Damuni and Reed, 1987). Decarboxylation of the BCKAs is catalyzed by E1, an $\alpha_2\beta_2$ tetramer that uses thiamin pyrophosphate (TPP) and Mg^{++} as cofactors. This E1 component transfers the branched chain acyl moiety to covalently bound lipoic acid on lysine 44 of the mature dihydrolipoyl branched chain acyltransferase (E2). Upon transfer of the acyl group to coenzyme A, the reduced lipoate is reoxidized by the flavoprotein, dihydrolipoamide dehydrogenase (E3) which functions as a homodimer. E3 is not unique to BCKD as it functions in a similar capacity with the other α-ketoacid dehydrogenase complexes as well as the glycine cleavage complex

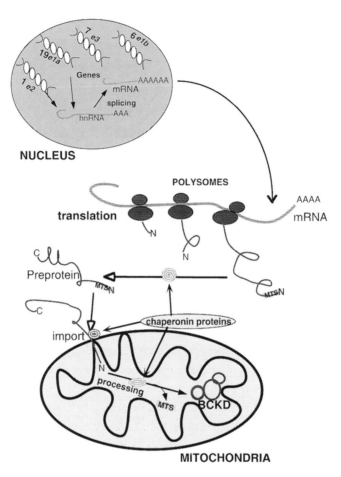

Figure 2. Pictorial representation of gene expression and branched chain α–ketoacid dehydrogenase complex assembly.

(Yeaman, 1989). It is thought that the complex forms with an exact stoichiometry of 24 E2 subunits as a core surrounded by 12 E1 tetramers and 6 E3 dimers (Reed and Hackert, 1990). E1, E2, and the regulatory components are substrate specific and unique to BCKD.

Genes for the BCKD subunits are found on separate chromosomes in the nucleus of human cells (see Tab. 2). After transcription, the hnRNA is converted to mRNA by the usual processes of capping, polyadenylation and splicing prior to exiting from the nucleus. In the cytosol, the mRNA is translated into protein on free ribosomes. Newly made protein must then find the mitochondria likely assisted by chaperon proteins (Gambill et al., 1993; Kiebler et al., 1993; Maccecchini et al., 1979; Manning-Krieg et al., 1991; Neupert and Pfanner, 1993). These precursor proteins have amino terminal peptides (mitochondrial targeting signal, MTS) which recognize mitochondrial receptors and direct their passage through the mitochondrial membranes (Glick and Schatz, 1991). Upon import, the MTS is proteolytically removed from the precursor protein trapping the mature protein within the matrix. Additional chaperon proteins are present within the mitochondrial matrix to aid entry and assist in folding and possibly location of these proteins (Craig et al., 1993; Craig et al., 1994; Wachter et al., 1994). The overall process is summarized in Figure 2. Factors involved in the regulation of gene expression, translation, mitochondrial import, and complex assembly are just beginning to be described. The molecular genetic basis for MSUD will be discussed in relation to these various processing steps.

Transcription

How stoichiometry is attained and maintained for the BCKD subunits in formation of the active complex remains an important question. Coordinated transcription of all the unique genes would provide one means by which this stoichiometry might be achieved. Since BCKD is constitutively expressed in all tissues, it is unlikely that a tissue-specific *trans* acting proteins will be found. Various lengths of 5' upstream DNA sequence are known for E1α, E1β, E2 and E3 (Chuang et al., 1993; Feigenbaum and Robinson, 1993; Johanning et al., 1992; Lau et al., 1992; Mitsubuchi et al., 1991a) and preliminary characterization of these regions was reported. Minimal lengths of DNA sequence which drive transcription have been determined by reporter gene assays. Within these regions are putative *cis* elements for a number of known transcription factors, but a common *cis* element unique to all four genes has not been found. A common *trans* factor protein(s) is being sought that binds to a region within all these genes and thus provides coordinated transcription.

One mechanism which could lead to MSUD would be a lack of gene expression. This would be detected by the antigenic absence of a protein subunit, or more specifically, the absence of mRNA for a subunit. Antigenic absence of protein was first demonstrated in 1985 for the E2 subunit by Western blot analysis (GM1366) (Danner et al., 1985). An absence of mRNA for E1α was demonstrated a few years later for one allele in an affected proband (Zhang et al., 1989), but the mechanism for this lack was not described. A homozygous lack of mRNA for E2 also has been reported in cells from studies of another proband (Litwer et al., 1989). Since no mRNA was detected by Rnase protection analysis it appears that neither allele is able to produce transcripts.

Parental cells in this family were unavailable for study, so confirmation was not possible from this type of analysis. When 5' promoter sequence was isolated from the proband cells and used in reporter gene assays, the sequence appeared to behave like wild-type sequence (D.J. Danner, unpublished). Nucleotide sequence analysis of ~700 bp in this region revealed two different sequences presumably from the two alleles. Both alleles also demonstrated nucleotide changes from reported sequence for wild-type alleles (Lau et al., 1992). When the changes from the patient alleles were engineered into the wild-type sequence, no effect on reporter gene activity was observed. Although RNase protection analysis should evaluate both nuclear and cytosolic RNA, it is possible that initial transcripts are produced and export from the nucleus is impaired. Studies have been reported describing a form of thalassemia that results from hnRNA processing defects (Loukopoulos, 1991). In other studies a block in hnRNA processing has been reported as human fibroblasts become senescent (Chang et al., 1991). Proteins which bind hnRNA are needed for proper processing of transcripts into mRNA (Dreyfus, 1986; Dreyfus and Matunis, 1993; Matunis et al., 1993; Pinol-Roma and Dreyfus, 1992) and some of these proteins are known to bind to intronic sequence (Charollais et al., 1994). It is known that the presence of introns determines the time of appearance of mRNA from the hnRNA (Charollais et al., 1994) and that nucleotide sequence within the intron is specifically recognized by some of the hnRNA binding proteins for processing of the hnRNA (Ghetti et al., 1992; Patton et al., 1991). Another laboratory has reported that GM1366 is homozygous for a point mutation in exon 6 (E163*), determined by ASO hybridization to genomic DNA fragments (Fisher et al., 1993). Our data suggests that two different alleles are present in GM1366 from nucleotide sequence analysis of the promoter regions and variations from wild-type alleles in the polypyrimidine tract in intron six. Point mutations within an exon do not usually result in complete absence of mRNA, however, with our increasing understanding of the involvement of hnRNA-binding proteins in the processing of transcripts, the combination of changes seen in this proband may produce the effect of decreased production of cytosolic mRNA. Further studies are required to establish the mechanism responsible for non-expression in this cell line.

This cell line was also used to test whether the absence of E2 was the sole reason for MSUD in this proband. Fibroblasts were transfected with plasmids harboring cDNA for wild-type preE2 and the *neo* gene for selection in G418 and it was determined if BCKD activity could be restored. After clonal selection of cells growing in G418, the clones were tested for BCKD activity. BKCD activity in the cells prior to transfection was less than 1% of that found in wild-type cells but after transfection BCKD activity was the same as wild-type cells (Litwer et al., 1989). Further, when tested for regulation by phosphorylation, the same activity state was observed in the clonal cells as found in the wild-type cells. This proves that the absence of E2 alone accounts for the inactive

BCKD. When all subunits are presented to the mitochondria within this mutant cell, the BCKD complex assembles and functions as in wild-type cells.

It is interesting to note that partial genomic deletion within the coding region of the e2 locus on chromosome 1p21-31 does not prevent mRNA formation. This region is rich in *Alu* repeat sequence (Jurka and Smith, 1988) and thus could be prone to deletions. It is reported for other genes that the presence of *Alu* repeats can lead to genomic deletion through base pairing between dispersed repeats. The explanation offered is that base pairing allows loops to form in the DNA structure which are easily broken and thus lost. Such an explanation was invoked for the loss of nearly 20 kb of DNA from within the e2 gene (Ellerine et al., 1993; Herring et al., 1992). This mechanism was inferred from analysis of both mRNA sequence and that of the genomic sequence from the cultured cells. mRNA contained bases from intron 6 linked to bases in exon 11. Obviously, there is still much to learn about transcript processing for these genes.

Other mechanisms for non-expression are known. At present, three alleles have been shown to involve exon skipping due to splicing errors (Herring et al., 1991; Herring et al., 1992; Mitsubuchi et al., 1991b). Again, ribonucleoproteins binding to hnRNA are known to affect splicing and exon selection (Mayeda et al., 1993). In one family the proband is a compound heterozygote with a maternal non-producing e2 allele and a paternal allele that produces transcripts lacking exon 2. This exon deletion results in a frame shift causing a termination codon after the first 12 codons in the MTS (Herring et al., 1991). In paternal cells the BCKD complex has only 15% activity compared to normal cells. Thus, it appears that the presence of the shorter mRNA or production of an abortive MTS may have a dominant effect and compromise BCKD activity. Despite the low BCKD activity, the father is phenotypically normal.

A second exon-skipping allele in another family also presents an interesting scenario (Herring et al., 1992). Again the proband is a compound heterozygote obtaining the *Alu*-based deletion gene described above from the father and a point mutation allele from the mother. This maternal allele has a G → A transition in the terminal base of the inframe exon 8. The wild-type codon is AA**G** and the mutant is AA**A**, both encoding lysine. By RT-PCR it was shown that two transcripts are found in the cultured cells from the affected proband. One mRNA lacks exon 8 while the second lacks exons 8–10. The mechanism for the multiple exon skipping likely relates to the lack of conformity to the most favored nucleotide sequence of the splice junction nucleotide sequences in this region. Using the Shapiro Senapathy method of calculation, values below 75 were found for the mutant exon 8/intron 8 junction, and the normally occurring junctions at the next two exons (Herring et al., 1992). These findings suggest that exon 8/intron 8 junction has a strong influence in subsequent removal of the next two introns.

Since exon 8 is inframe, the mRNA lacking this exon can make a protein that lacks 26 amino acids from the internal core binding domain of E2 (Griffin et al., 1988). When cloned, the cDNA was able to produce this protein *in vitro* and the protein is recognized by antibodies against native E2 (Ellerine et al., 1993). The truncated protein has been sought without success in various cells from the Caucasian patient including freshly isolated lymphocytes, and cultured fibroblasts and lymphoblasts. By yeast two-hybrid analysis it has been shown that the mature forms of this truncated protein will not interact with itself, supporting the importance of this region in E2 core formation (D.J. Danner, unpublished observations). The mRNA lacking exons 8–10 could encode a shorter protein, but the cloned cDNA has not been studied. A Japanese individual with MSUD has been reported with a G deletion at the exon 8/intron 8 boundary which also results in the absence of exon 8 in the mRNA. The shorter mRNA lacking exons 8–10 was not observed in the Japanese patient (Mitsubuchi et al., 1991b). In contrast to the Caucasian patient, the truncated protein from mRNA lacking exon 8 is antigenically present in cells from the Japanese patient. Two different mechanisms result in the absence of exon 8 and one is translationally active while the other is not. There is no obvious explanation for this discrepancy short of the genetic background being different for the two individuals.

Other conundrums exist for this Caucasian patient. Since the G → A change is silent at the amino acid level, if normal splicing occurred at some frequency, then a normal protein would be produced. In other examples of similar base substitutions in analogous positions within other genes, some normal spliced transcript have been found (Akli et al., 1990; Vidaud et al., 1989; Weil et al., 1989). Using readily available tissue from this patient (see above) and RT-PCR techniques, no full-length mRNAs were ever found (Ellerine et al., 1993; Herring et al., 1992). This offers further confusion since this patient also is "thiamin-responsive," and the mechanism for thiamin-responsive MSUD remains to be defined. Thiamin-responsive MSUD will be discussed later in this chapter specifically reflective of findings in this patient.

Translation/mRNA stability

Once the mRNA has been transferred to the cytoplasm it becomes available for translation. This process probably involves exchange of hnRNA binding proteins with mRNA-binding proteins, although an understanding and description of mRNA binding proteins is just beginning to be defined (Larson and Sells, 1987). The ability of mRNA to bind the translation protein complex depends in part on the nucleotide sequence surrounding the AUG initiation codon (Kozak, 1987, 1991). Nucleotide sequence surrounding the E1α subunit start methionine (GCCAAG**A**UG**G**) conforms very strongly with the Kozak consensus (GCC(A/G)CCAUGG) matching at six of the seven positions. When E1β and E2 mRNA sequence is examined, the matches are four and three respectively. Two methionine codons are present in the first six codons of E2 mRNA. It is not known for certain which methionine codon is used *in vivo* although *in vitro* transcription/translation experiments have demonstrated that when nucleotide sequence encoding both methionines is included in the transcripts a 30 fold increase in protein production results (D.J. Danner, unpublished). We do know that the second methionine can be used *in vivo* with production of a functional protein since reconstitution experiments in human cells were done with a cDNA construct that lacked the first methionine coding sequence (Litwer et al., 1989). Although not yet observed, point mutations in this region of the mRNA could reduce translation competence resulting in MSUD.

Translational competence does not only depend upon the nucleotide sequence at the initial AUG. Proteins that bind mRNA are not only those associated with translation. Proteins are known to bind the 3'UTR of mRNA as well as the polyadenylation tract and are shown to influence the translation ability of the mRNA (Berger et al., 1992a, b). The structures of the mRNAs for the α and β components of the E1 decarboxylase in BCKD are unremarkable. E1α has 392 nt in the 3'UTR with a typical polyadeynlation signal 24 nt from the addition site. E1β contains 156 nt in its 3'UTR and polyadenylation occurs 7 nt after the signal sequence. E2 is remarkably different in that 1850 nt are present in its 3'UTR and polyadenylation occurs some 371 nt beyond a consensus polyadenylation signal sequence (AATAAA). In many tissues a shorter mRNA for E2 is also detected by Northern blotting (Danner et al., 1989). This transcript is polyadenylated at a point 5' to the putative polyadenylation signal. In either case it is not known what signals polyadenylation of these transcripts. Again, although possible, no mutations in these regions of the mRNAs have been reported in MSUD patients.

From quantitative RNase protection analysis of mRNA in patient cells that antigenically lack a protein subunit of BCKD, it is suggested that steady-state mRNA concentrations may be influenced by this lack of protein (D.J. Danner, unpublished). Polyadenine tracts as tails on the 3'UTRs play an important role in mRNA stability (Bernstein et al., 1989). Half-lives of mRNAs for the subunits of BCKD have been found to be minimally different in two different cell types (McConnell et al., 1996). In DG75 cells, values for the three transcripts ranged from 12 to 17 h, while for HepG2 cells E1β and E2 showed half-lives of 21 h while E1α has a half-life of 11 h. The length of the polyadenine tract for these mRNAs was not estimated, nor was it determined whether the varied 3'UTRs play any role in half-lives. Further, it is not yet known if nutritional or hormonal conditions will influence these half-lives and whether this can impact on phenotypic expression of MSUD. Transcript instability, as yet, has not been implicated as a mechanism for MSUD.

Mitochondrial import

Assuming a protein is made from a transcript, it must now find its way to the mitochondria, be imported, and then associate with the other components of the BCKD complex. The journey for cytosolically made mitochondrial proteins after release from the ribosome is likely to involve chaperon proteins that help these proteins find the mitochondria while maintaining them in a form that allows transport across the two mitochondrial membranes (Becker and Craig, 1994; Georgopoulos and Welch, 1993). General mechanisms for the import process have been defined (Glover and Lindsay, 1992; Neupert, 1994). Within the mitochondrial matrix are additional chaperon proteins that bind the newly imported protein and pull the full protein into the matrix (Manning-Krieg et al., 1991; Stuart et al., 1994). A membrane potential across the inner membrane is the initial force used to bring the MTS into the matrix where it encounters the mitochondrial hsp70 chaperon. Current understanding suggests that ATP hydrolysis is then used by the mhsp70 to import the remaining protein structure (Glick, 1995; Ungermann et al., 1994; Wickner, 1994). During this process the MTS is proteolytically removed by a matrix protease (Stuart et al., 1994) helping to trap the protein within the matrix. The protein must then be folded into its final functional form, which may involve the addition of cofactors. Mitochondrial hsp60 aids in this folding function and may aid in complex assembly (Hallberg et al., 1993; Ostermann et al., 1989).

A few of these steps have been demonstrated for the proteins of the BCKD complex. All subunits are made as larger precursors, import is dependent upon a membrane potential and matrix ATP, and the mature protein has the MTS removed (Fig. 3) (Clarkson and Lindsay, 1991; D.J. Danner, unpublished). The use of mitochondrial chaperon proteins has been implied from studies with bacterial GroEL/ES (Wynn et al., 1992). By all accounts at this point there is nothing unusual about the processing of the subunits of the BCKD complex.

Mistargeting of proteins among organelles can result in a pathogenic phenotype for other human inherited disorders as shown for primary hyperoxaluria type 1 (Purdue et al., 1990). In individuals with one form of this disorder the alanine:glyoxylate aminotransferase which is normally targeted to the peroxisome now is found in the mitochondria. Some mammals localize this protein to the mitochondria by the presence of an MTS. Although nucleotide sequence in the gene is conserved from lower mammals through humans, humans have lost the AUG start codon ahead of the MTS and use a later AUG in the sequence. In some patients with this mistargeting a

Pro11 → Leu amino acid substitution appears to create a new MTS and thus account for the mistargeting. Others with mistargeting do not have this change in their alleles and amino acid substitutions further in the protein are implied to result in mistargeting (Danpure et al., 1993). No patients with MSUD have been identified as yet that result from alterations in targeting of the proteins to the mitochondria. It is known, however, that mistargeting of the BCKD proteins can occur. Evidence for this comes from reports of individuals with autoimmune disorders who make antibodies against proteins of the BCKD complex. Serum from patients with primary biliary cirrhosis and idiopathic dilated cardiomyopathy contain monospecific antibodies primarily for the E2 protein (Ansari et al., 1991; Van de Water et al., 1989). Immunostaining of diseased heart tissue in some cardiomyopathy patients clearly showed BCKD-E2 protein in the plasma membrane (Ansari et al., 1991). It therefore remains a possibility that some forms of MSUD may result from incorrect targeting of one or more of the BCKD subunits. MSUD patients have been studied by Western blot analysis (Danner et al., 1985), but generally these are done with mitochondrial proteins from cultured cells. If whole cell proteins are used for Western blots, it is possible to assess whether precursor proteins are accumulating without entering the mitochondria. In patients which antigenically lack a subunit in the mitochondria, no precursor accumulation has

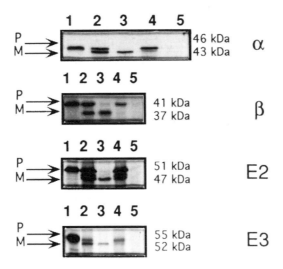

Figure 3. Autoradiograph of import and processing of *in* vitro-made subunits of the human branched chain α–ketoacid dehydrogenase into isolated mouse liver mitochondria. Lane 1 is translation reaction product. Lane 2 is translation product plus mitochondria. Lane 3 is as 2 but treated briefly with trypsin to digest proteins not contained within the mitochondrial matrix. Lane 4 is as 2 except mitochondria were first treated with rhodamine 123 to disrupt the membrane potential. Lane 4 and lane 5 were treated with trypsin.

been seen for those examined. Cell fractionation studies have not been done to determine if the missing protein has been placed in another subcellular compartment.

Other possibilities for MSUD could result from an inability of BCKD-subunits to interact with the chaperon proteins. Amino acids changes within regions of BCKD-subunits necessary for specific interaction with chaperons could result in an inability to find the mitochondria, or be efficiently imported into the matrix. Mutations of this type have yet to be described for MSUD.

BCKD complex assembly

Mechanisms for assembly of the α–ketoacid dehydrogenase complexes are not well understood. Lipoate must be covalently bound to specific lysine residues of E2 components of all complexes in an enzyme-catalyzed reaction. Lipoyltransferase has been recently characterized and is localized to the mitochondria in mammals (Fujiwara et al., 1994; Morris et al., 1995). It remains to be determined when during import and assembly the addition of lipoate takes place. Point mutations within these E2 proteins could alter the recognition peptide substrate for the lipoyltransferase and without lipoate the E2 is catalytically inactive. Point mutations of this type are not known in MSUD.

Thiamin pyrophosphate (TPP) is the dissociable cofactor that forms the active site in the E1 component of these complexes. Recent reports on the pyruvate dehydrogenase proteins demonstrate interactive sites for TPP with both the α and β subunits and due to the high degree of similarity in protein structure between PD-E1 and BCKD-E1 it is anticipated the same will be true for BCKD-E1. Point mutations that affect the binding site for TPP in either the α or β subunit could compromise activity of E1 and might also be a mechanism to explain thiamin-responsive MSUD discussed below. Specific point mutations in BCKD-E1 components and their relationship to MSUD are discussed in detail in the chapter by Indo and Matsuda in this volume.

In addition to binding cofactors, the individual proteins must find one another and form the large catalytically active complex. Carboxyl-terminal domains within BCKD-E2 are suggested as binding sites for self-association and mid-protein domains are implicated for binding to the other subunits (Griffin et al., 1988). Engineered and naturally occurring amino acid changes within the E2 protein are being tested in various systems to more specifically address how the changes affect complex assembly and catalytic function. For example, we have shown by the yeast two-hybrid analysis that truncation of as little as 61 amino acids from the C-terminal end of E2 prevents interaction of the truncated protein with itself or the full-length E2 (D.J. Danner, unpublished). Critical amino acids within this region are being defined by this method of analysis using site-directed mutagenesis to change single amino acids in BCKD-E2. Interaction of BCKD-E2 with E1α and E1β is also evaluated in this same way. As more point mutations in E2 are defined from studies of MSUD patients, a clearer understanding of critical residues will be possible. The best example of amino acid change that affects protein interaction comes from mutations in E1α that prohibit interaction with E1β (Chuang et al., 1994). These are discussed in greater in the chapter by Indo and Matsuda, this volume.

Relationship of genotype to phenotype

Although it is reasonable to assume that expression of the MSUD phenotype would vary depending on a specific alteration within any specific gene of the complex, correlations of this type are difficult with the advent of early intervention and diet therapy which mitigate phenotypic expression. Newborn screening programs and prenatal monitoring in families with known affected individuals has allowed identification of new MSUD cases prenatally or within the first few days after birth. Synthetic formulas without BCAAs enable the pediatrician and dietitian to selectively add these essential amino acids to the diet in concentrations that do not elevate plasma or tissue concentrations beyond that needed for formation of new protein. Diets are specifically tailored to each individual to maximize growth and development (Elsas and Acosta, 1987; Elsas and Danner, 1989). Careful monitoring of plasma amino acid concentrations has minimized complications that would result from elevated values without these constraints. It would be unethical to withhold this treatment and allow a mutant-specific phenotype to develop in different individuals solely to define more clearly genotype-phenotype relationships. It should be noted that even within the Mennonite population where a single mutation accounts for the majority of affected individuals, a consistent phenotype is not reported. As stated above, patients with well controlled diets develop within normal parameters and lead healthy, productive lives and are able to reproduce. A correlation that is possible under these conditions is the dietary restraints necessary in one genotype *versus* another. These comparisons also are difficult since the genetic background in each individual would influence their overall metabolism.

One phenotype is interesting to discuss, the so-called "thiamin-responsive MSUD." First described in a 1971 report (Scriver et al., 1971), this patient showed a marked and immediate reduction in plasma BCAAs when her protein-restricted diet was supplemented with pharmacologic thiamin. Other reports have followed and mixed conclusions have been reached as to the effect of this supplementation (Danner et al., 1978b; Fernhoff et al., 1985; Pueschel et al., 1979) Contributing to the confusion is a follow-up report on this first patient, suggesting that removal of thiamin supplementation was without effect (Scriver et al., 1985). A specific mechanism for the response has not emerged and might reflect that different mechanisms are needed to explain different cases. An obvious explanation would be an altered affinity of the E1 $\alpha_2\beta_2$ tetramer for TPP and this was reported as findings from studies on cells cultured from the original patient (Chuang et al., 1982). Our original report suggested the thiamin response was time and concentration dependent, requiring at least 3 weeks of supplementation before a noted lowering of plasma BCAAs and effect on BCKD activity was seen (Danner et al., 1978). Supplementation of a normal individual's diet with pharmacologic thiamin again for 3 weeks, resulted in an increase in BCKD activity in the liver (Danner et al., 1975). At that time it was not known that BCKD activity in the cells is varied with the phosphorylation state of E1α within the complex so the results need to be viewed with that caveat. It should be noted however, that the BCKD activity measured in the studies was in the liver cells and the activity state of BCKD in liver is usually $> 90\%$ active (Veerkamp and Wagenmakers, 1987). We further showed that TPP could stabilize BCKD against proteolytic turnover (Danner et al., 1978a, 1980; Heffelfinger et al., 1984). Four additional patients were studied in a carefully controlled 2-month-long clinical and biochemical protocol. Individuals were maintained on their protein restricted diet with thiamin supplement for 1 month. This same diet was then supplemented with pharmacologic thiamin for a second month. Three of the four individuals responded by lowered plasma BCAA concentrations and increased BCKD activity, comparing values at the end of the first and second month (Fernhoff et al., 1985). Compiling all our results, we postulated that thiamin-responsive MSUD could be due to a slowing of turnover of BCKD when the BCKD complex is saturated with TPP. Saturation would require the cellular concentration of TPP to increase in a time-dependent manner (Elsas and Danner, 1982). Direct quantitation of cellular TPP concentrations before and after this diet supplementation with thiamin have not been made.

These concepts were again brought into question with the finding that at least two of the patients reported to be "thiamin-responsive" antigenically lack the E2 core protein of the BCKD complex (Chuang, 1989; Ellerine et al., 1993). One of these patients is the first one described by Scriver et al., which now casts into doubt the altered TPP binding site hypothesis (Chuang et al., 1982; Scriver et al., 1971). The other patient has also been discussed above as the compound heterozygote lacking E2 due to a gene deletion allele and a splicing defect allele (Ellerine et al., 1993; Herring et al., 1992). This individual was evaluated by whole-body leucine oxidation methodology (Elsas et al., 1993) which quantifies the production of $^{13}CO_2$ from 1-^{13}C-leucine after oral loading with the isotope-tagged amino acid. When her protein restricted diet was supplemented with ~200 mg thiamin/day she was able to oxidize leucine at near 25% the rate of age-matched BCKD-normal controls. The thiamin supplement was stopped for 3 weeks and then leucine oxidation was not detectable. During this period of unsupplemented diet was the only time in her life that she developed acidosis from her inherited defect in BCKD. The study was not repeated for ethical reasons.

The genotype of this patient leaves possible the stabilization hypothesis and introduces yet other hypotheses. Although one allele produces no protein due to a 20 kb genomic deletion the other allele could produce a functional E2 if normal splicing occurs. Using cultured cells or freshly isolated lymphocytes from this individual it was never possible to demonstrate that any full-length mRNA was produced (Ellerine et al., 1993). We are left with the possibility that thiamin supplementation could stabilize a small amount of E2 produced in a tissue specific manner. It is also conceivable that thiamin supplementation enables production of full-length mRNA by affecting the splicing, again in a tissue-specific way. Alternatively, the excess TPP may in some way allow the E1 decarboxylase to function by itself and thus account for the leucine oxidation when on the vitamin supplementation. It is interesting to note that the Japanese patient who also produces a transcript that lacks exon 8 due to a G deletion at the same junction is reported to be thiamin unresponsive. The Caucasian patient has developed normally and is now 12 years old, her IQ test scores are above 120, and she is socially and academically functioning well in a standard school setting. Our quest to explain the thiamin response in her continues since it appears that understanding this situation could greatly benefit other individuals with MSUD. Results from studies on this patient suggest that all tissues need not have a functional BCKD complex and therefore gene therapy for MSUD might be a reasonable alternative therapy.

Conclusion

The inherited human disorder MSUD presents a continuing challenge. Mutations are known to occur in each of the genes of the core complex. No patient is known who is a compound heterozygote with mutations in two different subunits. It is unlikely that such an individual exists since heterozygotes for a single mutant allele are functionally normal. Most mutations are exclusive within a family, so broad-based molecular genetic testing methods are not applicable with the exception of the Mennonite population, which harbors a common mutation.

Excellent management of these patients is possible through protein-restricted diets. Although patients must remain on these diets throughout their lives, it appears from the present data available that a normal life span can be expected. Newborn screening programs have been useful in identifying individuals at risk but diligent pediatric care can also quickly identify individuals who have missed the screening.

Results from the thiamin-responsive patient described above suggest that it is possible to consider gene therapy in treatment of MSUD. Data suggest that gene replacement need only be in selected tissue, as yet to be determined. Individuals treated with gene therapy would no longer need protein-restricted diets or face the complications of ketoacidosis. The challenge will be to identify which tissue should be targeted, the best delivery system for selective targeting, and how often the treatment needs to be repeated throughout the patient's life-span.

Acknowledgement

The authors acknowledge grant support from the National Institutes of Health, DK 38320. Thanks are especially given to the graduate students and post-doctoral fellows whose efforts have contributed greatly to these studies. The understanding we now have of the molecular genetics of maple syrup urine disease would not have been possible without these students and fellows.

References

Aki, K., Ogawa, K. Shirai, A. and Ichihara, A. (1967) Transaminases of branched chain amino acids. 3. Purification and properties of the mitochondrial enzyme from hog heart and comparison with the supernatant enzyme. *J. Biochem.* 62: 610–617.

Aki, K., Ogawa, K. and Ichihara, A. (1968) Transaminases of branched chain amino acids. IV. Purification and properties of two enzymes from rat liver. *Biochim. Biophys. Acta* 159: 276–284.

Akli, S., Chelly, J., Mezard, C., Gandy, S., Kahn, A. and Poenaru, L. (1990) A "G" to "A" mutation at position-1 of a 5' splice site in a late infantile form of Tay-Sachs disease. *J. Biol. Chem.* 265: 7324–7330.

Ansari, A.A., Wang, Y.-C., Danner, D.J., Gravanis, M.B., Mayne, A., Neckelmann, N., Sell, K.W. and Herskowitz, A. (1991) Abnormal expression of histocompatibility and mitochondrial antigens by cardiac tissue from patients with myocarditis and dilated cardiomyopathy. *Am. J. Path.* 139: 337–354.

Antonozzi, I., Dominici, R., Andreoli, M. and Monaco, F. (1980) Neonatal screening in Italy for congenital hypothyroidism and metabolic disorders: hyperphenylalaninemia, maple syrup urine disease and homocystinuria. *J. Endocrinol. Invest.* 3: 357–363.

Becker, J., and Craig, E.A. (1994) Heat-shock proteins as molecular chaperones. *Eur. J. Biochem.* 219: 11–23.

Berger, L.C., Bag, J. and Sells, B.H. (1992a) Identification of proteins associating with poly(A)-binding-protein mRNA. *Eur. J. Biochem.* 204: 733–743.

Berger, L.C., Bag, J. and Sells, B.H. (1992b) Translation of poly(A)-binding protein mRNA is regulated by growth conditions. *Biochem. Cell Biol.* 70: 770–778.

Bernstein, P., Peltz, S.W. and Ross, J. (1989) The poly(A)-poly(A)-binding protein complex is a major determinant of mRNA stability *in vitro*. *Mol. Cell. Biol.* 9: 659–670.

Berry, G.T., Heidenreich, R., Kaplan, P., Levine, F., Mazur, A., Palmieri, M.J., Yudkoff, M. and Segal, S. (1991) Branched-chain amino acid-free parenteral nutrition in the treatment of acute metabolic decompensation in patients with maple syrup urine disease. *N. Eng. J. Med.* 324: 175–179.

Chang, C.D., Phillips, P., Lipson, K.E., Cristofalo, V.J. and Baserga, R. (1991) Senescent human fibroblasts have a post-transcriptional block in the expression of the proliferating cell nuclear antigen gene. *J. Biol. Chem.* 266: 8663–8666.

Charollais, R.H., Surmacz E., and Baserga, R. (1994) Introns determine the time of appearance of PCNA mRNA in 3T3 cells stimulated by growth factors. *Biochem. Biophys. Res. Comm.* 201: 841–847.

Chemke, J., andLevin, S. (1975) Maple syrup urine disease. Two cases in Israel. *Isr. J. Med. Sci.* 11: 809–816.

Chuang, D.T., Ku, L.S. and Cox, R.P. (1982) Thiamin-responsive maple-syrup-urine disease: Decreased affinity of the mutant branched-chain alpha-keto acid dehydrogenase for alpha-ketoisovalerate and thiamin pyrophosphate. *Proc. Natl. Acad. Sci. USA* 79: 3300–3304.

Chuang, D.T. (1989) Molecular studies of mammalian branched chain alpha-keto acid dehydrogenase complexes: domain structures, expression and inborn errors. *Ann. N.Y. Acad. Sci.* 573: 137–154.

Chuang, J.L., Cox, R.P. and Chuang, D.T. (1993) Characterization of the promoter-regulatory region and structural organization of E1alpha (BCKDHA) of human branched-chain alpha-keto acid dehydrogenase complex. *J. Biol. Chem.* 268: 8309–8316.

Chuang, J.L., Fisher, C.R., Cox, R.P., and Chuang, D.T., (1994) Molecular basis of maple syrup urine disease: Novel mutations at the E1α locus that impair E1(α2β2) assembly of decrease steady-state E1α mRNA levels of branched chain α-ketoacid dehydrogenase complex. *Am. J. Hum. Genet.* 55: 297–304.

Clarkson, G.H.D., and Lindsay, J.G. (1991) Immunology, biosynthesis and *in vivo* assembly of the branched-chain 2-oxoacid dehydrogenase complex from bovine kidney. *Eur. J. Biochem.* 196: 95–100.

Craig, E.A., Gambill, B.D. and Nelson, R.J. (1993) Heat shock proteins: Molecular chaperones of protein biogenesis. *Microbiol. Rev.* 57: 402–414.

Craig, E.A., Weissman, J.S. and Horwich, A.L. (1994) Heat shock proteins and molecular chaperones: mediators of protein conformation and turnover. *Cell* 78: 365–372.

Damuni, Z., Merryfield, M.L., Humphreys J.S., and Reed, L.J. (1984) Purification and properties of branched-chain alpha-ketoacid dehydrogenase phosphatase from bovine kidney. *Proc. Natl. Acad. Sci. USA* 81: 4335–4338.

Damuni, Z., and Reed, L.J. (1987) Purification and properties of the catalytic subunit of the branched-chain alpha-ketoacid dehydrogenase phosphatase from bovine kidney mitochondria. *J. Biol. Chem.* 262: 5129–5132.

Dancis, J., Levitz, M., Miller S. and Westall, R.G., (1959) Maple syrup urine disease. *Brit. Med. J.* 1: 91–93.

Dancis, J., Hutzler, J., Snyderman, S.E. and Cox, R.P. (1972) Enzyme activity in classical and variant forms of maple syrup urine disease. *J. Pediatr.* 81: 312–320.

Danner, D.J., Davidson, E.D. and Elsas, L.J. (1975) Thiamine increases the specific activity of human liver branched chain alpha-ketoacid dehydrogenase. *Nature* 254: 529–530.

Danner, D.J., Lemmon, S.K. and Elsas, L.J. (1978a) Substrate specificity and stabilization by thiamine pyrophosphate of rat liver branched chain alpha-ketoacid dehydrogenase. *Biochem. Med.* 19: 27–38.

Danner, D.J., Wheeler, F.B., Lemmon, S.K. and Elsas, L.J. (1978b) *In vivo* and *in vitro* response of human branched chain alpha-ketoacid dehydrogenase to thiamine and thiamine pyrophosphate. *Pediat. Res.* 12: 235–238.

Danner, D.J., Lemmon, S.K., Besharse, J.C. and Elsas, L.J. (1979) Purification and characterization of branched chain alpha-ketoacid dehydrogenase from bovine liver mitochondria. *J. Biol. Chem.* 254: 5522–5526.

Danner, D.J., Lemmon, S.K. and Elsas, L.J. (1980) Stabilization of mammalian liver branched chain alpha-ketoacid dehydrogenase by thiamin pyrophosphate. *Arch. Biochem. Biophys.* 202: 23–28.

Danner, D.J., Armstrong, N., Heffelfinger, S.C., Sewell, E.T., Priest, J.H. and Elsas, L.J. (1985) Absence of branched chain acyl-transferase as a cause of maple syrup urine disease. *J. Clin. Invest.* 75: 858–860.

Danner, D.J., and Elsas, L.J. (1989) Disorders of branched chain amino acid and keto acid metabolism. *In*: C. R. Scriver, A. L. Beaudet, W. S. Sly and D. Valle (eds): *The Metabolic Basis of Inherited Diseas.* McGraw-Hill Book Company, New York, pp 671–692.

Danner, D.J., Litwer, S., Herring, W.J. and Pruckler, J. (1989) Construction and nucleotide sequence of a cDNA encoding the full-length preprotein for human branched chain acyltransferase. *J. Biol. Chem.* 264: 7742–7746.

Danpure, C.J., Purdue, P.E., Fryer, P., Griffiths, S., Allsop, J., Lumb, M.J., Guttridge, K.M., Jennings, P.R., Scheinman, J.I., Mauer, S.M. and Davidson, N.O. (1993) Enzymological and mutational analysis of a complex primary hyperoxaluria type I phenotype involving alanine:glyoxylate aminotransferase peroxisome-to-mitochondrion mistargeting and intraperoxisomal aggregation. *Am. J. Hum. Genet.* 53: 417–432.

Dastur, D.K., Manghani, D.K., Joshi, M.K. and Adavi, S.V. (1966) Maple syrup urine disease in Indian baby: branched chain amino and ketoaciduria. *Indian J. Med. Res.* 54 (10): 915–922.

Dreyfus, G. (1986) Structure and function of nuclear and cytoplasmic ribonucleoprotein particles. *Ann. Rev. Cell Biol.* 2: 459–498.

Dreyfus, G., and Matunis, M.J. (1993) hnRNPproteins and the biogenesis of mRNA. *Ann. Rev. Biochem.* 62: 289–321.

Ellerine, N.P., Herring, W.J., Elsas, L.J., McKean, M.C., Klein, P.D. and Danner, D.J. (1993) Thiamin-responsive maple syrup urine disease in a patient antigenically missing dihydrolipoamide acyltransferase. *Biochem. Med. Metab. Biol.* 49: 363–374.

Elsas, L.J., Pask, B.A., Wheeler, F.B., Perl, D.P. and Trusler, S. (1972) classical maple syrup urine disease: Cofactor resistance. *Metabolism* 21: 929–944.

Elsas, L.J., Priest, J.H., Wheeler F.B., Danner, D.J. and Pask, B.A. (1974) Maple syrup urine disease: Coenzyme function and prenatal monitoring. *Metabolism* 23: 569–579.

Elsas, L.J., and Danner, D.J. (1982) The role of thiamine in maple syrup urine disease. *Ann. N.Y. Acad. Sci.* 378: 404–.

Elsas, L.J., and Acosta, P.E. (1987) Nutritional management of inherited metabolic disorders. *In*: M.E. Shils and V. Young (eds): *Modern Nutrition in Health and Disease,* Seventh Edition. Lea and Febiger, Philadelphia.

Elsas, L. J., and Danner, D.J. (1989) Recent advances in maple syrup urine disease. *Rivista di Pediatria Preventiva e Sociale* 39: 177–193.

Elsas, L.J., Ellerine, N.P and Klein, P.D. (1993) Practical methods to estimate whole body leucine oxidation in maple syrup urine disease. *Pediatr. Res.* 33: 445–451.

Feigenbaum, A.S., and Robinson, B.H. (1993) The structure of the human dihydrolipoamide dehydrogenase gene (DLD) and its upstream elements. *Genomics* 17: 376–381.

Felber, S.R., Sperl, W., Chemelli, A, Murr C., and Wendel, U. (1993) Maple syrup urine disease: metabolic decompensation monitored by proton magnetic resonance imaging and spectroscopy. *Ann. Neurol.* 33: 396–401.

Fernhoff, P.M., Fitzmaurice N., Milner, J., McEwen C.T., Dembure, P.P., Brown, A.L., Wright, L., Acosta, P.B. and Elsas, L.J. (1982) Coordinated system for comprehenxive newborn metabolic screening. *South. Med. J.* 75: 529–532.

Fernhoff, P.M., Lubitz, D., Danner, D.J., Dembure, P.P., Schwarz, H.P., Hillman, R., Bier, D.M. and Elsas. L.J. (1985) Thiamine responsive maple syrup urine disease. *Pediat. Res.* 19: 1011–1016.

Fisher, C.W., Fisher C.R., Chuang, J.L., Lau, K.S., Chuang, D.T. and Cox, R.P. (1993) Occurrence of a 2-bp (AT) deletion allele and a nonsense (G-to-T) mutant allele at the E2 (DBT) locus of six patients with maple syrup urine disease: Multiple-exon skipping as a secondary effect of the mutations. *Am. J. Hum. Genet.* 52: 414–424.

Fujiwara, K., Okamura, K. Ikeda, and Motokawa, Y. (1994) Purification and characterization of lipoyl-AMP:N epsilon-lysine lipoyltransferase from bovine liver mitochondria. *J. Biol. Chem.* 269: 16605–16609.

Gambill, B.D., Voos, W., Kang, P.J., Miao B., Langer, T., Craig, E.A. and Pfanner, N. (1993) A dual role for mitochondrial heat shock protein 70 in membrane translocation of preproteins. *J. Cell Biol.* 123: 109–117.

Garrod, A.E. (1908) Inborn errors of metabolism. *Lancet* 2: 1, 73, 142, 214.

Garrod, A.E. (1923) *Inborn Errors of Metabolism*, Second Edition. Oxford University, London.

Gazzola, G.C., Dall'Asta, V. and Guidotti, G.G. (1980) The transport of neutral amino acids in cultured human fibroblasts. *J. Biol. Chem.* 255: 929–936.

Georgopoulos, C., and Welch, W.J. (1993) Role of the major heat shock proteins as molecular chaperones. *Annu. Rev. Biol.* 9: 601–634.

Ghetti, A., Pinol-Roma, S., Michael, W.M., Morandi, C. and Dreyfus, G. (1992) HnRNP I, the polypyrimidine tract-binding protein: distinct nuclear localization and association with hnRNAs. *Nucleic Acids Res.* 20: 3671–3678.

Glick, B. and Schatz, G. (1991) Import of proteins into mitochondria. *Ann. Rev. Genet.* 25: 21–44.

Glick, B.S. (1995) Can Hsp70 proteins act as force-generating motors? *Cell* 80: 11–14.

Glover, L.A. and Lindsay, J.G. (1992) Targeting proteins to mitochondria: a current overview. *Biochem. J.* 284: 609–620.

Gonzalez-Rios, M. del C., Chuang, D.T., Cox, R.P., Schmidt, K., Knopf, K. and Packman, S. (1985) A distinct variant of intermediate maple syrup urine disease. *Clin. Genet.* 27: 153–159.

Gouyon, J.B., Desgres J. and Mousson, C. (1994) Removal of branched-chain amino acids by peritoneal dialysis, continuous arteriovenous hemofiltration, and continuous arteriovenous hemodialysis in rabbits: implications for maple syrup urine disease treatment. *Pediatr. Res.* 35: 357–361.

Griffin, T.A., Lau, K.S. and Chuang, D.T. (1988) Characterization and conservation of the inner E2 core domain struture of branched chain alpha-keto acid dehydrogenase complex from bovine liver. Construction of a cDNA encoding the entire transacylase (E2b) precursor. *J. Biol. Chem.* 263: 14008–14014.

Hallberg, E.M., Shu, Y. and Hallberg, R.L. (1993) Loss of mitochondrial hsp60 function: Nonequivalent effects on matrix-targeted and intermembrane-targeted proteins. *Mol. Cell. Biol.* 13: 3050–3057.

Harper, P.A.W., Healy, P.J. and Dennis, J.A. (1986) Ultrastructural findings in maple syrup urine disease in Poll Hereford calves. *Acta Neuropathol.* 71: 316–320.

Harper, P.A.W., Healy, P.J. and Dennis, J.A. (1990) Animal model of human disease. Maple syrup urine disease (branched chain ketoaciduria). *Am. J. Path.* 136: 1445–1447.

Harris, R.A., Powell, S.M., Paxton, R., Gillim, S.E. and Nagae, H. (1985) Physiological covalent regulation of rat liver branched-chain alpha-ketoacid dehydrogenase. *Arch. Biochem. Biophys.* 243: 542–555.

Harris, R.A., Popov, K.M., Zhao, Y. and Shimomura, Y. (1994) Regulation of branched-chain amino acid metabolism. *J. Nutrition* 124: 1499S-1502S.

Heffelfinger, S.C., Sewell, E.T., Elsas, L.J. and Danner, D.J. (1984) Direct physical evidence for stabilization of branched chain ketoacid dehydrogenase by thiamin pyrophosphate. *Am. J. Hum. Genet.* 36: 802–807.

Herring, W.J., Litwer, S., Weber, J.L. and Danner, D.J. (1991) Molecular genetic basis of maple syrup urine disease in a family with two defective alleles for branched chain acyltransferase and localization of the gene to human chromosome 1. *Am. J. Hum. Genet.* 48: 342–350.

Herring, W.J., McKean, M., Dracopoli, N. and Danner, D.J. (1992) Branched chain acyltransferase absence due to an ALU-based genomic deletion allele and an exon skipping allele in a compound heterozygote proband expressing maple syrup urine disease. *Biochim. Biophys. Acta* 1138: 236–242.

Hilliges, C., Awiszus, D. and Wendel, U. (1993) Intellectual performance of children with maple syrup urine disease. *Eur. J. Pediatr.* 152: 144–147.

Hutson, S.M., and Rannels, S.L. (1985) Characterization of a mitochondrial transport system for branched chain alpha-ketoacids. *J. Biol. Chem.* 260: 14189-.

Johanning, G.L., Morris, J.I., Madhusudhan, K.T., Samols, D. and Patel, M.S. (1992) Characterization of the transcriptional regulatory region of the human dihydrolipoamide dehydrogenase gene. *Proc. Natl. Acad. Sci. USA* 89: 10964–10968.

Jones, S.M., and Yeaman, S.J. (1986) Oxidative decarboxylation of 4-methylthio-2-oxobutyrate by branched-chain 2-oxo acid dehydrogenase complex *Biochem. J.* 237: 621–623.

Jurka, J., and Smith, T. (1988) A fundamental division in the alu family of repeated sequences. *Proc. Natl. Acad. Sci. USA* 85: 4775–4778.

Kam-pui, F., Kit, C. Wun and Ying, C. Pui (1985) Maple syrup urine disease in Chinese. *Chinese Med. J.* 99: 119–120.

Kamei, A., Takashima, S., Chan, F. and Becker, L.E. (1992) Abnormal dendritic development in maple syrup urine disease. *Pediatr. Neurol.* 8: 145–147.

Kaplan, P., Mazur, A., M. Field, Berlin, J.A., Berry, G.T., Heidenreich, R., Yadkoff, M. and Segal, S. (1991) Intellectual outcome in children with maple syrup urine disease. *J. Pediatr.* 119: 46–50.

Kiebler, M., Becker, K., Pfanner, N. and Neupert, W. (1993) Mitochondrial protein import: Specific recognition and membrane translocation of preproteins. *J. Membrane Biol.* 135: 191–207.

Koch, R., Levy, H.L., Matalon, R., Rouse, B., Hanley, W. and Azen, C. (1993) The North American collaborative study of maternal phenylketonuria. Status report (1993) *Am. J. Dis. Child.* 147: 1224–1230.

Kozak, M. (1987) At least six nucleotides preceding the AUG initiator codon enhance translation in mammalian cells. *J. Mol. Biol.* 196: 947–950.

Kozak, M. (1991) Structural features in eukaryotic mRNAs that modulate the initiation of translation. *J. Biol. Chem.* 266: 19867–19870.

Langenbeck, U. (1986) Two different forms of maple syrup urine disease in a single family. *Hum. Genet.* 72: 279.

Larson, D.E. and Sells, B.H. (1987) The function of proteins that interact with mRNA. *Mol. Cell. Biochem.* 74: 5–15.

Lau, K.S., Herring, W.J., Chuang, J.L., McKean, M., Danner, D.J., Cox, R.P. and Chuang, D.T. (1992) Structure of the gene encoding dihydrolipoyl transacylase (E2) component of human branched chain alpha-keto acid dehydrogenase complex and characterization of an E2 pseudogene. *J. Biol. Chem.* 267: 24090–24096.

Litwer, S., Herring, W.J. and Danner, D.J. (1989) Reversion of maple syrup urine disease phenotype of impaired branched chain alpha-ketoacid dehydrogenase complex activity in fibroblasts from an affected child. *J. Biol. Chem.* 264: 14597–14600.

Loukopoulos, D. (1991) Thalassemia: genotypes and phenotypes. *Ann. Hematol.* 62: 85–94.

Maccecchini, M.-L., Rubin, Y., Blobel, G. and Schatz, G. (1979) Import of proteins into mitochondria: precursor forms of the extramitochondrially made F1-ATPase subunits in yeast. *Proc. Natl. Acad. Sci. USA* 79: 343–347.

Manning-Krieg, U.C., Scherer, P.E. and Schatz, G. (1991) Sequential action of mitochondrial chaperones in protein import into the matrix. *EMBO J.* 10: 3273–3280.

Matunis, E.L., Matunis, M.J. and Dreyfus, G. (1993) Association of individual hnRNP proteins and snRNPs with nascent transcripts. *J. Cell Biol.* 121: 219–228.

Mayeda, A., Helfman, D.M. and Krainer, A.R. (1993) Modulation of exon skipping and inclusion by heterogeneous nuclear ribonucleaoprotiens A1 and premRNA splicing factor SF2/ASF. *Mol. Cell. Biol.* 13: 2993–3001.

McConnell, B.B., McKean, M.C. and Danner, D.J. (1996) Influence of subunit transcript and protein levels on formation of a mitochondrial multienzyme complex. *J. Cell Biochem.* 60; in press.

McKusick, V.A., Francomano, C.A., Antonarakis, S.E. and Pearson, P.L. (1994) *Mendelian Inheritance in Man*, Eleventh Edition, Vol. 2. Johns Hopkins University Press, Baltimore.

Menkes, J.H. (1959) Maple syrup disease. Isolation and identification of organic acids in the urine. *Pediatrics* 23: 348–353.

Menkes, J.H., Hurst, P.L. and Craig, J.M. (1954) A new syndrome: Progressive familial infantile cerebral dysfunction associated with an unusual urinary substance. *Pediatrics* 14: 462–467.

Merinero, B., del Valle, J.A., Garcia, M.J., Garcia Miguel, M.J., Barrio, M.I., Garcia Hortelano, J., Morales, E., Gonzalez, F., Garcia Aparicio, J., Saez Perez, E. et al. (1983) Three Patients with maple syrup urine disease. *An. Esp. Pediatr.* 19: 393–400.

Mitsubuchi, H., Nobukuni, Y., Endo, F. and Matsuda, I. (1991a) Structural organization and chromosomal localization of the gene for the E1beta subunit of human branched chain alpha-keto acid dehydrogenase. *J. Biol. Chem.* 266: 14686–14691.

Mitsubuchi, H., Nobukuni, Y., Akaboshi, I., Indo, Y., Endo, F. and Matsuda, I. (1991b) Maple syrup urine disease caused by a partial deletion in the inner E2 core domain of the branched chain alpha-keto acid dehydrogenase complex due to aberrant splicing. A single base deletion at a 5'-splice donor site of an intron of the E2 gene disrupts the consensus sequence in this region. *J. Clin. Invest.* 87: 1207–1211.

Morris, T.W., Reed, K.E. and Cronan, J.E., Jr. (1995) Lipoic acid metabolism in *Escherichia coli*: the lplA and lipB genes define redundant pathways for ligation of lipoyl groups to apoprotein. *J. Bacteriology* 177: 1–10.

Muller, K., Kahn, T. and Wendel, U. (1993) Is demyelination a feature of maple syrup urine disease? *Pediatr. Neurol.* 9: 375–382.

Neupert, W. and Pfanner, N. (1993) Roles of molecular chaperones inprotein targeting to mitochondria. *Phil. Trans. R. Soc.* 339: 355–361.

Neupert, W. (1994) Transport of proteins across mitochondrial membranes. *J. Clin. Invest.* 72: 251–261.

Ostermann, J., Horwich, A.L., Neupert, W. and Hartl, F.-U. (1989) Protein folding in mitochondria requires complex formation with hsp60 and ATP hydrolysis. *Nature* 341: 125–130.

Patton, J. G., Mayer, S.A., Tempst, P. and Nadal-Ginard, B. (1991) Characterization and molecular cloning of polypyrimidine tract-binding protein: a component of a complex necessary for premRNA splicing. *Genes Dev.* 5: 1237–1251.

Paxton, R., Scislowski, P.W.D., Davis, E.J. and Harris, R.A. (1986) Role of branched-chain 2-oxoacid dehydrogenase and pyruvate dehydrogenase in 2-oxobutyrate metabolism. *Biochem. J.* 234: 295–303.

Peinemann, F. and Danner, D.J. (1994) Maple syrup urine disease 1954 to 1993. *J. Inherit. Metab. Dis.* 17: 3–15.

Pettit, F.H., Yeaman, S.J. and Reed, L.J. (1978) Purification and characterization of branched chain alpha-ketoacid dehydrogenase complex of bovine kidney. *Proc. Natl. Acad. Sci. USA* 75: 4881–4886.

Pinol-Roma, S. and Dreyfus, G. (1992) Shuttling of pre-mRNA binding proteins between nucleus and cytoplasm. *Nature* 355: 730–732.

Popa, W., Apostolescu, I. and Popescu, G. (1970) Clinico-metabolic aspects of maple syrup urine disease and manifestation of some new trophometabolic relationships. *Pediatria* 19: 51–56.

Pueschel, S.M., Bresnan, M.J., Shih, V.E. and Levy, H.L. (1979) Thiamine-responsive intermittent branched-chain ketoaciduria. *J. Pediatr.* 94: 628–631.

Purdue, P.E., Takada, Y. and Danpure, C.J. (1990) Identification of mutations associated with peroxisome-to-mitochondrion mistargeting of alanine/glyoxylate aminotransferase in primary hyperoxaluria Type 1. *J. Cell Biol.* 111: 2341–2351.

Raven, E.J. (1969) Progressive encephalopathy after a symptom-free period in the newborn infant, due to disorders of amino acid metaboism. Illustrated with case histories of patients with hyperglycinemia and maple syrup urine disease. *Ned. Tijdschr. Geneeskd.* 113: 1850–1853.

Reed, L.J. and Hackert, M.L. (1990) Structure-function relationships in dihydrolipoamide acyltransferases. *J. Biol. Chem.* 265: 8971–8974.

Riviello, J.J., Rezvani, I., DiGeorge, A.M. and Foley, C.M. (1991) Cerebral edema causing death in children with maple syrup urine disease. *J. Pediatr.* 119: 42–45.

Rokkones, T. (1970) Maple syrup urine disease. *Tidsskr. Nor. Laegeforen* 90: 239–242.

Scriver, C.R., Clow, C.L., Mackenzie, S. and Delvin, E. (1971) Thiamine-responsive maple-syrup-urine disease. *Lancet* i: 310–312.

Scriver, C.R., Clow, C.L. and George, H. (1985) So-called thiamin-responsive maple syrup urine disease: 15-year follow-up of the original patient. *J. Pediatr.* 107: 763–765.

Sherwood, R.A., Titheradge, A.C. and Richards, D.A. (1990) Measurement of plasma and urine amino acids by high-performance liquid chromatography with electrochemical detection using phenylisothiocyanate derivatives. *J. Chromatogr.* 528: 293–303.

Snyderman, S.E., Norton, P.M., Roitman, E. and Holt, L.E. (1964) Maple syrup urine disease, with particular reference to dietotherapy. *J. Pediatr.* 34: 454–472.

Stuart, R.A., Cyr, D.M., Craig, E.A. and Neupert, W., (1994) Mitochondrial molecular chaperones: their role in protein translocation. *TIBS* 19: 87–92.

Taccone, A., Schiaffino, M.C., Cerone, R., Fondelli, M.P. and Romano, C. (1992) Computed tomography in maple syrup urine disease. *Eur. J. Radiol.* 14: 207–212.

Tanaka, K., West-Dull, A., Hine, D.G., Lynn, T.B. and Lowe, T. (1980) Gas-chromatographic method of analysis for urinary organic acids. II. Description of the procedure and its application to diagnosis of patients with organic acidurias. *Clin. Chem.* 26: 1847–1853.

Thomas, F. (1992) Dietary management of inborn errors of amino acid metabolism with protein-modified diets. *J. Child. Neurol.* 7: S92–111.

Thompson, G.N., Francis, D.E., Kirby, D.M. and Compton, R. (1991) Pregnancy in phenylketonuria: dietary treatment aimed at normalising maternal plasma phenylalanine concentration. *Arch. Dis. Child.* 66: 1346–1349.

Treacy, E., Clow, C.L., Reade, T.R., Chitayat, D., Mamer, O.A. and Scriver, C.R. (1992) Maple syrup urine disease: Interrelationships between branched chain amino-, oxo-, and hydroxyacids; implications for treatment; association with CNS dysmyelination. *J. Inherit. Metab. Dis.* 15: 121–135.

Ungermann, C., Neupert, W. and Cyr, D.M. (1994) The role of hsp70 in conferring unidirectionality on protein translocation into mitochondria. *Science* 266: 1250–1253.

Van Calcar, S.C., Harding, C.O., Davidson, S.R., Barness, L.A. and Wolff, J.A. (1992) Case reports of successful pregnancy in women with maple syrup urine disease and propionic acidemia. *Am. J. Med. Genet.* 44: 641–646.

Van de Water, J., Cooper, A., Surh, C.D., Coppel, R., Danner, D.J., Ansari, A., Dickson, R. and Gershwin, M.E. (1989) Detection of autoantibodies to recombinant mitochondrial proteins in patients with primary biliary cirrhosis. *N. Eng. J. Med.* 320: 1377–1380.

Veerkamp, J.H., and Wagenmakers, A.J.M. (1987) Postnatal development of the actual and total activity of the branched-chain 2-oxoacid dehydrogenase complex in rat tissues. *Int. J. Biochem.* 19: 205–207.

Vidaud, M., Gattoni, R., Stevenin, J., Vidaud, D., Amselem, S., Chibani, J., Rosa, J. and Goossens, M. (1989) A 5' splice-region G → C mutation in exon 1 of the human beta-globin gene inhibits pre-mRNA splicing: A mechanism for beta+-thalassemia. *Proc. Natl. Acad. Sci. USA* 86: 1041–1045.

Wachter, C., Schatz, G. and Glick, B.S. (1994) Protein import into mitochondria: the requirement for external ATP is precursor-specific whereas intramitochondrial ATP is universally needed for translocation into the matrix. *Mol. Biol. Cell.* 5: 465–474.

Weil, D., D'Alessio, M., Ramirez, F., de Wet, W., Cole, W.G., Chan, D. and Bateman, J.F. (1989) A base substitution in the exon of a collagen gene causes alternative splicing and generates a structurally abnormal polypeptide in a patient with Ehlers-Danlos syndrome type VII. *EMBO J.* 8: 1705–1710.

Wendel, U., Becker, K., Przyrembel, H., Bulla, M., Manegold, C., Mench-Hoinowski, A. and Langenbeck, U. (1980) Peritoneal dialysis in maple syrup urine disease: Studies on branched-chain amino and ketoacids. *Eur. J. Pediatr.* 129: 57–63.

Westall, R.G., Dancis, J. and Miller, S. (1957) Maple syrup urine disease. *Am. J. Dis. Child.* 94: 571–572.

Wickner, W.T. (1994) How ATP drives proteins across membranes. *Science* 266: 1197–1198.

Wynn, R.M., Davie, J.R., Cox, R.P. and Chuang, D.T. (1992) Chaperonins GroEL and GroES promote assembly of heterotetramers (alpha2beta2) of mammalian mitochondrial branched chain alpha-ketoacid decarboxylase in *Escherichia coli. J. Biol. Chem.* 267: 12400–12403.

Yadav, G.C., and Reavey, P.C. (1988) Aminoacidopathies: a review of 3 years' experience of investigations in a Kuwait hospital. *J. Inherit. Metab. Dis.* 11: 277–284.

Yeaman, S.J. (1989) The 2-oxo acid dehydrogenase complexes: recent advances. *Biochem. J.* 257: 625–632.

Zhang, B., Edenberg, H.J., Crabb, D.W. and Harris, R.A (1989) Evidence for both a regulatory mutation and a structural mutation in a family with maple syrup urine disease. *J. Clin. Invest.* 83: 1425–1429.

Structure and chromosomal localization of the human 2-oxoglutarate dehydrogenase gene

K. Koike

Department of Pathological Biochemistry, Atomic Disease Institute, Nagasaki University, School of Medicine, Sakamoto 1-12-4, Nagasaki 852, Japan

Summary. In this review the molecular cloning, structural analysis and determination of chromosomal localization of the human 2-oxoglutarate dehydrogenase (OGDH) gene are discussed. Genomic clones covering the entire sequence of the gene encoding human OGDH were isolated by screening leukocyte and placenta genomic libraries with the human OGDH cDNA as a probe. The human OGDH gene contains 22 exons spanning approximately 85 kb. Primer extension of the OGDH poly(A)$^+$RNA from HeLa cells revealed a major transcription start point at a thymine base 55-bp upstream of the ATG start codon. The 5'-flanking region of the human OGDH gene lacked canonical TATA or CAAT boxes. Using DNAs from human × rodent somatic cell hybrids that segregate human chromosomes in conjunction with fluorescence *in situ* hybridization, the human OGDH gene was assigned to chromosome 7p13 and 7p14.

Introduction

The 2-oxo acid dehydrogenase multienzyme complexes that catalyze a thiamin pyrophosphate (TPP)-, CoA-, NAD$^+$- and FAD-dependent and lipoic acid-mediated oxidative decarboxylation of 2-oxo acids have been isolated from prokaryotic and eukaryotic cells as functional units with molecular weights in the millions and with distinct molecular shapes, as previously reviewed (Reed, 1974; Koike and Koike, 1976; Yeaman, 1987). Three complexes with different substrate specificities have been purified, one specific for pyruvate, a second for 2-oxoglutarate and a third for the branched-chain 2-oxo acids. Each complex is organized about a central core, dihydro-lipoamide acyltransferase (E2) with covalently bound lipoic acid essential cofactor, to which multiple copies of TPP-dependent and substrate-specific 2-oxo acid decarboxylase-dehydro-genase (E1) and dihydrolipoamide dehydrogenase (E3), a flavoprotein that is the only common component of three complexes, are noncovalently bound by self-assembling. These three compo-nent enzymes, acting in sequence, catalyze the overall oxidative decarboxylation of 2-oxo acids.

The 2-oxoglutarate dehydrogenase (OGDH) complex has been isolated from porcine heart muscle in highly purified state as a multienzyme complex with molecular weight of 2.7 million (Hirashima et al., 1971). The OGDH complex was further separated into three component enzymes and reconstituted by self-assembling of the isolated component enzymes (Koike et al.,

1971; Tanaka et al., 1971). It is now apparent that the OGDH complex is composed of a dihy-drolipoamide succinyltransferase core (E2o; EC 2.3.1.61; $M_r = 1\,000\,000$), multiple copies of 2-oxoglutarate: lipoamide oxidoreductae (decarboxylating and acceptor-succinylating, (OGDH or E1o; EC 1.2.4.2; $M_r = 216\,000$) and a common component dihydrolipoamide dehydrogenase (E3; $M_r = 110\,000$) (Koike et al., 1971; Tanaka et al., 1974; Moriyasu et al., 1986). E1o catalyzes the TPP-dependent irreversible decarboxylation and the subsequent reductive succinylation of the lipoyl moiety which is covalently bound to E2o (Tanaka et al., 1971; Petit et al., 1973; Koike et al., 1974).

Three component enzymes of the OGDH complex are nuclear coded, transferred into mito-chondria, and apparently assembled into organized functional multienzyme complex. We recently cloned the human OGDH cDNA from fetal liver cDNA library (Koike et al., 1992). The cDNA contains a 3006-bp open reading frame encoding a 40-amino acid leader peptide and a mature 962-amino acid enzyme protein ($M_r = 108\,878$). Thus, the human OGDH cDNA encodes a precursor protein including a leader peptide essential for its transport into mitochondria and subsequent processing. To obtain further insights into the regulation of its expression, I have undertaken to isolate genomic OGDH clones using the human OGDH cDNA as a probe, and characterize the OGDH gene structure (Koike, 1995). In addition, the chromosomal localization of the human OGDH gene was determined.

Isolation and characterization of human OGDH genomic clones

Three human leukocyte and one placenta genomic libraries in λ phage vectors and one leukocyte cosmid library were screened by standard procedures (Sambrook et al., 1989) using a full length human OGDH cDNA, its 5'-end region (a 760-bp EcoRI-BamHI fragment) and its 3'-end region (a mixture of three SmaI-SmaI, SmaI-SmaI and SmaI-EcoRI fragments with the total length of 1.9 kb). A 385-bp Sau3AI-BstPI fragment corresponding to the 5'-end region of a genomic clone, λHGOG 1 (see Fig. 1C), was also used as a probe. These probes were labeled with digoxigenin (DIG) (Höltke et al., 1992) or [α-^{32}P]ddCTP by random priming method (Feinberg et al., 1983).

Sau3AI-generated leukocyte libraries constructed in λEMBL3 and λEMBL3 SP6/T7 vectors (Clontech, Palo Alto, California, USA) yielded three clones (designated λHGOG 2, 8 and 9) and one clone (λHGOG 10), respectively. Lymphocyte and placenta libraries in λDASH and λFIXII vectors (Stratagene, La Jolla, California, USA), respectively, yielded six clones (λHGOG 1, 3–7) and one clone (λHGOG 11). These positive clones were purified and subjected to restriction enzyme mapping, Southern blot hybridization analyses and sequencing analyses. As shown in

Table 1. Sizes of exons and introns and splice junction sequences of the human OGDH gene

Exon[a] No.	Size (bp)	Boundary sequence[b] 5'-splice donor	Intron[a] Size (kb)	Boundary sequence[b] 3'-splice acceptor
1	277	GTA CAT AAG V - H - K 34	gtaaggctcg --- 24.0 --- ttgtctttag	TCA TGG GAC S - W - D 35
2	192	GCA TAT CAG A - Y - Q 98	gtaaggcggg --- 2.0 --- ccactcatag	ATA CGA GGG I - R - G 99
3	103	AAA CTT G K - L - 133	gtgagggtct --- 19.0 --- ctccctgcag	GG TTC TAT G - F - Y 133
4	116	CGG CTG GAG R - L - E 171	gtaagagcag --- 7.0 --- tatggaatag	ATG GCC TAC M - A - Y 172
5	155	TCC ACC AG S - T - R 223	gtatgggtct --- 0.47 ---ccctgtccag	G TTT GAG - F - E 223
6	147	CCA CAC AG P - H - R 272	gtacagccaa --- 0.62 --- tacgttccag	A GGG CGG - G - R 272
7	91	GCT GAT GAG A - D - E 302	gtgaggcacc --- 0.7 ---accccatcag	GGC TCC GGA G - S - G 303
8	180	GGG AAA AAG G - K - K 342	gtaaggccca --- 6.0 --- gtgtttacag	GTC ATG TCC V - M - S 343
9	129	AAC AAC CAG N - N - Q 385	gtacctcaca --- >8.0 --- cctcctgcag	ATC GGC TTC I - G - F 386
10	180	GTC GAT TTG V - D - L 445	gtgagtgact --- 0.41 --- ctctgagcag	GTG TGT TAC V - C - Y 446
11	153	GAG TAT GAG E - Y - E 496	gtacgtccct --- 2.01 --- aaccctccag	GAG GAA ATT E - E - I 497
12	103	TGG CCT G W - P - 531	gtgagtgaag --- 0.3 --- cttctcctag	GC TTC TTC G - F - F 531
13	129	CAT GGA G H - G - 574	gtaacacgct --- 0.8 --- gtctctgcag	GG CTG AGC G - L - S 574
14	151	ACA TTC AG T - F - S 624	gtaacgttct --- 0.27 --- cctgtctcag	C CAC CGC - H - R 624
15	128	GTG CTG G V - L - 667	gtgagtgcct --- 0.1 --- tcccttgtag	GC TTT GAA G - F - E 667
16	179	GAG GGC ATG E - G - M 726	gtgagcctct --- 0.41 --- gtggttccag	GGT CCA GAA G - P - E 727

Table 1. (continued)

Exon[a] No.	Size (bp)	Boundary sequence[b] 5'-splice donor	Intron[a] Size (kb)	Boundary sequence[b] 3'-splice acceptor
17	72	GTC CTG CCA gtgagtaata --- 1.88 --- cctgctgcag V - L - P 750		GAC CTT AAA D - L - K 751
18	129 ^	CGG AAG CCG gtcagtaaca --- 1.28 ---tgtatttcag R - K - P 793		TTA ATT ATC L - I - I 794
19	73	CTT CCA G gtgggtgtga --- 5.0 --- tctcttgtag L - P - 818		GA ACC CAC G - T - H 818
20	164	ATT GAG CAG gtgagggcag --- 0.19 ---tgtctcctag I - E - Q 872		CTG TCG CCA L - S - P 873
21	155	CCC GTC TG gtaaggcttc --- 0.14 --- catctctctag P - V - W 924		G TAT GCC - Y - A 924
22	1,114	TTTATTTATTTTGGTT(A)$_n$		

[a]See Figure 1A; [b]the exonic nucleotide sequences are shown in upper case letters, while intronic nucleoide sequences are in lower case. The exon size in bp and the estimated size of the introns in kb are indicated. The letters below the nucleotide sequence indicate the amino acid residue flanking the exon boundaries.

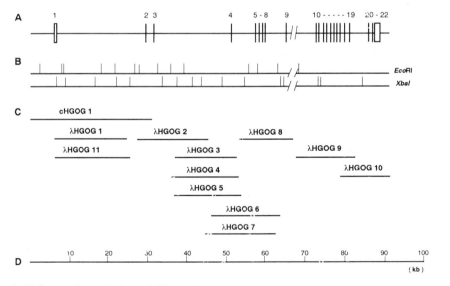

Figure 1. (A) Structural organization and (B) restriction map of the human OGDH gene. Solid bars indicate the locations of the 22 exons and open bars are untranslated exonic sequences. (C) The 12 overlapping clones were vertically aligned according to restriction mapping, Southern blot hybridization, partial nucleotide sequence and (D) the scale. The scale (D) designates nucleotide positions in kb relative to ATG codon.

Figure 1, the results revealed that the phage clones overlapped with each other and contained the entire nucleotide sequence of the human OGDH cDNA. λHGOG 1 and 11 contained the 5'-end region of the cDNA, including the ATG start codon, but lacked a further upstream region. Therefore, a *Mbo*I-generated leukocyte library in cosmid pWE15 vector (Stratagene) was screened and one positive clone (cHGOG 1) containing 6-kb upstream of the start codon was obtained. The human OGDH gene spanned approximately 85 kb (Fig. 1A). The *Eco*RI and *Xba*I restriction map and overlapping clones are shown in Figures 1B and C, respectively.

Structural organization of the human OGDH gene

DNA fragments containing exons and their surrounding regions were sequenced. Comparison of the nucleotide sequence of the human OGDH gene with its cDNA indicated that the gene consisted of 22 exons interrupted by 21 introns. The location of these exons and introns are shown in Figure 1A. The sequence of the exon-intron boundaries and the size of the exons and introns are summarized in Table 1. The sizes of both exon and intron regions vary considerably, and exons range in size between 72 bp (exon 17) and 1 114 bp (exon 22), while introns range between 105 bp (intron 15) and 24 kb (intron 1). The sequences at the exon-intron boundaries were in good agreement with the consensus proposed by Mount (1992) and the GT/AG rule of Breathnach et al. (1981). Exon 1 contained a transcriptional start point, the ATG start codon and the sequence encoding a leader peptide of 40 amino acid residues of the N-terminal region (see Fig. 3). Although exon 22 is the longest exon, it encoded only 18 amino acid residues of the C-terminal region, consisting mostly of a long 3'-untranslated region. In exon 22, a potential polyadenylation signal was not detected. The human OGDH gene utilyzes TPP as a coenzyme like other 2-oxo acid dehydrogenases. Hawkins et al. (1989) proposed that a putative TPP-binding sequence motif in the 2-oxo acid dehydrogenases consists of 30 amino acid residues beginning with -GQG- and ending -NN-. This motif was identified in exon 9.

Determination of transcriptional start point

Primer extension analysis of the OGDH poly(A)$^+$RNA from HeLa cells revealed a major transcriptional start point at a thymine (+1 in Figs 2 and 3). The transcriptional start point is located 55-bp upstream from the ATG start codon in exon 1 as shown in Figure 3. This is consistent with common structural properties of housekeeping gene encoded mitochondrial enzyme. A computer analysis search of the nucleotide sequence of the 5'-flanking region of the human OGDH gene did not reveal canonical TATA or CAAT boxes. An inverted GC box (Sp1-binding site) was found at position nucleotide –401 to –396, and two Ap1-binding sites at positions nucleotide –620

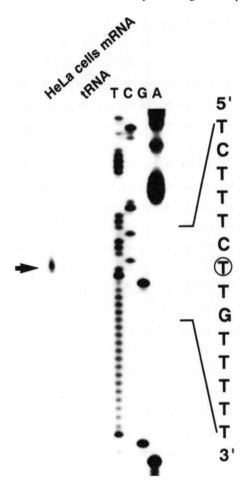

Figure 2. Determination of the transcriptional start point for the human OGDH gene by primer extension analysis. Lanes: HeLa cells poly(A)+RNA (mRNA); yeast transfer RNA. Primer-extended fragment is indicated by arrowhead. The four sequencing reaction ladder (T, C, G and A) served as a chain length marker to precisely locate the transcriptional start point. The start point of human OGDH mRNA is circled. Analysis was carried out using HeLa cells poly(A)+RNA according to the described method (Sambrook et al., 1989). An oligodeoxyribonucleotide primer (30-mer, 5'-CTT AGC AGC ACA AGT CCT TAA ATG AAA CAT-3') complimentary to nucleotide 56 to 85 of exon 1 (see Fig. 3) was synthesized, radiolabeled at the 5'-end using [γ-³²P]ATP and separated by electrophoresis in 15% polyacrylamide/8 M urea denaturing gel. The end-labeled primer was hybridized to poly(A)+RNA (10 µg) and was extended in the presence of reverse transcriptase (15 units). The products were separated in 6% polyacrylamide/8 M urea denaturing gel and autoradiographed. A 1.8-kb *Pst*I-*Eco*RI fragment containing exon 1 served as a template for nucleotide sequencing using the same unradiolabeled oligodeoxyribonucleotide primer as reported (Sanger et al., 1977). The sequencing reaction products were run in parallel with the primer extended fragments in a sequencing gel.

```
tctaattgtttttatatatatgtaatatgtaaatttaaAGAACAtcgtatgtttatcatct   -670
                                      GRE
tagcccttaaactagatttaatggttttttgttttgttttgttttgagaCAGAGTCTcac    -610
                                                  Ap1
tctgttgctaggctggagtacagtggtgcaatctcggctcactgcagcctccacctacca    -550

ggttcaagcaattctcctgcctcagcctcccaagtatctgggtctacaggcatgcaccac    -490

cacacccagctaatttttgtattttttagtagagatgtggtttcaccatgttggccaggat   -430

ggtcttgatctcctgacctcgtgatccaCCCGCCttggcctcccaaagtgctgggattac    -370
                            Sp1
aggtgtgagccacCGAGTCTGgcctagatttaaatttttaagtgaagatggtaatatctg    -310
             Ap1
caactgtgcactcaccttccttgggcacgcacaggcatcctgtgttagtaggtgcaagag    -250

tgtgtttcagtttcacacagtcacagctctgttgtgctctcaggcacatgaactggtggc    -190

actgaacaggtagcactcttgtgaatacatttataggccagggtctttctcctcaacaat   -130

tagagttgaacatgttttgttttttgtttttgtttttctagtagcttttaaaaaaaaatta   -70

tcctttctctagagcatcaggatctagtagccttgtctcctaaatactcatttttaaaac   -10
                    Sau3AI
ctttctttcTTGTTTTTTTTTTTTTTTTTTTTTTGTACAGGCAGTTGTGAAAAACTTCAGGAC +51
          +1
             ==================== Exon 1 ====================
AAAAATGTTTCATTTAAGGACTTGTGCTGCTAAGTTGAGGCCATTGACGGCTTCCCAGAC    +111
    M  F  H  L  R  T  C  A  A  K  L  R  P  L  T  A  S  Q  T     -22
=============================================================
TGTTAAGACATTTTCACAAAACAGACCAGCAGCAGCTAGGACATTTCAACAGATTCGGTG    +171
  V  K  T  F  S  Q  N  R  P  A  A  A  R  T  F  Q  Q  I  R  C    -2
=============================================================
CTATTCTGCACCTGTTGCTGCTGAGCCCTTTCTCAGTGGGACTAGTTCGAACTATGTGGA    +231
  T  S  A  P  V  A  A  E  P  F  L  S  G  T  S  S  N  Y  V  E    +19
=============================================================
GGAGATGTACTGTGCTTGGCTGGAAAACCCCAAAAGTGTACATAAGgtaaggctcgcagg    +291
  E  M  Y  C  A  W  L  E  N  P  K  S  V  H  K                  +34
=============================================================
gctgtgggtctgcctcatgttggtgtgttggcctgttcctgactggtcaccaggtaagtg   +351
                                            BstPI
```

Figure 3. The nucleotide sequence of the 5'-flanking region of the human OGDH gene including exon 1. The sequences of the 5'-flanking region, untranslated exon and part of the first intron are in lower case letters, and translated exonic sequence is in upper case. Bases are numbered on the right beginning with the transcriptional start point, base T (boldface, designated +1). The deduced amino acid sequence of the translated exonic sequence is shown below the nucleotide sequence in upper case with the single-letter code and numbered relative to the N-terminus. The sequence of the putative regulatory elements, inverted GC-box (Sp1-binding site), Ap1-binding site and GRE are indicated in boldface and underlined. Exon 1 (underlined with double dashed line) includes 55 bp of untranslated region (lower case), 120 bp encoding a leader peptide and 102 bp encoding the mature protein ending with the 5'-donor site, gt (boldface). The ATG translation start codon and N-terminal amino acid are in boldface. The complementary sequence from the ATG start codon (+56) to AAG (+85) used for primer extension analysis is underlined. The Sau3AI and BstPI sites are in boldface and underlined.

Table 2. Segregation of the human OGDH gene in human × rodent somatic cell hybrids

Hybrid	OGDH gene[a]	_Number of human chromosome present in rodent cell line[b]_																							
		1	2	3	4	5	6	**7**	8	9	10	11	12	13	14	15	16	17	18	19	20	21	22	X	Y
212	–	–	–	–	–	Dq	–	–	–	–	–	–	–	–	–	–	–	–	–	–	–	–	–	–	+
324	–	–	–	–	–	–	–	–	–	–	–	–	–	–	–	–	–	–	+	–	–	–	–	–	–
423	–	–	–	+	–	–	–	–	–	–	–	–	–	–	–	–	–	–	–	–	–	–	–	–	–
507	–	–	–	+	–	+	–	–	–	–	–	–	–	+	–	(+)	–	–	–	(–)	–	(–)	–	–	+
683	–	–	–	–	–	+	–	–	–	–	–	–	+	–	+	–	–	–	–	+	–	+	+	–	–
734	–	–	–	–	–	+	–	–	–	+	–	–	–	–	–	–	–	+	–	–	–	–	–	–	–
750	–	–	–	–	–	D	–	–	–	–	–	–	–	+	+	+	–	–	–	+	–	–	–	–	–
756	+	–	–	–	–	D	(+)	+	–	–	–	–	+	(+)	(–)	–	–	–	–	+	+	+	–	–	+
803	–	–	–	–	+	+	–	–	+	–	–	(–)	–	–	–	–	–	–	–	–	–	+	+	–	–
811	–	–	–	–	–	–	–	–	+	–	–	–	–	–	–	–	(+)	(+)	–	–	–	–	–	–	–
852	–	–	+	–	–	–	–	–	–	–	–	–	–	–	–	–	–	–	–	–	–	–	–	–	–
854	–	–	–	–	–	+	–	–	–	–	–	–	–	–	–	–	–	–	–	–	–	–	–	–	–
860	–	–	–	(–)	–	+	+	–	–	–	(–)	–	–	–	–	–	–	–	(–)	–	+	–	–	–	–
862	–	–	–	–	–	+	–	–	+	–	–	–	–	–	–	–	–	–	–	–	–	–	–	–	–
867	–	(+)	–	–	–	+	–	–	–	–	–	–	–	+	+	–	–	–	+	+	–	–	–	–	–
904	–	–	–	–	–	D	+	–	–	–	–	–	–	+	–	–	(–)	–	–	–	+	–	–	–	+
909	–	–	–	–	–	D	+	–	+	–	–	–	–	+	–	–	–	–	–	–	+	–	–	+	–
937	–	+	–	–	–	+	–	–	–	–	–	–	–	+	–	–	+	–	+	–	+	–	–	–	–
940	–	–	–	–	–	+	–	–	–	–	–	–	–	–	–	–	–	–	–	+	–	–	–	–	–
967	–	–	–	–	–	+	–	–	+	–	–	–	–	–	–	–	+	–	–	–	–	–	–	–	–
968	–	–	–	–	–	+	–	–	–	+	–	–	–	+	–	–	–	–	–	–	–	–	–	(+)	–
983	–	–	–	–	–	+	–	–	–	–	(–)	–	–	–	–	–	–	–	–	–	–	–	–	–	–
1006	+	–	–	–	(+)	+	–	+	+	–	–	–	–	+	–	+	–	–	–	+	–	+	–	–	–
1049	–	–	–	–	–	+	–	–	–	–	–	+	–	–	–	–	–	–	–	–	–	–	–	–	–
1079	–	–	–	+	–	+	–	–	–	–	–	–	–	–	(–)	+	–	–	–	–	–	–	–	–	–
1099	–	(+)	–	–	–	D	–	–	–	–	–	–	–	+	–	–	–	–	–	(–)	–	+	+	–	–

OGDH/Chromosome

Concordant

		1	2	3	4	5	6	**7**	8	9	10	11	12	13	14	15	16	17	18	19	20	21	22	X	Y
+/+		0	0	0	1	2	1	**2**	1	0	0	0	1	2	1	1	0	0	0	2	1	2	0	0	1
–/–		21	23	20	23	4	21	**24**	20	21	22	22	21	20	18	22	21	22	20	19	22	19	20	21	21

Discordant

		1	2	3	4	5	6	**7**	8	9	10	11	12	13	14	15	16	17	18	19	20	21	22	X	Y
+/–		2	2	2	1	0	1	**0**	1	2	2	2	1	0	1	1	2	2	2	0	1	0	2	2	1
–/+		3	1	4	1	20	3	**0**	4	3	2	2	3	4	6	2	3	2	4	5	2	5	4	3	3
% Discordancy		19	12	23	8	76	15	**0**	19	19	15	15	15	15	27	12	19	15	23	19	12	19	23	19	15

[a] +, strong hybridization; –, no hybridization to the cDNA probe in genomic Southern blot analysis.
[b] Intact chromosome: +, present in > 75%; (+) in 45–74%; (–), in 5–44%; –, absent; (D) contains deletion(s); (Dq) multiple deletions in 5q.

to –613 and –356 to –349. A glucocorticoid response element (GRE) was found at position nucleotide –692 to –687. In exon 22 AT-rich regions (ATTTAT, 45-bp upstream, and ATTTATTTATTTT, 4-bp upstream of the poly A tail) were detected instead of typical polyadenylation signal (ATTAAA).

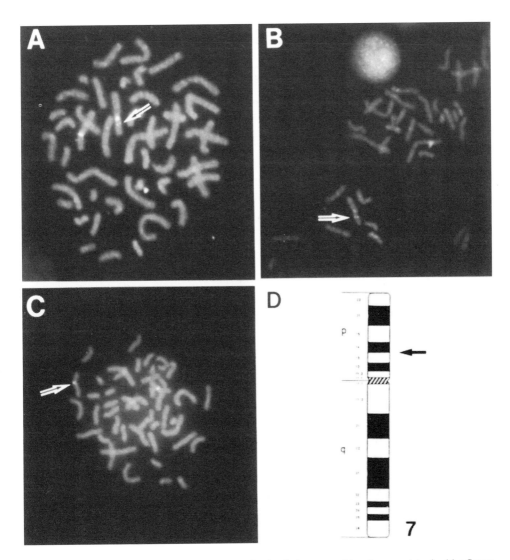

Fig. 4. Regional mapping of the human OGDH gene. Regional chromosomal location was detemined by fluorescence *in situ* hybridization, as reported by Richter et al. (1990). A λ phage clone (λHGOG 9) DNA was biotinylated by nick translation and hybridized to metaphase chromosome spreads. To confirm the location of the λ phage probe a biotinylated probe which is specific for the centromere of chromosome 7 was cohybridized. (A) Fluorescence *in situ* hybridization of biotinylated λHGOG 9 clone DNA and a probe specific for the centromere of chromosome 7 to a human chromosome spread. The hybridization signal of λ phage probe is indicated by an arrow. The hybridization signals of biotinylated λ phage probe alone (B) and the chromosome 7-specific probe (C) are indicated by the arrows, respectively. (D) Idiogram of human chromosome 7 with the map position of the human OGDH gene denoted by an arrow.

Chromosomal localization of the human OGDH gene

Chromosomal mapping of the human OGDH gene was first determined by Southern blot hybridization of chromosome panels contained *Bam*HI digested DNAs from human, 26 human × rodent somatic cell hybrids and hamster (BIOS, New Haven, CT, USA) with a radiolabeled 640-bp *Pst*I-*Pst*I fragment of 3'-untranslated region of the human OGDH cDNA corresponding to a part of exon 22, as summarized in Table 2. The results indicate that this gene is localized in chromosome 7. Regional chromosomal localization was determined by fluorescence *in situ* hybridization. Measurements of 10 specifically labeled chromosome 7 showed that the fluorescence labeled λ phage clone (λHGOG 9) DNA is located 30% distant from the centromere to the telomere of chromosome arm 7p, an area that corresponds to the boundary between bands 7p13 and 7p14, as shown in Figures 4A and B. In a recent report (Szabo et al., 1994), the OGDH (E1o) gene locus was mapped to chromosome 7p11.2-7p13 using a pair of human × rodent hybrids panels. I have now located it more precisely at the boundary between bands 7p13 and 7p14 of chromosome 7.

The gene loci of the two other component enzymes, E2o and E3, of the human OGDH complex were then assigned to chromosome 14 at q24.2-24.3 (Nakano et al., 1993) and chromosome 7 at q31-q32 (Scherer et al., 1991), respectively. Regulation of expression of the genes encoding these three component enzymes of the OGDH complex remains to be elucidated. Of particular interest is how the expression of these three genes distributed on three loci on two chromosome is coordinated to produce multiple copies of three component enzymes of the complex which are subsequently targeted to the mitochondrial compartments.

Note

The DDBJ, EMBL and GenBank accession numbers are: D32056, D32057, D32058, D32059, D32060, D32061, D32062, D32063 and D32064. An error has been found in the nucleotide and amino acid sequences of the human OGDH cDNA, and the correction is available from DDBJ, EMBL and GenBank (D90499).

Acknowledgement

I thank Dr. Lester J. Reed for critical reading of this manuscript, Drs. Kiyoshi Fukui and Masatugu Horiuchi for technical advice and helpful discussions, and Dr. Sachiko Matsuo for providing HeLa cells. This work was supported in part by a grant-in-aid for scientific research from Ministry of Education, Science and Culture of Japan and by a grant from the Vitamin B Research Committee.

References

Breathnach, R. and Chambon, P. (1981) Organization and expression of eucaryotic split genes coding for proteins. *Ann. Rev. Biochem.* 50: 349–383.

Feinberg, A.P. and Vogelstein, B.V. (1983) A technique for radiolabeling DNA restriction endonuclease fragments to high specific activity. *Anal. Biochem.* 132: 6–13.

Hawkins, C.F., Borges, A. and Perham, R.N. (1989) A common structural motif in thiamin pyrophosphate-binding enzymes. *FEBS Lett.* 255: 77–82.

Hirashima, M., Hayakawa, T. and Koike, M. (1967) Mammalian α-keto acid dehydrogenase complexes: II. An improved procedure for the preparation of 2-oxoglutarate dehydrogenase complex from pig heart muscle. *J. Biol. Chem.* 242: 902–907.

Höltke, H.J., Sanger, G., Kessler, C. and Schmitz, G. (1992) Sensitive chemiluminescent detection of digoxigenin-labeled nucleic acids: A fast and simple protocol and its applications. *Bio Techniques* 12: 104–113.

Koike, K., Tanaka, N., Hamada, M., Otsuka, K.-I., Suematsu, T. and Koike, M. (1971) Resolution and reconstitution of pig heart 2-oxoglutarate dehydrogenase complex. *J. Biochem.* 69: 1143–1147.

Koike, K., Hamada, M., Tanaka, N., Otsuka, K.-I., Ogasahara, K. and Koike, M. (1974) Properties and subunit composition of the pig heart 2-oxoglutarate dehydrogenase. *J. Biol. Chem.* 249: 3836–3842.

Koike, M. and Koike, K. (1976) Structure, assembly and function of mammalian α-keto acid dehydrogenase complexes. *Adv. Biophys.* 9: 182–227.

Koike, K., Urata, Y. and Goto, S. (1992) Cloning and nucleotide sequence of the cDNA encoding human 2-oxoglutarate dehydrogenase (lipoamide). *Proc. Natl. Acad. Sci. USA* 89: 1963–1967.

Koike, K. (1995) The gene encoding human 2-oxoglutarate dehydrogenase: structural organization and mapping to chromosome 7p13-p14. *Gene* 159: 261–266.

Lichter, P., Tang, C.C., Call, K., Hermanson, G., Evans, G.A., Housman, E.D. and Word, D.C. (1990) High resolution mapping of human chromosome 11 by *in situ* hybridization with cosmid clones. *Science* 247: 64–69.

Moriyasu, M., Koike, K., Urata, Y., Tsuji, A. and Koike, M. (1986) Rapid and simple isolation procedure for three component enzymes of pig heart 2-oxoglutarate dehydrogenase complex. *J. Nutr. Sci. Vitaminol.* 32: 33–40.

Mount, S.M. (1992) A catalogue of splice junction sequences. *Nucleic Acids Res.* 10: 459–472.

Nakano, K., Matsuda, S., Sakamoto, T., Takase, C., Nakagawa, S., Ohta, S., Ariyama, T., Inagawa, J., Abe, T. and Miyata, T. (1993) Human dihydrolipoamide succinyltransferase: cDNA cloning and localization on chromosome 14q24.2-q24.3. *Biochim. Biophys. Acta* 1216: 360–368.

Petit, F.H., Hamilton, L., Munk, P., Namihira, G., Eley, M.H., Willms, C.R. and Reed, L.J. (1973) α-Keto acid dehydrogenase complexes. *J. Biol. Chem.* 248: 5282–5290.

Reed, L.J. (1974) Multienzyme complexes. *Acc. Chem. Res.* 7: 40–60.

Sambrook, J., Fritsch, E.F. and Maniatis, T. (1989) *Molecular Cloning: A Laboratory Manual*, Second edition. Cold Spring Harbor Laboratory Press, Cold Spring Harbor, NY.

Sanger, F., Nicklen, S. and Coulson, A.R. (1977) DNA sequencing with chain-terminating inhibitors. *Proc. Natl. Acad. Sci. USA* 74: 5463–5467.

Scherer, S.W., Otulakowski, G., Robinson, B.H. and Tsui, L.-C. (1991) Localization of the human dihydrolipoamide dehydrogenase gene (DLD) to 7q31-> q32. *Cytogen. Cell Gen.* 56: 176–177.

Szabo, P., Cai, X., Ali, G. and Blass, J.P. (1994) Localization of the gene (OGDH) coding for the E1k component of the α-ketoglutarate dehydrogenase complex to chromosome 7p13-p11.2. *Genomics* 20: 324–326.

Tanaka, N., Koike, K., Hamada, M., Otsuka, K.-I., Suematsu, T. and Koike, M. (1971) Mammalian α-keto acid dehydrogenase complexes: VII. Resolution and reconstitution of the pig heart 2-oxoglutarate dehydrogenase complex. *J. Biol. Chem.* 247: 4043–4049.

Tanaka, N., Koike, K., Otsuka, K.-I., Hamada, M., Ogasahara, K. and Koike, M. (1974) Mammalian α-keto acid dehydrogenase complexes: VIII. Properties and subunit composition of the pig heart lipoate succinyltransferase. *J. Biol. Chem.* 249: 191–198.

Yeaman, S.J. (1989) The 2-oxo acid dehydrogenase complexes: recent advances. *Biochem. J.* 257: 625–632.

Alpha-Keto Acid Dehydrogenase Complexes
M.S. Patel, T.E. Roche and R.A. Harris (eds)
© 1996 Birkhäuser Verlag Basel/Switzerland

Pyruvate dehydrogenase complex as an autoantigen in primary biliary cirrhosis

S.J. Yeaman and A.G. Diamond[1]

Departments of Biochemistry and Genetics and Immunology[1], The Medical School, University of Newcastle upon Tyne, Newcastle upon Tyne NE2 4HH, UK

Introduction

Primary biliary cirrhosis (PBC) is a chronic cholestatic liver disease which is characterised by progressive inflammatory obliteration of the intrahepatic bile ducts, leading to fibrosis, liver cell damage and ultimately liver failure (Sherlock and Dooley, 1993). A variety of experimental evidence has indicated an autoimmune basis to this disease. Over 30 years ago it was recognised by Walker et al. (1965) that PBC-specific autoantibodies are present in the sera of patients and that these autoantibodies can be detected by indirect immunofluorescence. Subsequently the autoantigens involved were shown to be associated with the inner mitochondrial membrane (Berg et al., 1967). A variety of other conditions are also associated with the presence of autoantibodies to mitochondrial antigens but a subset of these antigens, termed M2, have been defined as specific to PBC (Berg et al., 1982). Furthermore a variety of other autoantibodies have been reported to be associated with PBC but they are not as disease-specific or are found in a lower proportion of patients (Berg and Klein, 1989).

Immunoblotting of mitochondrial extracts with PBC sera revealed that M2 consists of several distinct polypeptides which have been classified by Lindenborn-Fotinos et al. (1985) according to their apparent molecular mass on SDS-PAGE. The predominant autoantigen, termed M2a, is recognised by more than 95% of patients with PBC and migrates on SDS-PAGE with an apparent M_r of approximately 74 000.

Identification of the M2a autoantigen as the E2 component of pyruvate dehydrogenase complex

Availability of high titer sera against the M2a autoantigen allowed the screening of an expression cDNA library from rat liver. Using this approach, Gershwin et al. (1987) identified a partial cDNA clone of 1370 base pairs, the nucleotide sequencing of which indicated that it encoded a 456 amino acid residue peptide apparently unrelated to any known protein. Further inspection of this predicted protein sequence by Fussey et al. (1988) and analysis of a related sequence from a human placental library (Coppel et al., 1988) indicated that this polypeptide corresponded to the E2 component of the pyruvate dehydrogenase complex (E2p).

Direct experimental evidence in support of this was provided by immunoblotting experiments whereby it was demonstrated that PBC sera react specifically with E2p and protein X of pyruvate dehydrogenase purified from bovine heart. Using purified E2p+X, it was demonstrated that autoantibodies (IgG) against these polypeptides were present in 38 of 40 (95%) of PBC patients sera studied (Yeaman et al., 1988). Subsequently a similar percentage of patients were shown to have IgM autoantibodies against these polypeptides (Mutimer et al., 1989). A similar finding was reported by Van de Water et al. (1988a), in which they reported that 29 out of 30 PBC sera studied had antibodies which reacted in immunoblotting against a fusion protein containing recombinant rat E2p. Furthermore, this fusion protein immunosorbs all the reactivity in PBC sera against the M2a antigen in beef heart mitochondrial extracts (Van de Water et al., 1988a).

It has also been demonstrated by Van de Water et al. (1988b) that the autoantibodies inhibit the activity *in vitro* of E2p, although this is unlikely to be of any physiological or pathological significance. An interesting aside is that when experimental animals are immunised with recombinant human E2p they produce antibodies to the protein but show no symptoms of PBC (Krams et al., 1989). It is also noteworthy that the titer of autoantibodies in patients' sera is incredibly high, with in some cases a positive reaction being detected by ELISA at a sera dilution of 10^{-8} (Heseltine et al., 1990).

Other components of the α-keto acid dehydrogenase complexes as autoantigens

Subsequent to the identification of E2p and protein X as autoantigens, a variety of studies using purified protein and also recombinant polypeptides were carried out to investigate the possible role as autoantigens of other components of pyruvate dehydrogenase and of the other α-keto acid dehydrogenase complexes.

Immunoblotting of 40 PBC sera against purified polypeptides demonstrated that autoantibodies against the E2 components of the 2-oxoglutarate and branched-chain 2-oxo acid dehydrogenase complexes (E2o and E2b respectively) are also found in 73 and 63% of PBC sera respectively (Fussey et al., 1988). Separate studies have confirmed this finding, with 71% of sera containing antibodies against native E2b (Fregeau et al., 1989) and 37% against recombinant human E2b, as shown by Surh et al. (1989). Furthermore 39% of patients' sera were found by Fregeau et al. (1990a) to react against purified E2o and inhibit complex activity.

It has also been demonstrated that E1pα and E1pβ are autoantigens in this disease. This was initially demonstrated by Fussey et al. (1989a), who reported that 41% of sera reacted with E1Pα and 7% with E1Pβ. Subsequently Fregeau et al. (1990b) demonstrated reactivity against E1Pα in 66% of their patients' sera. There are also reports that E3 is an autoantigen in PBC (Maeda et al., 1991), but this remains somewhat contentious in that other evidence indicates that it is not an autoantigen (Adami et al., 1991) and that apparent reactivity in immunoblotting is due to a proteolytic fragment of E2p co-migrating with E3 upon SDS-PAGE (Fussey et al., 1991).

The vast majority of work investigating reactivity of autoantibodies against components of the mammalian complexes has involved use of recombinant antigens expressed in *E.coli* or else antigens purified from animal tissues. Recently however Palmer et al. (1993) have compared reactivity of PBC sera against E2p+X prepared from human and bovine heart. Reactivity against the antigens from different species was identical, both in terms of titer and affinity of the autoantibodies, indicating that the reactivity is truly of an autoimmune nature.

In addition to the reactivity against components of the mammalian complexes, the sera of PBC patients also contain antibodies reacting against E2p and E2o of *E. coli* (Fussey et al., 1989b,1990) and indeed a bacterial aetiology has been proposed for the disease (Stemerowicz et al., 1988). Arguing against this possibility however is the observation that a distinct species of antibodies recognise the bacterial proteins and that their titer is significantly lower than the autoantibodies targetted against the mammalian complexes (Fussey et al., 1991). Flannery et al. (1989) have detected some cross-reactivity of the antibodies for the mammalian and bacterial antigens. E2p from *S. cerevisiae* is also recognised by antibodies in patients' sera (Fussey et al., 1989b) and in this case it appears that this is due to cross-reactivity with the autoantibodies against the mammalian proteins (Fussey et al., 1991).

With regards to the autoantibodies against the mammalian antigens, it appears that several species are present with different specificities. Antibodies recognising E2p cross-react with protein X and *vice versa* (Surh et al., 1989b) but do not recognise E2o or E2b as shown in immunoblotting studies by Fussey et al. (1991). A second species of autoantibodies has been reported to recognise both E2o and E2b (Fussey et al., 1991), although others have reported that autoantibodies against E2o are specific for that polypeptide (Fregeau et al., 1990a). Antibodies specific for the α and β subunits respectively of Elp are also present (Fussey et al., 1991).

An unrelated observation is that in halothane hepatitis in which susceptible individuals have been exposed to this anaesthetic, autoantibodies are generated which recognise E2p (Christen et al., 1993). This is thought to be due to molecular mimicry, in which the autoantibodies generated in response to the elicited trifluoroacetylated proteins cross-react with the lipoylated region of E2p (Christen et al., 1994).

In summary, the sera of PBC patients contain antibodies against six or more components of the α-keto acid dehydrogenase complexes. The most commonly occurring autoantibody is against E2p (and protein X) and is present in high titer in the sera of approximately 95% of PBC sera (Bassendine et al., 1989). Indeed the presence of autoantibodies is essentially disease-specific, leading to the use of ELISAs as a diagnostic technique for the disease (Mutimer et al., 1989; Van de Water et al., 1989).

Nature of the antigenic determinants on E2p

A key feature of the structure of the different E2s of the α-keto acid dehydrogenase complexes is their highly segmented nature, consisting of several distinct domains linked by flexible regions of polypeptide (reviewed by Perham, 1991). The presence of structurally independent domains allows studies to be carried out using limited proteolysis or sub-cloning and expression to investigate the function of different regions of the polypeptide. Using this approach, it was possible initially to demonstrate that the bulk of the reactivity of the autoantibodies in PBC were targetted against the lipoyl domains of the E2 polypeptides. This has been demonstrated using both recombinant protein (Van de Water et al., 1988a) and also native protein (Fussey et al., 1989b). All experimental data to date indicates that the immunodominant epitopes are located within the lipoyl domains, with some evidence for the inner of the two lipoyl domains on E2p being the dominant epitope (Surh et al., 1990a, b).

Within the lipoyl domain itself there is also the question of the role of the lipoic acid residue in recognition by the autoantibodies. Again current data are consistent with the possibility that the lipoic acid plays a key role in effective recognition and that the epitope consists of elements of both the polypeptide and the lipoic acid. Work for this originally came from use of E2p from *E.coli*, a polypeptide with significant homology to its mammalian counterpart. Using a variety of preparations in which the E2p was either non-lipoylated, lipoylated or modified with a lipoic acid pre-cursor, namely octanoic acid, it was demonstrated that maximum reactivity was observed against the lipoylated peptide, with the octanoylated polypeptide being recognised less effectively but more so than the naked polypeptide (Fussey et al., 1990). This work has also been carried out subsequently using isolated recombinant inner lipoyl domains from the human E2p. Again using a combination of ELISAs, immunoblotting and affinity studies, it has been demonstrated that the titer and affinity of the autoantibodies is higher against the lipoylated than non-lipoylated form of the domain (Quinn et al., 1993).

This work is supported broadly by studies using synthetic peptides, but one apparent discrepancy is that in one report using these peptides essentially no reactivity was observed against the naked polypeptide, indicating that the presence of lipoic acid is absolutely essential for recognition (Tuaillon et al., 1992). Within the native polypeptides, although lipoic acid is important, the polypeptide lacking this cofactor is still recognised by the autoantibodies. These findings are consistent with a conformationally-dependent epitope being recognised by the antibodies, with elements contributed by the polypeptide and by the lipoic acid co-factor.

To date, there is no information available concerning the epitopes on the other components of the complexes which act as autoantigens. Whilst it might be expected that the lipoyl domains of E2b and E2o contain the immunodominant epitope, it will be of particular interest to determine whether the thiamine pyrophosphate binding site(s) of E1p are involved in the autoimmune process.

Possible cell surface expression of E2p in PBC

An intriguing nature of this autoimmune disease is the tissue specificity of the damage. Equally, how can autoantibodies targetted against intracellular antigens "protected" by three membranes cause a pathogenic effect? There are a variety of possible explanations of this, one of which might be localised expression of an antigen on the surface of the target cell. Indeed, there is now some very clear evidence in support of this.

Two groups have reported the presence of molecules on the surface of biliary epithelial cells which are recognised by antibodies to E2p. Using one mouse monoclonal antibody and a human combinatorial antibody against E2p, reactivity was seen with molecules expressed on the luminal surface of biliary epithelial cells in PBC patients. However, using seven other mouse monoclonals against the same antigen, no such reactivity could be detected (Van de Water et al., 1993). Similarly, using a rabbit anti-E2p antibody, high intensity non-mitochondrial staining has been observed localised to biliary epithelial cells from PBC patients, but not controls (Joplin et al., 1991, 1994). Confocal microscopy confirmed that this molecule was associated with the cell membrane of cultured biliary epithelial cells derived from PBC patients (Joplin et al., 1992). Recently this has been extended by Joplin et al. (1995) to include work with affinity-purified autoantibodies from PBC patients.

As yet these antigens have not been fully characterised but it is an exciting possibility that they are a form of E2p or an immunologically cross-reactive species. If they are a variant of E2p we need to understand the structure and derivation of the molecules at the cell surface. Do they contain the lipoic acid moiety so important for reactivity with autoantibodies? What causes their appearance at the cell surface? Could they arise through changes in leader sequences normally responsible for targetting these molecules to the mitochondrial membranes, or from some other change within the biliary epithelial cell? Are any such changes causally related to the disease process? If not E2p, are these molecules the product of other cellular genes – or of an as yet unidentified infectious agent that might be the trigger for the induction of PBC? Clearly, under-standing the nature of these cell surface molecules will be a crucial advance.

T-Cell responses in PBC

Although there is some evidence that autoantibodies may be involved in the pathogenesis of the disease, it remains likely that tissue damage in PBC is mediated primarily by T-cells. This idea is supported by the presence of a marked activated T-cell infiltrate in the portal tracts, co-incident with biliary epithelial cell damage (Eggink et al., 1992; Van den Oord et al,1984; Yamada et al,1986). Furthermore there is a marked elevation of HLA class I and class II expression on bile duct epithelia and hepatocytes in the vicinity of the T-cell infiltrate, probably under the influence of cytokines secreted by these T-cells (Van den Oord et al., 1986; Spengler et al., 1988). All of this is consistent with the possibility that it is T-cell mediated damage which occurs in PBC.

Despite advances made in the identification of the antigens involved in the B-cell response, rather less is known about the possible T-cell antigens in PBC. The presence of high levels of circulating IgG against components of pyruvate dehydrogenase is certainly consistent with T-cell help from CD4[+] ("Helper") T-cells reactive with the same molecules, leading to the production of antibodies of this isotype. Whether T-cells with the same specificity are involved in damage to the biliary epithelia remains to be established. If this is indeed the case, it would provide a solution to the problem described above, of accessibility of biliary cell antigen needed to cause cell destruction. Because the complexes recognised by T-cells consist of an antigen-derived peptide bound to an MHC molecule, it would only be necessary to postulate the presence of peptide fragments of E2p (or other subunits) at the biliary epithelial cell surface. Why this would be unique to these cells or to biliary epithelial cells from individuals with PBC would, however, remain unclear.

Several studies have now shown the existence of T-cells in this disease directed against components of the pyruvate dehydrogenase complex. CD4[+] T-cell clones grown in the presence of the antigens, from cells infiltrating the livers of PBC patients have been shown to produce interleukin 2 in response to stimulation by E2p or E2b (Van de Water et al., 1991). Nine from a total of 115 T-cell clones grown in the absence of specific antigen, from liver-infiltrating cells of PBC patients, responded to intact pyruvate dehydrogenase complex (Lohr et al., 1993), suggesting a major role for this molecule as a T-cell antigen.

Recent studies using peripheral blood lymphocytes from PBC patients have demonstrated a proliferative response to preparations of E2p and X. This response is apparently disease-specific (Jones et al., 1995). By way of contrast, a non-disease-specific response to E1 was observed, with no response to E3. A similar disease-specific response to E2p was found by a second group; these workers, however, noted that the response to E1p was restricted to PBC patients (Van de Water et al., 1995). Despite these discrepancies, both these studies are consistent with the possibility that E2 and/or X play a key role in generating responses from T-cells. One problem common to both these studies is their use of bovine, rather than human, antigen preparations. Subsequently, using E2p and X, E1p and E3 preparations from human heart, our laboratory has seen responses essentially identical to those to the equivalent bovine antigens (Jones et al., manuscript in preparation). Peripheral blood responses to recombinant inner (amino acids 120–233) or outer (aa 1–98) lipoyl domains of human E2p are also found in the majority of PBC patients (Van de Water et al., 1995). In a few patients responses to both domains were observed. The same study showed a variety of patterns of responses among liver-derived T-cell clones. Several clones responded to bovine E2p but to neither human domain, suggesting either specificity for the bovine antigen or the presence of epitopes elsewhere in the human molecule. However, the majority reacted with one of the human lipoyl domains but not the intact bovine E2p. These cells could be specific for the human form of the molecule, or it may be that the epitopes that they recognise were generated following proteolysis of the isolated domains but not the intact subunit. The patterns of reactivity observed imply that there are several important T-cell epitopes within the E2p subunit; with at least one in each lipoyl domain and in the intervening 22 amino acid segment.

Further work is needed to clarify the location of the T-cell epitopes, their variability between individuals with different MHC haplotypes, the processing pathways that generate the peptides and whether the presence of the lipoyl cofactor, so important for antibody recognition, has any influence on this process. Ultimately, the crucial questions of how these autoreactive T-cells are generated, whether they initiate the disease process, their role in pathogenesis and, perhaps, how their activity can be downregulated to abrogate the disease process will have to be answered.

One further caveat to these T-cell studies is their concentration, mainly for technical reasons, on the "helper" CD4[+] population. In the tissue, the infiltrating T-cells most closely associated with the biliary epithelial cells are predominantly of the "killer" CD8[+] subset (Si et al,1984; Yamada et al., 1986). T-cells of this phenotype have been shown to be isolatable from the spleen of PBC patients which have cytotoxic activity against autologous biliary epithelial cells (Onishi et al., 1993). Although no information is presently available, it is tempting to speculate that these T-cells might also recognise E2 antigens and, perhaps, receive "help" from the CD4[+] populations described above. If this is indeed the case, then once again we are left with the question of precisely what is different about the expression of E2 molecules in biliary epithelial cells that renders them susceptible to this form of immune attack?

Conclusions

The identification of a role for the α-keto acid dehydrogenases as antigens in human autoimmune liver disease adds a fascinating new dimension to the biology of the complexes. Not only has knowledge of the structure and function of the complexes been applied to gain some under-standing of the autoimmune response in PBC, the availability of high titer PBC sera has allowed successful cloning of rat, and subsequently human, cDNA encoding E2p.

In many ways identification of E2p and other components as autoantigens has raised more questions that it has answered. It has provided however a good example of how basic scientific knowledge can further knowledge of an important clinical condition. Further application of a biochemical and molecular biological approach, particularly to the complex as a T-cell antigen, should provide further significant insight.

Acknowledgements
Work in the authors' laboratories is funded by the Wellcome Trust. We thank all our co-workers and collaborators, past and present, for their contributions to work emanating from our laboratories.

References

Adami, P., Berrez, J.M. and Latruffe, N. (1991) Mitochondrial pyruvate dehydrogenase complex subunits as autoantigens in human primary biliary cirrhosis. *Biochem. Int.* 23: 429–437.

Bassendine, M.F., Fussey, S.P.M., Mutimer, D.J., James, O.F.W. and Yeaman, S.J. (1989) Identification and characterisation of four M2 mitochondrial autoantigens in primary biliary cirrhosis. *Seminars in Liver Disease* 9: 124–131.

Berg, P.A., Doniach, D. and Roitt, I.M. (1967) Mitochondrial antibodies in primary biliary cirrhosis. *J. Exp. Med.* 126: 277–291.

Berg, P.A., Lindenborn-Fotinos, J., Klein, R. and Kloppel, W. (1982) ATPase-associated antigen (M2): marker antigen for serological diagnosis of primary biliary cirrhosis. *Lancet* 2: 1423–1426.

Berg, P.A. and Klein, R. (1989) Heterogeneity of antimitochondrial antibodies. *Seminars in Liver Disease* 9: 103–116.

Christen, U., Jeno, P. and Gut, J. (1993) Halothane metabolism: The dihydrolipoamide acetyltransferase subunit of the pyruvate dehydrogenase complex molecularly mimics trifluoroacetyl-protein adducts. *Biochemistry* 32: 1492–1499.

Christen, U., Quinn, J., Yeaman, S.J., Kenna, J.G., Clarke, J.B., Gandolfi, A.J. and Gut, J. (1994) Identification of the dihydrolipoamide acetyltransferase subunit of the human pyruvate dehydrogenase complex as an autoantigen in halothane hepatitis. *Eur. J. Biochem.* 223: 1035–1047.

Coppel, R.L., McNeilage, L.J., Surh, C.D., Van de Water, J., Spithill, T.W., Whittingham, S. and Gershwin, M.E. (1988) Primary structure of the human M2 mitochondrial autoantigen of primary biliary cirrhosis: Dihydrolipoamide acetyltransferase. *Proc. Natl. Acad. Sci. USA* 85: 7317–7321.

Eggink, H.F., Houthoff, H.J., Huitema, S., Gips, C.H. and Poppema, S. (1982) Cellular and humoral immune reactions in chronic active liver disease.I. Lymphocyte subsets in liver biopsies of patients with untreated idiopathic autoimmune hepatitis, chronic active hepatitis B and primary biliary cirrhosis. *Clin. Exp. Immunol.* 50: 17–24.

Flannery, G.R., Burroughs, A.K., Butler, P., Chelliah, J., Hamilton-Miller, J., Brumfitt, W. and Baum, H. (1989) Antimitochondrial antibodies in primary biliary cirrhosis recognise both specific peptides and shared epitopes of the M2 family of antigens. *Hepatology* 10: 370–374.

Fregeau, D.R., Davis, P.A., Danner, D.J., Ansari, A., Coppel, R.L., Dickson, E.R. and Gershwin, M.E. (1989) Antimitochondrial antibodies of primary biliary cirrhosis recognise dihydrolipoamide acyltransferase and inhibit enzyme function of the branched chain α-ketoacid dehydrogenase complex. *J. Immunol.* 142: 3815–3820.

Fregeau, D.R., Roche, T.E., Davis, P.A., Coppel, R. and Gershwin, M.E. (1990a) Primary biliary cirrhosis: Inhibition of pyruvate dehydrogenase complex activity by autoantibodies specific for E1α, a non-lipoic acid containing mitochondrial enzyme. *J. Immunol.* 144: 1671–1676.

Fregeau, D.R., Prindiville, T., Coppel, R.L., Kaplan, M., Dickson, E.R. and Gershwin, M.E. (1990b) Inhibition of α-ketoglutarate dehydrogenase activity by a distinct population of autoantibodies recognising dihydrolipoamide succinyltransferase in primary biliary cirrhosis. *Hepatology* 11: 975–981.

Fussey, S.P.M., Guest, J.R., James, O.F.W., Bassendine, M.F. and Yeaman, S.J. (1988) Identification and analysis of the major M2 autoantigens in primary biliary cirrhosis. *Proc. Natl. Acad. Sci. USA* 85: 8654–8658.

Fussey, S.P.M., Bassendine, M.F., James, O.F.W. and Yeaman, S.J. (1989a) Characterisation of the reactivity of autoantibodies in primary biliary cirrhosis. *FEBS Lett.* 246: 49–53.

Fussey, S.P.M., Bassendine, M.F., Fittes, D., Turner, I.B., James, O.F.W. and Yeaman, S.J. (1989b) The E1α and β subunits of the pyruvate dehydrogenase complex are M2'd' and M2'e' autoantigens in primary biliary cirrhosis. *Clin. Sci.* 77: 365–368.

Fussey, S.P.M., Ali, S.T., Guest, J.R., James, O.F.W., Bassendine, M.F. and Yeaman, S.J. (1990) Reactivity of primary biliary cirrhosis sera with *Escherichia coli* dihydrolipoamide acetyltransferase (E2p): Characterization of the main immunogenic region. *Proc. Natl. Acad. Sci. USA* 87: 3987–3991.

Fussey, S.P.M., Lindsay, J.G., Fuller, C., Perham, R.N., Dale, S., James, O.F.W., Bassendine, M.F. and Yeaman, S.J. (1991) Autoantibodies in primary biliary cirrhosis: Analysis of reactivity against eukaryotic and prokaryotic 2-oxo acid dehydrogenase complexes. *Hepatology* 13: 467–474.

Gershwin, M.E., Mackay, I.R., Sturgess, A. and Coppel, R. (1987) Identification and specificity of a cDNA encoding the 70 kD mitochondrial antigen recognised in primary biliary cirrhosis. *J. Immunol.* 138: 3525–3531.

Heseltine, L., Turner, I.B., Fussey, S.P.M., Kelly, P.J., James, O.F.W., Yeaman, S.J. and Bassendine, M.F. (1990) Primary biliary cirrhosis: Quantitation of autoantibodies to purified mitochondrial enzymes and correlation with disease progression. *Gastroenterology* 99: 1786–1792.

Jones, D.E.J., Palmer, J.M., James, O.F.W., Yeaman, S.J., Bassendine, M.F. and Diamond, A.G. (1995) T-cell responses to the components of pyruvate dehydrogenase complex in primary biliary cirrhosis. *Hepatology* 21: 995–1002.

Joplin, R., Lindsay, J.G., Hubscher, S.G., Johnson, G.D., Shaw, J.C., Strain, A.J. and Neuberger, J.M. (1991) Distribution of dihydrolipoamide acetyltransferase (E2) in the liver and portal lymph nodes of patients with primary biliary cirrhosis: an immunohistochemical study. *Hepatology* 14: 442–447.

Joplin, R., Lindsay, J.G., Johnson, G.D., Strain, A. and Neuberger, J.M. (1992) Membrane dihydrolipoamide acetyltransferase (E2) on human biliary epithelial cells in primary biliary cirrhosis. *Lancet* 339: 93–94.

Joplin, R., Johnson, G.D, Matthews, J.B., Hamburger, J., Lindsay, J.G., Hubscher, S.G., Strain, A.J. and Neuberger, J.M. (1994) Distribution of pyruvate dehydrogenase dihydrolipoamide acetyltransferase (PDC-E2) and another mitochondrial marker in salivary gland and biliary epithelium from patients with primary biliary cirrhosis. *Hepatology* 19: 1375–1380.

Joplin, R., Wallace, L.L., Johnson, G.D., Lindsay, J.G., Yeaman, S.J., Palmer, J.M., Strain, A.J. and Neuberger, J.M. (1995) Sub-cellular localization of pyruvate dehydrogenase dihydrolipoamide acetyltransferase in human intrahepatic biliary epithelial cells. *J. Pathol.* 176: 381–390.

Krams, S.M., Surh, C.D., Coppel, R.L., Ansari, A., Ruebner, B. and Gershwin, M.E. (1989) Immunisation of experimental animals with dihydrolipoamide acetyltransferase, as a purified recombinant polypeptide, generates mitochondrial antibodies but not primary biliary cirrhosis. *Hepatology* 9: 411–416.

Lindenborn-Fotinos, J., Baum, H. and Berg, P.A. (1985) Mitochondrial antibodies in primary biliary cirrhosis: species and nonspecies specific determinants of M2 antigen. *Hepatology* 5: 763–769.

Lohr, H., Fleischer, B., Gerken, G., Yeaman, S.J., Meyer zum Buschenfelde, K.-H. and Manns, M. (1993) Autoreactive liver-infiltrating T-cells in primary biliary cirrhosis recognize inner mitochondrial epitopes and the pyruvate dehydrogenase complex. *J. Hepatology* 18: 322–327.

Maeda, T., Loveland, B.E., Rowley, M.J. and Mackay, I.R. (1991) Autoantibody against dihydrolipoamide dehydrogenase, the E3 subunit of the 2-oxoacid dehydrogenase complexes: significance for primary biliary cirrhosis. *Hepatology* 14: 994–999.

Mutimer, D.J., Fussey, S.P.M., Yeaman, S.J., Kelly, P.J., James, O.F.W. and Bassendine, M.F. (1989) Frequency of IgG and IgM autoantibodies to four specific M2 mitochondrial autoantigens in primary biliary cirrhosis. *Hepatology* 10: 403–407.

Onishi, S., Saibara, T., Nakata, S., Maeda, T., Iwasaki, S., Iwamura, S., Miyazaki, M., Yamamoto, Y. and Enzan, H. (1993) Cytotoxic activity of spleen-derived T lymphocytes against autologous biliary epithelial cells in autopsy patients with primary biliary cirrhosis. *Liver* 13: 188–192.

Palmer, J.M., Bassendine, M.F., James, O.F.W. and Yeaman, S.J. (1993) Human pyruvate dehydrogenase complex as an autoantigen in primary biliary cirrhosis. *Clin. Sci.* 85: 289–293.

Perham, R.N. (1991) Domains, motifs and linkers in 2-oxo acid dehydrogenase multienzyme complexes: A paradigm in the design of a multifunctional protein. *Biochemistry* 30: 8501–8512.

Quinn, J., Diamond, A.G., Palmer, J.M., Bassendine, M.F., James, O.F.W. and Yeaman, S.J. (1993) Lipoylated and unlipoylated domains of human PDC-E2 as autoantigens in primary biliary cirrhosis: Significance of lipoate attachment. *Hepatology* 18: 1384–1391.

Sherlock, S. and Dooley, J. (1993) *Diseases of the Liver and Biliary System*, Ninth Edition. Blackwell Scientific Publications, Oxford, pp 236–250.

Si, L., Whiteside, T.L., Schade, R.R., Starzl, T.E. and Van Thiel, D.H. (1984) T-lymphocyte subsets in liver tissues of patients with primary biliary cirrhosis (PBC), patients with primary sclerosing cholangitis (PSC), and normal controls. *J. Clin. Immunol.* 4: 262–272.

Spengler, U., Pape, G.R., Hoffmann, R.M., Johnson, J.P., Eisenburg, J., Paumgartner, G. and Riethmuller, G. (1988) Differential expression of MHC class II subregion products on bile duct epithelial cells and hepatocytes in patients with primary biliary cirrhosis. *Hepatology* 8: 459–462.

Stemerowicz, R., Moller, B., Rodloff, A., Freudenhan, M., Hopf, U., Wittenbrink, C., Reinhardt, R. and Galanos, C. (1988) Are antimitochondrial antibodies in primary biliary cirrhosis induced by r (rough)-mutants of enterobacteriaceae? *Lancet* ii: 1166–1169.

Surh, C.D., Danner, D.J., Ahmed, A., Coppel, R.L., Mackay, I.R., Dickson, E.R. and Gershwin, M.E. (1989a) Reactivity of primary biliary cirrhosis sera with a human fetal liver cDNA clone of branched-chain α-keto acid dehydrogenase dihydrolipoamide acyltransferase, the 52 kD mitochondrial autoantigen. *Hepatology* 9: 63–68.

Surh, C.D., Roche, T.E., Danner, D.J., Ansari, A., Coppel, R.L., Prindiville, T., Dickson, R.E. and Gershwin, M.E. (1989b) Antimitochondrial autoantibodies in primary biliary cirrhosis recognise cross-reactive epitope(s) on protein X and dihydrolipoamide acetyltransferase of pyruvate dehydrogenase complex. *Hepatology* 10: 127–133.

Surh, C.D., Ahmed-Ansari, A. and Gershwin, M.E. (1990a) Comparative epitope mapping of murine monoclonal and human autoantibodies to human PDH-E2, the major mitochondrial autoantigen of primary biliary cirrhosis. *J. Immunol.* 144: 2647–2652.

Surh, C.D., Coppel, R. and Gershwin, M.E. (1990b) Structural requirement for autoreactivity on human pyruvate dehydrogenase-E2, the major autoantigen of primary biliary cirrhosis. *J. Immunol.* 144: 3367–3374.

Tuaillon, N., Andre, C., Briand, J.-P., Penner, E. and Muller, S. (1992) A lipoyl synthetic octadecapeptide of dihydrolipoamide acetyltransferase specifically recognised by anti-M2 autoantibodies in primary biliary cirrhosis. *J. Immunol.* 148: 445–450.

Van de Water, J., Gershwin, M.E., Leung, P., Ansari, A. and Coppel, R.L. (1988a) The autoepitope of the 74-kD mitochondrial autoantigen of primary biliary cirrhosis corresponds to the functional site of dihydrolipoamide acetyltransferase. *J. Exp. Med.* 167: 1791–1799.

Van de Water, J., Fregeau, D., Davis, P., Ansari, A., Danner, D., Leung, P., Coppel R. and Gershwin M.E. (1988b) Autoantibodies of primary biliary cirrhosis recognise dihydrolipoamide acetyltransferase and inhibit enzyme function. *J. Immunol.* 141: 2321–2324.

Van de Water, J., Cooper, A., Surh, C.D., Coppel, R., Danner, D., Ansari, A., Dickson, R. and Gershwin, M.E. (1989) Detection of autoantibodies to recombinant mitochondrial proteins in patients with primary biliary cirrhosis. *N. Eng. J. Med.* 320: 1377–1380.

Van de Water, J., Ansari, A.A., Surh, C.D., Coppel, R., Roche, T., Bonkovsky, H., Kaplan, M. and Gershwin, M.E. (1991) Evidence for the targeting by 2-oxo-dehydrogenase enzymes in the T-cell response of primary biliary cirrhosis. *J. Immunol.* 146: 89–94.

Van de Water, J., Turchany, J., Leung, P.S.C., Lake, J., Munoz, S., Surh, C.D., Coppel, R., Ansari, A., Nakanuma, Y. and Gershwin, M.E. (1993) Molecular mimicry in primary biliary cirrhosis: Evidence for biliary epithelial expression of a molecule cross-reactive with pyruvate dehydrogenase complex-E2. *J. Clin. Invest.* 91: 2653–2664.

Van de Water, J., Ansari, A., Prindiville, T., Coppel, R.L., Ricalton, N., Kotzin, B.L., Liu, S., Roche, T.E., Krams, S.M., Munoz, S. and Gershwin, M.E. (1995) Heterogeneity of autoreactive T-cell clones specific for the E2 component of the pyruvate dehydrogenase complex in primary biliary cirrhosis. *J. Exp. Med.* 181: 723–733.

Van den Oord, J.J., Fevery, J., de Groote, J. and Desmet, V.J. (1984) Immunohistochemical characterization of inflammatory infiltrates in primary biliary cirrhosis. *Liver* 4: 264–274.

Van den Oord, J.J., Sciot, R. and Desmet, V.J. (1986) Expression of MHC products by normal and abnormal bile duct epithelium. *J. Hepatology* 3: 310–317.

Walker, J.G., Doniach, D., Roitt, I.M. and Sherlock, S. (1965) Serological tests in diagnosis of primary biliary cirrhosis. *Lancet* 1: 827–831.

Yamada, G., Hyodo, I., Tobe, K., Mizuno, M., Nishihara, T., Kobayashi, T. and Nagashima, H. (1986) Ultrastructural immunocytochemical analysis of lymphocytes infiltrating bile duct epithelia in primary biliary cirrhosis. *Hepatology* 6: 385–391.

Yeaman, S.J., Fussey, S.P.M., James, O.F.W., Danner, D.J., Mutimer, D.J. and Bassendine, M.F. (1988) Primary biliary cirrhosis: identification of two major M2 mitochondrial autoantigens. *Lancet* 1: 1067–1070.

Subject index

(The page number refers to the first page of the chapter in which the keyword occurs)

Interface between Chemistry and Biochemistry

Edited by
P. Jollès, *Univ. René Descartes, Paris, France*
H. Jörnvall, *Karolinska Institute, Stockholm, Sweden*

1995. 312 pages. Hardcover • ISBN 3-7643-5081-4
(EXS 73)

The increasing importance of the interface between chemistry and biology is probably the largest change in chemistry in the past 15 years. More and more organic chemists are working on problems dealing with biology. Once considered to be at the very outside edge of either field, interfacial research is poised to move into the mainstream of both disciplines. This merging of two types of approach has resulted in a vigorous research discipline with unprecedented potential to address important biological and chemical problems. A series of examples are developed in this book.

Some analytical aspects are discussed first as the fundamental concepts are not only chemical, but chemistry has provided biochemistry with powerful tools of analysis.

Physico-chemical aspects are devoted to spectrometric studies of nucleic acids as well as lipids, lipases and membrane proteins (receptors). Three chapters are included in the section dealing with enzymes. The part devoted to metalloproteins is mainly directed toward zinc metallochemistry and NMR structural work on zinc proteins.

Chemists have been able to bring to biology their characteristic approach of synthesizing new molecules; three chapters are devoted to peptides, sugar compounds and biocatalysts. Two chapters discuss new active compounds (antibacterial peptides, catalytic antibodies), which are the result of collaboration between chemists and biochemists.

From the Contents:
Chemistry at interfaces and in transport • Enzyme function in organic solvents • Chemistry and biochemistry • Analysis of proteins and nucleic acids • UV and nucleic acids • Synthesis of active compounds • Metalloproteins

Birkhäuser Verlag • Basel • Boston • Berlin

S. Papa, *Institute of Medical Biochemistry and Chemistry, University of Bari, Italy*
J.M. Tager, *E.C. Slater Institute, University of Amsterdam, The Netherlands (Eds)*

Biochemistry of Cell Membranes

A Compendium of Selected Topics

1995. 376 pages. Hardcover. ISBN 3-7643-5056-3 (MCBU)

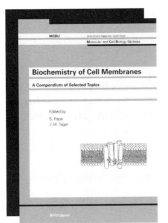

This book consists of a series of reviews on selected topics within the rapidly and vastly expanding field of membrane biology. Its aim is to highlight the most significant and important advances that have been made in recent years in understanding the structure, dynamics and functions of cell membranes.

Areas covered in this monograph include

- Signal Transduction
- Membrane Traffic: Protein and Lipids
- Bioenergetics: Energy Transfer and Membrane Transport
- Cellular Ion Homeostasis
- Growth Factors and Adhesion Molecules
- Structural Analysis of Membrane Proteins
- Membranes and Disease

Biochemistry of Cell Membranes should serve as a benchmark for indicating the most important lines for future research in these areas.

Birkhäuser Verlag • Basel • Boston • Berlin

Analysis of Free Radicals in Biological Systems

Edited by
A.E. Favier, *CHU Albert Michallon, Grenoble, France*
J. Cadet, *CEA, Grenoble, France*
B. Kalyanaraman, *Medical College of Wisconsin, Milwaukee, WI, USA*
M. Fontecave, J.L. Pierre, *LEDSS II, St. Martin d'Hères, France*

1995. 312 pages. Hardcover. ISBN 3-7643-5137-3

The main aim of this book is to provide a comprehensive survey on recent methodological aspects of the measurement of damage within cellular targets, information which may be used as an indicator of oxidative stress.

In the introductory chapters, emphasis is placed on the chemical properties of reactive oxygen species and their role in the induction of cellular modifications together with their links to various diseases. The central part of the book is devoted to the description of selected methods aimed at monitoring the production of free radicals in cellular systems. In addition, several assays are provided to assess the chemical damage induced by reactive oxygen species in critical cellular-targets in vitro and in humans in vivo.

Thus both practical aspects and general considerations, including discussions on the applications and limitations of the assays, are critically reviewed. One of the major features of the book is the description of new experimental methods. These include the measurement of oxydized DNA bases and nucleosides, new techniques for the determination of LDL oxidation using spin-trap agents, and the use of salicylate as an indicator of oxidative stress. In addition, more classical though significantly improved techniques devoted to the measurement of hydroperoxides and aldehydes are described.

This book will serve a large scientific community including biologists, chemists, and clinicians working on the chemical and biological effects of oxidative stress. It may also be of interest to investigators in the fields of drug, cosmetic and new food research.

Birkhäuser Verlag • Basel • Boston • Berlin